Cambridge
International AS and A Level Mathematics

Mechanics

Sophie Goldie
Series Editor: Roger Porkess

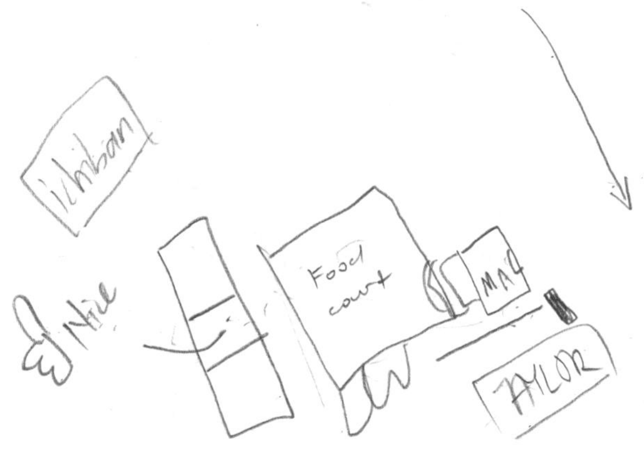

HODDER EDUCATION
AN HACHETTE UK COMPANY

Questions from the Cambridge International Examinations AS and A Level Mathematics papers are reproduced by permission of University of Cambridge International Examinations.

Questions from the MEI AS and A Level Mathematics papers are reproduced by permission of OCR.

We are grateful to the following companies, institutions and individuals who have given permission to reproduce photographs in this book.

Photo credits: page 2 © Mathematics in Education and Industry; page 6 © Radu Razvan – Fotolia; page 22 © Iain Masterton / Alamy; page 24 © photoclicks – Fotolia; page 40 © DOD Photo / Alamy; page 44 © Dr Jeremy Burgess / Science Photo Library; page 45 ©Jonathan Pope / http://commons.wikimedia.org/wiki/File:Olympic_Curling,_Vancouver_2010_crop_sweeping.jpg/http://creativecommons.org/licenses/by/2.0/deed.en/16thJan2011; page 47 © imagehit – Fotolia; page 53 © NASA / Goddard Space Flight Center / Arizona State University; page 59 © Lebrecht Music and Arts Photo Library / Alamy; page 60 © Mehau Kulyk / Science Photo Library; page 64 © Dmitry Lobanov – Fotolia; page 67 © Imagestate Media (John Foxx); page 85 l © Dean Moriarty – Fotolia; page 85 c © Masson – Fotolia; page 85 r © Marzanna Syncerz – Fotolia; page 99 ©Tifonimages – Fotolia; page 115 © Kathrin39 – Fotolia; page 138 © SHOUT / Alamy; page 153 © Mathematics in Education and Industry; page 154 © Image Asset Management Ltd. / SuperStock; page 175 © Steve Mann – Fotolia; page 184 © photobyjimshane – Fotolia; page 253 © Millbrook Proving Ground Ltd; page 264 © cube197 – Fotolia; page 266 l © M.Rosenwirth – Fotolia; page 266 c ©Michael Steele / Getty Images; page 266 r © NickR – Fotolia; page 280 © Steeve ROCHE – Fotolia; page 295 © Lovrencg – Fotolia; page 308 © blueee – Fotolia

Photo credits for CD material: Exercise 14B question 11 © Monkey Business – Fotolia

l = left, c = centre, r = right

Orders: please contact Bookpoint Ltd, 130 Milton Park, Abingdon, Oxon OX14 4SB. Telephone: (44) 01235 827720. Fax: (44) 01235 400454. Lines are open 9.00–5.00, Monday to Saturday, with a 24-hour message answering service. Visit our website at www.hoddereducation.co.uk

Much of the material in this book was published originally as part of the MEI Structured Mathematics series. It has been carefully adapted for the Cambridge International AS and A Level Mathematics syllabus.

The original MEI author team for Mechanics comprised John Berry, Pat Bryden, Ted Graham, David Holland, Cliff Pavelin and Roger Porkess.

First published in 2012 by
Hodder Education, an Hachette UK company,
338 Euston Road
London NW1 3BH

Cover photo by © Imagestate Media (John Foxx)
Illustrations by Pantek Media, Maidstone, Kent
Typeset in 10.5pt Minion by Pantek Media, Maidstone, Kent
Printed in Dubai

A catalogue record for this title is available from the British Library

Contents

Key to symbols in this book

This symbol means that you may want to discuss a point with your teacher. If you are working on your own there are answers in the back of the book. It is important, however, that you have a go at answering the questions before looking up the answers if you are to understand the mathematics fully.

This is a warning sign. It is used where a common mistake, misunderstanding or tricky point is being described.

This is the ICT icon. It indicates where you could use a graphic calculator or a computer. Graphical calculators and computers are not permitted in any of the examinations for the Cambridge International AS and A Level Mathematics 9709 syllabus, however, so these activities are optional.

This symbol and a dotted line down the right-hand side of the page indicates material which is beyond the syllabus but which is included for completeness.

Introduction

This is one of a series of books for the University of Cambridge International Examinations syllabus for Cambridge International AS and A Level Mathematics 9709. There are fifteen chapters in this book; the first nine cover Mechanics 1 and the remaining six Mechanics 2. The series also includes two books for pure mathematics and one for statistics.

These books are based on the highly successful series for the Mathematics in Education and Industry (MEI) syllabus in the UK but they have been redesigned for Cambridge International students; where appropriate new material has been written and the exercises contain many past Cambridge examination questions. An overview of the units making up the Cambridge international syllabus is given in the diagram on the next page.

Throughout the series the emphasis is on understanding the mathematics as well as routine calculations. The various exercises provide plenty of scope for practising basic techniques; they also contain many typical examination questions.

In the examinations of the Cambridge International AS and A Level Mathematics 9709 syllabus the value of g is taken to be $10\,\mathrm{m\,s^{-2}}$ and this convention is used in this book; however, in a few questions readers are introduced to a more accurate value, typically $9.8\,\mathrm{m\,s^{-2}}$.

An important feature of this series is the electronic support. There is an accompanying disc containing two types of Personal Tutor presentation: examination-style questions, in which the solutions are written out, step by step, with an accompanying verbal explanation, and test yourself questions; these are multiple-choice with explanations of the mistakes that lead to the wrong answers as well as full solutions for the correct ones. In addition, extensive online support is available via the MEI website, www.mei.org.uk.

The books are written on the assumption that students have covered and understood the work in the Cambridge IGCSE® syllabus. There are places where the books show how the ideas can be taken further or where fundamental underpinning work is explored and such work is marked as 'Extension'.

The original MEI author team would like to thank Sophie Goldie who has carried out the extensive task of presenting their work in a suitable form for Cambridge international students and for her original contributions. They would also like to thank University of Cambridge International Examinations for their detailed advice in preparing the books and for permission to use many past examination questions.

Roger Porkess
Series Editor

The Cambridge International AS and A Level Mathematics syllabus

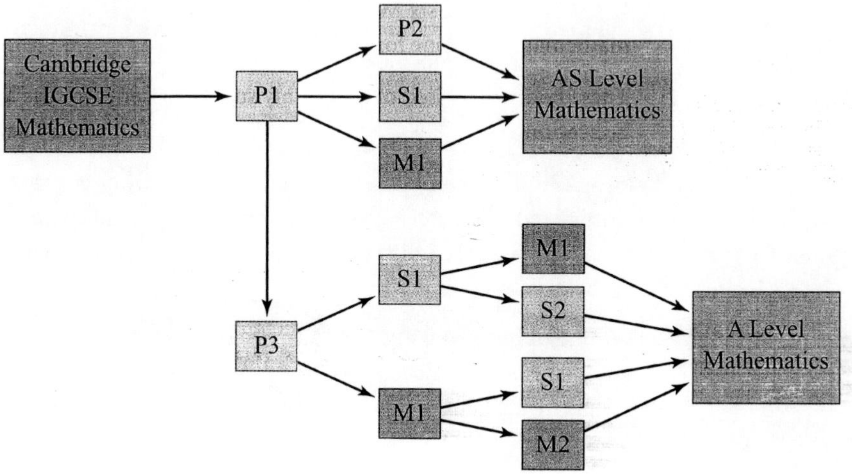

Mechanics 1

M1

Motion in a straight line

The whole burden of philosophy seems to consist in this – from the phenomena of motions to investigate the forces of nature.

Isaac Newton

The language of motion

Throw a small object such as a marble straight up in the air and think about the words you could use to describe its motion from the instant just after it leaves your hand to the instant just before it hits the floor. Some of your words might involve the idea of direction. Other words might be to do with the position of the marble, its speed or whether it is slowing down or speeding up. Underlying many of these is time.

Direction

The marble moves as it does because of the gravitational pull of the earth. We understand directional words such as up and down because we experience this pull towards the centre of the earth all the time. The *vertical* direction is along the line towards or away from the centre of the earth.

In mathematics a quantity which has only size, or magnitude, is called a *scalar.* One which has both magnitude and a direction in space is called a *vector.*

Distance, position and displacement

The total *distance* travelled by the marble at any time does not depend on its direction. It is a scalar quantity.

Position and displacement are two vectors related to distance: they have direction as well as magnitude. Here their direction is up or down and you decide which of these is positive. When up is taken to be positive, down is negative.

The *position* of the marble is then its distance above a fixed origin, for example the distance above the place it first left your hand.

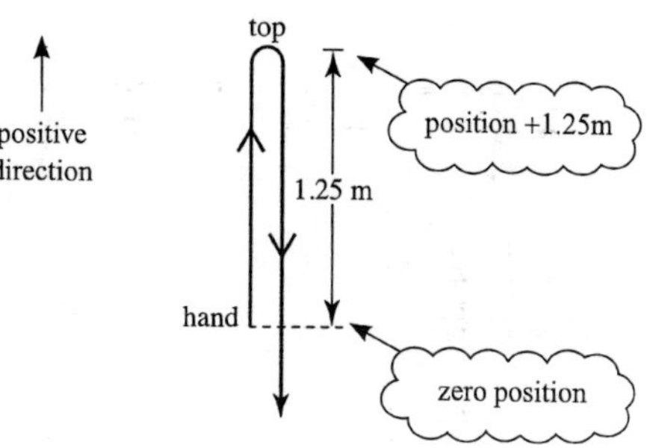

Figure 1.1

When it reaches the top, the marble might have travelled a distance of 1.25 m. Relative to your hand its position is then 1.25 m upwards or +1.25 m.

At the instant it returns to the same level as your hand it will have travelled a total distance of 2.5 m. Its *position*, however, is zero upwards.

A position is always referred to a fixed origin but a *displacement* can be measured from any position. When the marble returns to the level of your hand, its displacement is zero relative to your hand but −1.25 m relative to the top.

? What are the positions of the particles A, B and C in the diagram below?

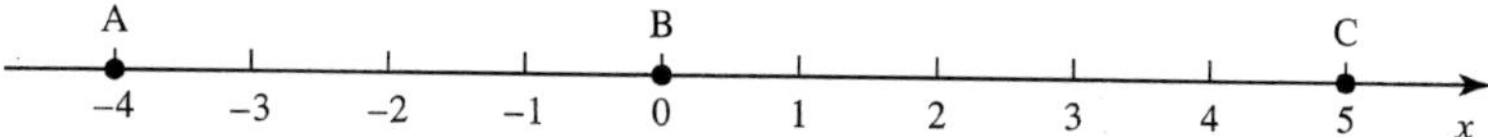

Figure 1.2

What is the displacement of B

(i) relative to A **(ii)** relative to C?

Diagrams and graphs

In mathematics, it is important to use words precisely, even though they might be used more loosely in everyday life. In addition, a picture in the form of a diagram or graph can often be used to show the information more clearly.

Figure 1.3 is a *diagram* showing the direction of motion of the marble and relevant distances. The direction of motion is indicated by an arrow. Figure 1.4 is a *graph* showing the position above the level of your hand against the time. Notice that it is *not* the path of the marble.

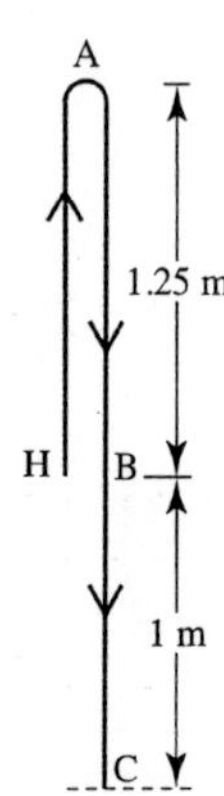

Figure 1.3

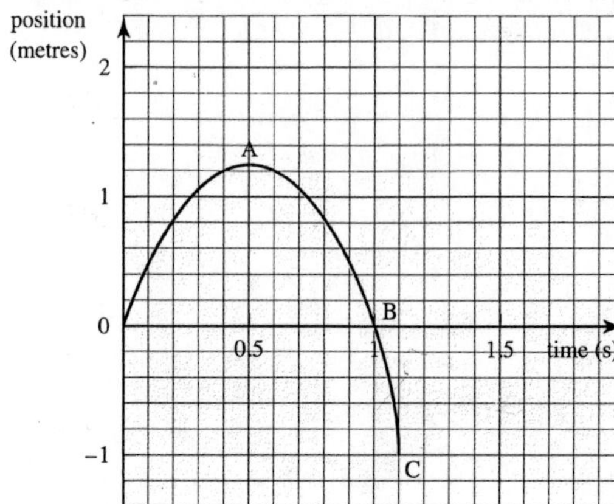

Figure 1.4

? The graph in figure 1.4 shows that the position is negative after one second (point B). What does this negative position mean?

Note

When drawing a graph it is very important to specify your axes carefully. Graphs showing motion usually have time along the horizontal axis. Then you have to decide where the origin is and which direction is positive on the vertical axis. In this graph the origin is at hand level and upwards is positive. The time is measured from the instant the marble leaves your hand.

Notation and units

As with most mathematics, you will see in this book that certain letters are commonly used to denote certain quantities. This makes things easier to follow. Here the letters used are:

- s, h, x, y and z for position
- t for time measured from a starting instant
- u and v for velocity
- a for acceleration.

The S.I. (Système International d'Unités) unit for *distance* is the metre (m), that for *time* is the second (s) and that for *mass* the kilogram (kg). Other units follow from these so speed is measured in metres per second, written $\mathrm{m\,s^{-1}}$.

EXERCISE 1A

1 When the origin for the motion of the marble (see figure 1.3) is on the ground, what is its position

(i) when it leaves your hand?

(ii) at the top?

2 A boy throws a ball vertically upwards so that its position y m at time t is as shown in the graph.

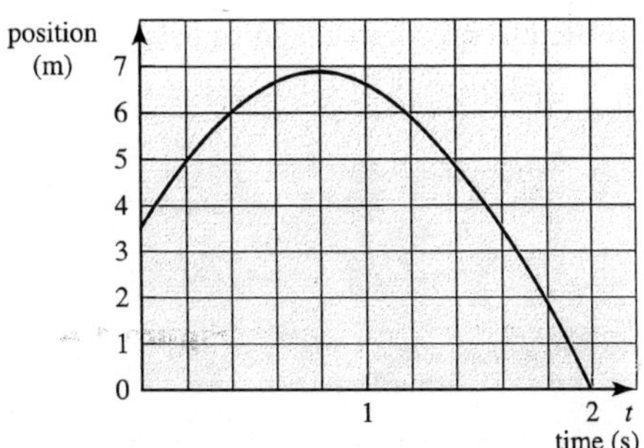

(i) Write down the position of the ball at times $t = 0$, 0.4, 0.8, 1.2, 1.6 and 2.

(ii) Calculate the displacement of the ball relative to its starting position at these times.

(iii) What is the total distance travelled

(a) during the first 0.8 s **(b)** during the 2 s of the motion?

3 The position of a particle moving along a straight horizontal groove is given by $x = 2 + t(t - 3)$ for $0 \leqslant t \leqslant 5$ where x is measured in metres and t in seconds.

(i) What is the position of the particle at times $t = 0$, 1, 1.5, 2, 3, 4 and 5?

(ii) Draw a diagram to show the path of the particle, marking its position at these times.

(iii) Find the displacement of the particle relative to its initial position at $t = 5$.

(iv) Calculate the total distance travelled during the motion.

4 For each of the following situations sketch a graph of position against time. Show clearly the origin and the positive direction.

(i) A stone is dropped from a bridge which is 40 m above a river.

(ii) A parachutist jumps from a helicopter which is hovering at 2000 m. She opens her parachute after 10 s of free fall.

(iii) A bungee jumper on the end of an elastic string jumps from a high bridge.

5 The diagram is a sketch of the position–time graph for a fairground ride.

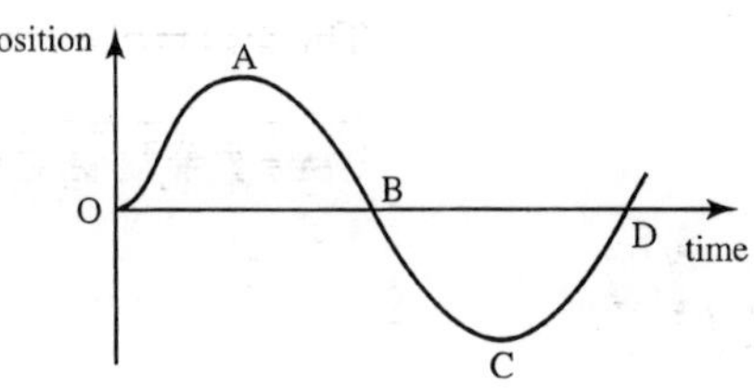

(i) Describe the motion, stating in particular what happens at O, A, B, C and D.

(ii) What type of ride is this?

Speed and velocity

Speed is a scalar quantity and does not involve direction. *Velocity* is the vector related to speed; its magnitude is the speed but it also has a direction. When an object is moving in the negative direction, its velocity is negative.

Amy has to post a letter on her way to college. The post box is 500 m east of her house and the college is 2.5 km to the west. Amy cycles at a steady speed of $10\,\mathrm{m\,s^{-1}}$ and takes 10 s at the post box to find the letter and post it.

Figure 1.5 shows Amy's journey using east as the positive direction. The distance of 2.5 km has been changed to metres so that the units are consistent.

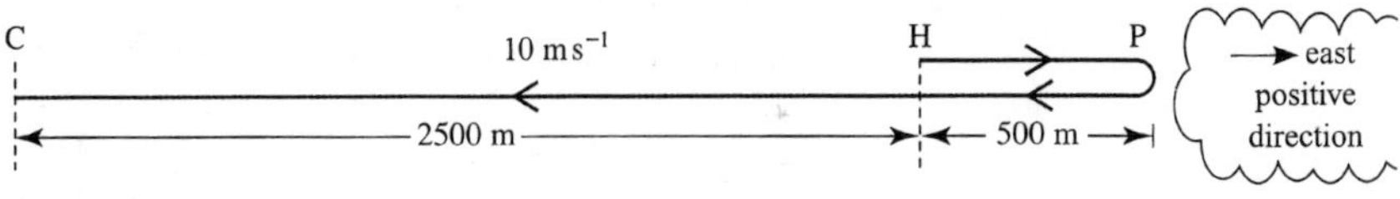

Figure 1.5

After she leaves the post box Amy is travelling west so her velocity is negative. It is $-10\,\mathrm{m\,s^{-1}}$.

The distances and times for the three parts of Amy's journey are:

	Distance	Time
Home to post box	500 m	$\frac{500}{10} = 50$ s
At post box	0 m	10 s
Post box to college	3000 m	$\frac{3000}{10} = 300$ s

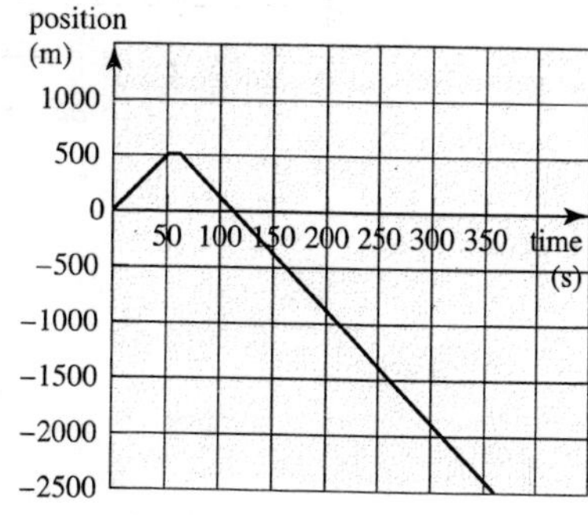

Figure 1.6

These can be used to draw the position–time graph using home as the origin, as in figure 1.6.

Calculate the gradient of the three portions of this graph. What conclusions can you draw?

The velocity is the rate at which the position changes.

- Velocity is represented by the gradient of the position–time graph.

Figure 1.7 is the velocity–time graph.

Note

By drawing the graphs below each other with the same horizontal scales, you can see how they correspond to each other.

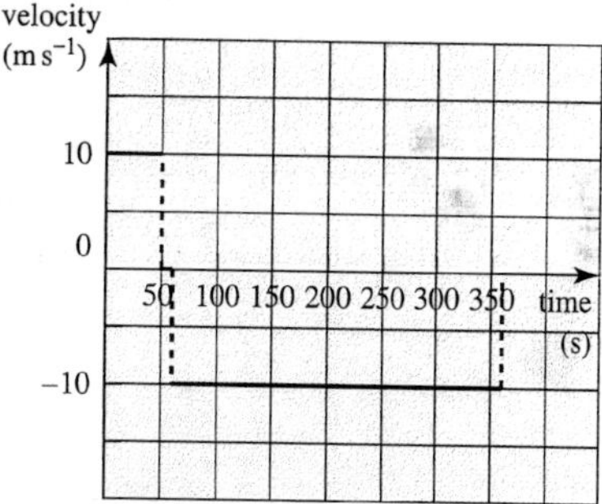

Figure 1.7

Distance–time graphs

Figure 1.8 is the distance–time graph of Amy's journey. It differs from the position–time graph because it shows how far she travels irrespective of her direction. There are no negative values.

The gradient of this graph represents Amy's speed rather than her velocity.

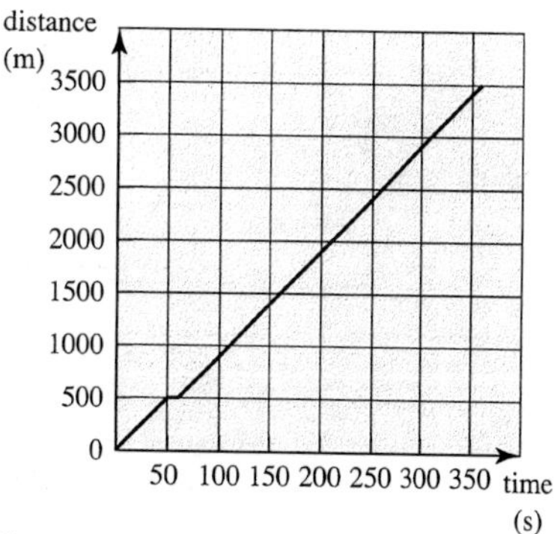

Figure 1.8

? It has been assumed that Amy starts and stops instantaneously. What would more realistic graphs look like? Would it make a lot of difference to the answers if you tried to be more realistic?

Average speed and average velocity

You can find Amy's average speed on her way to college by using the definition

- $\text{average speed} = \dfrac{\text{total distance travelled}}{\text{total time taken}}$

When the distance is in metres and the time in seconds, speed is found by dividing metres by seconds and is written as m s^{-1}. So Amy's average speed is

$$\frac{3500\,\text{m}}{360\,\text{s}} = 9.72\,\text{m s}^{-1}$$

Amy's average velocity is different. Her displacement from start to finish is −2500 m so

The college is in the negative direction.

- $\text{average velocity} = \dfrac{\text{displacement}}{\text{time taken}}$

$$= \frac{-2500}{360}$$

$$= -6.94\,\text{m s}^{-1}$$

If Amy had taken the same time to go straight from home to college at a steady speed, this steady speed would have been $6.94\,\text{m s}^{-1}$.

Velocity at an instant

The position–time graph for a marble thrown straight up into the air at $5\,\mathrm{m\,s^{-1}}$ is curved because the velocity is continually changing.

The velocity is represented by the gradient of the position–time graph. When a position–time graph is curved like this you can find *the velocity at an instant* of time by drawing a tangent as in figure 1.9.

The velocity at P is approximately

$$\frac{0.6}{0.25} = 2.4\,\mathrm{m\,s^{-1}}$$

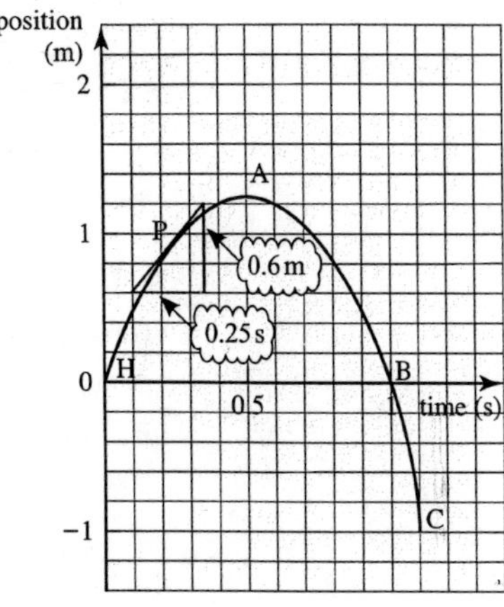

Figure 1.9

The velocity–time graph is shown in figure 1.10.

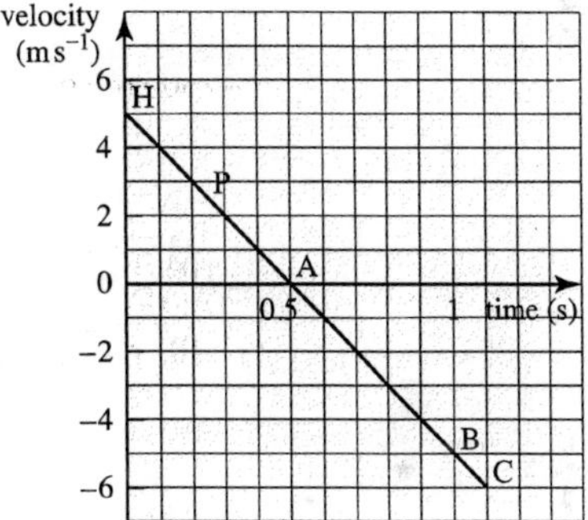

Figure 1.10

? What is the velocity at H, A, B and C? The speed of the marble increases after it reaches the top. What happens to the velocity?

At the point A, the velocity and gradient of the position–time graph are zero. We say the marble is *instantaneously at rest.* The velocity at H is positive because the marble is moving in the positive direction (upwards). The velocity at B and at C is negative because the marble is moving in the negative direction (downwards).

EXERCISE 1B

1 Draw a speed–time graph for Amy's journey on page 6.

2 The distance–time graph shows the relationship between distance travelled and time for a person who leaves home at 9.00 am, walks to a bus stop and catches a bus into town.

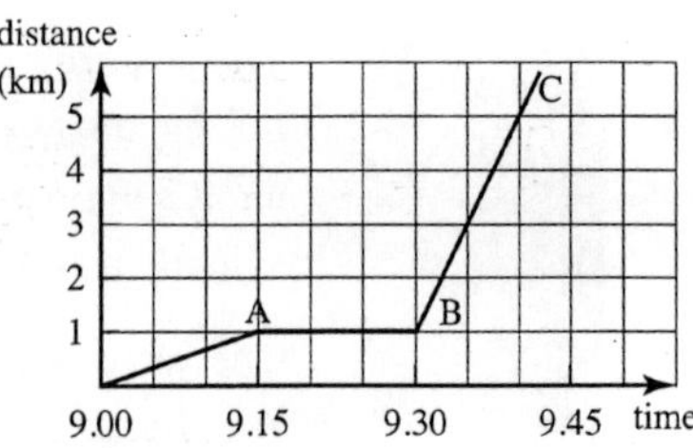

(i) Describe what is happening during the time from A to B.

(ii) The section BC is much steeper than OA; what does this tell you about the motion?

(iii) Draw the speed–time graph for the person.

(iv) What simplifications have been made in drawing these graphs?

3 For each of the following journeys find

(a) the initial and final positions

(b) the total displacement

(c) the total distance travelled

(d) the velocity and speed for each part of the journey

(e) the average velocity for the whole journey

(f) the average speed for the whole journey.

(i)

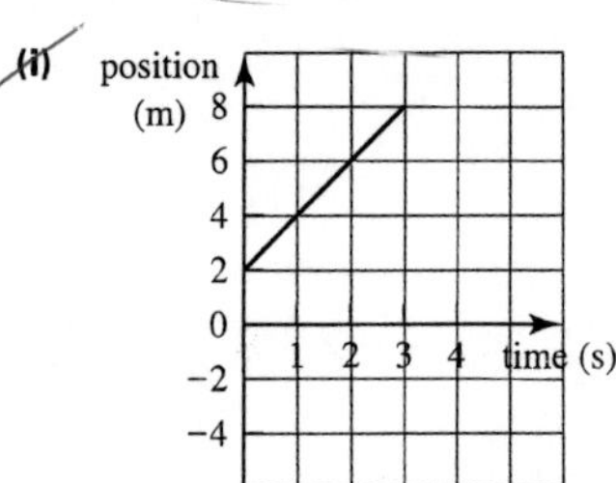

(ii)

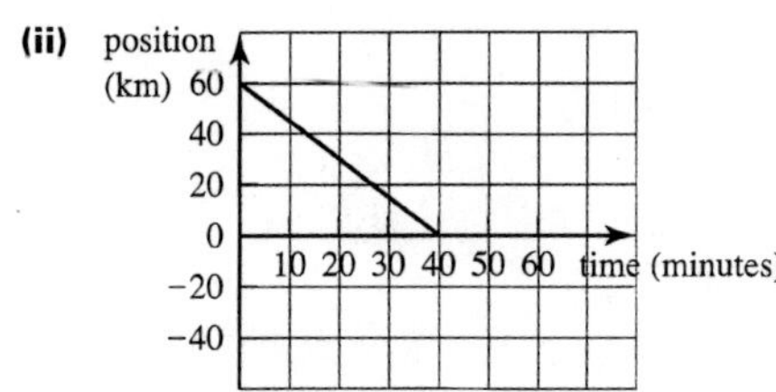

(iii)

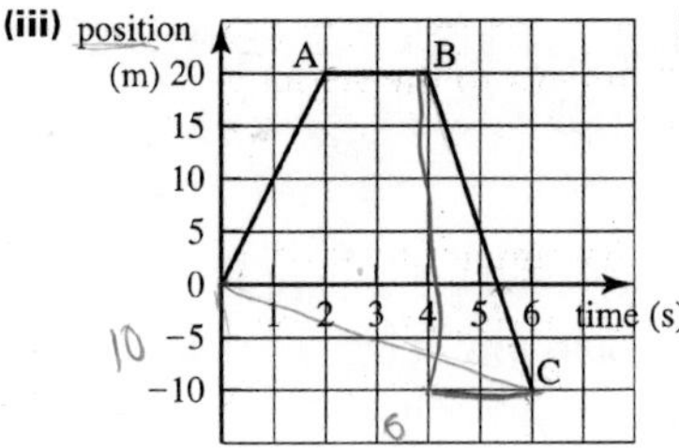

(iv)

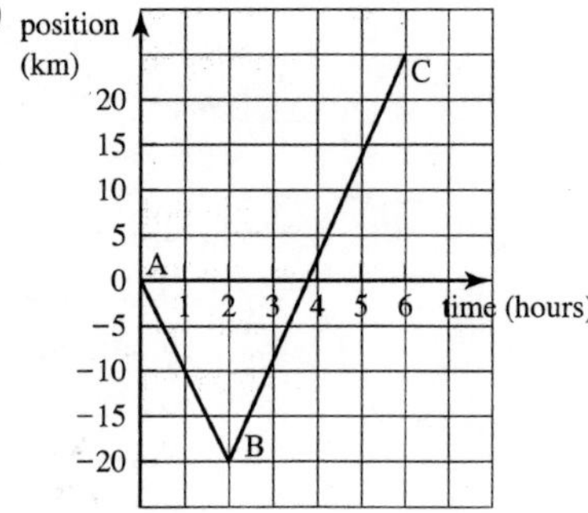

4 A plane flies from London to Toronto, a distance of 5700 km, at an average speed of $1280\,\text{km}\,\text{h}^{-1}$. It returns at an average speed of $1200\,\text{km}\,\text{h}^{-1}$. Find the average speed for the round trip.

Acceleration

In everyday language, the word 'accelerate' is usually used when an object speeds up and 'decelerate' when it slows down. The idea of deceleration is sometimes used in a similar way by mathematicians but in mathematics the word *acceleration* is used whenever there is a change in velocity, whether an object is speeding up, slowing down or changing direction. Acceleration is *the rate at which the velocity changes.*

Over a period of time

- $\text{average acceleration} = \dfrac{\text{change in velocity}}{\text{time taken}}$

Acceleration is represented by the gradient of a velocity–time graph. It is a vector and can take different signs in a similar way to velocity. This is illustrated by Tom's cycle journey which is shown in figure 1.11.

Tom turns on to the main road at $4\,\text{m s}^{-1}$, accelerates uniformly, maintains a constant speed and then slows down uniformly to stop when he reaches home.

Between A and B, Tom's velocity increases by $(10 - 4) = 6\,\text{m s}^{-1}$ in 6 seconds, that is by 1 metre per second every second.

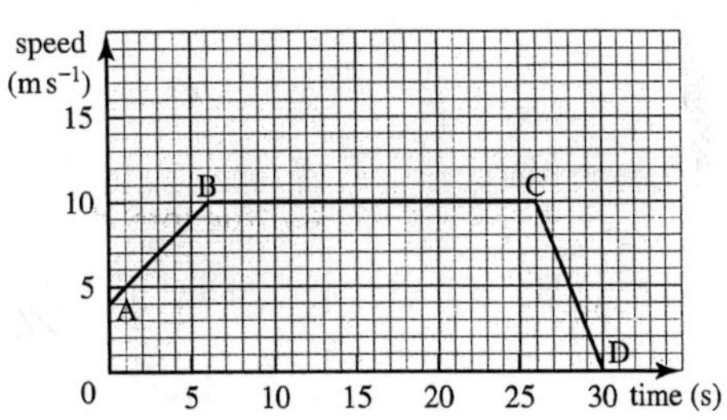

Figure 1.11

This acceleration is written as $1\,\text{m s}^{-2}$ (one metre per second squared) and is the gradient of AB.

From B to C $\quad$ acceleration $= 0\,\text{m s}^{-2}$ ← There is no change in velocity.

From C to D $\quad$ acceleration $= \dfrac{(0 - 10)}{(30 - 26)} = -2.5\,\text{m s}^{-2}$

From C to D, Tom is slowing down while still moving in the positive direction towards home, so his acceleration, the gradient of the graph, is negative.

The sign of acceleration

Think again about the marble thrown up into the air with a speed of $5\,\text{m s}^{-1}$.

Figure 1.12 represents the velocity when *upwards* is taken as the positive direction and shows that the velocity *decreases* from $+5\,\text{m s}^{-1}$ to $5\,\text{m s}^{-1}$ in 1 second.

This means that the gradient, and hence the acceleration, is *negative.* It is $-10\,\text{m s}^{-2}$. (You might recognise the number 10 as an approximation to *g*. See Chapter 2 page 28.)

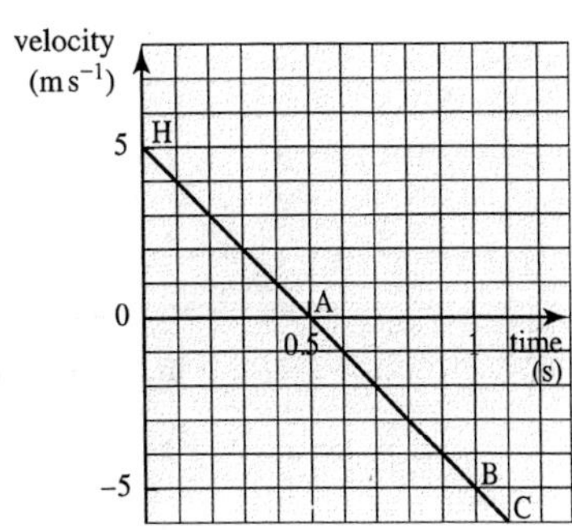

Figure 1.12

A car accelerates away from a set of traffic lights. It accelerates to a maximum speed and at that instant starts to slow down to stop at a second set of lights. Which of the graphs below could represent

(i) the distance–time graph

(ii) the velocity–time graph

(iii) the acceleration–time graph of its motion?

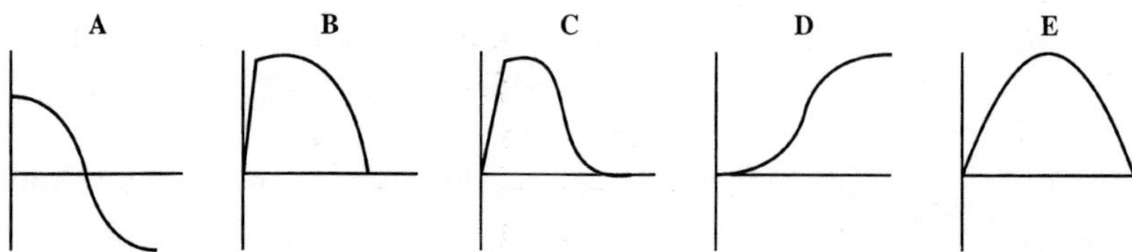

Figure 1.13

EXERCISE 1C

1 (i) Calculate the acceleration for each part of the following journey.

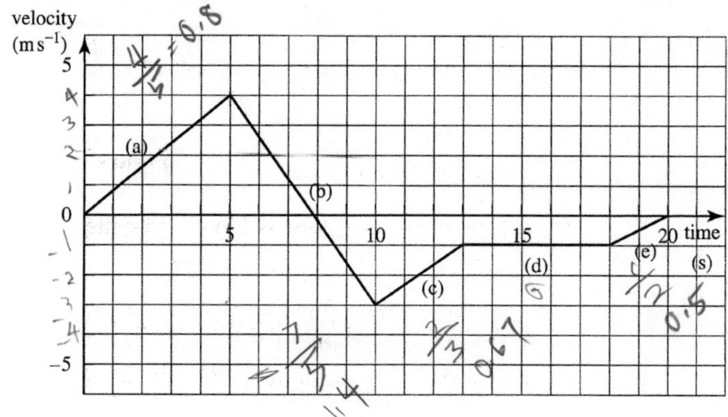

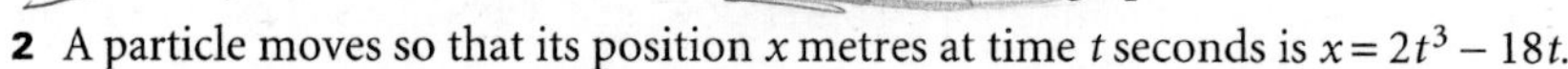

(ii) Use your results to sketch an acceleration–time graph.

2 A particle moves so that its position x metres at time t seconds is $x = 2t^3 - 18t$.

(i) Calculate the position of the particle at times $t = 0, 1, 2, 3$ and 4.

(ii) Draw a diagram showing the position of the particle at these times.

(iii) Sketch a graph of the position against time.

(iv) State the times when the particle is at the origin and describe the direction in which it is moving at those times.

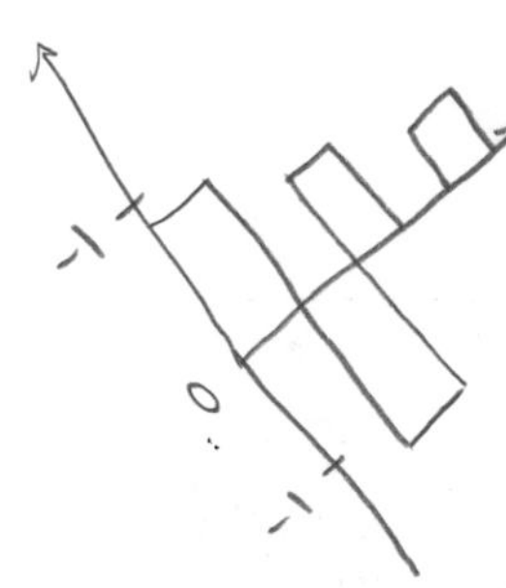

3 A train takes 45 minutes to complete its 24 kilometre trip. It stops for 1 minute at each of 7 stations during the trip.

(i) Calculate the average speed of the train.

(ii) What would be the average speed if the stop at each station was reduced to 20 seconds?

4 When Louise is planning car journeys she reckons that she can cover distances along main roads at roughly $100\,\mathrm{km\,h^{-1}}$ and those in towns at $30\,\mathrm{km\,h^{-1}}$.

(i) Find her average speed for each of the following journeys.

(a) 20 km on main roads and then 10 km in a town

(b) 150 km on main roads and then 2 km in a town

(c) 20 km on main roads and then 20 km in a town

(ii) In what circumstances would her average speed be $65\,\mathrm{km\,h^{-1}}$?

5 A lift travels up and down between the ground floor (G) and the roof garden (R) of a hotel. It starts from rest, takes 5 s to increase its speed uniformly to $2\,\mathrm{m\,s^{-1}}$, maintains this speed for 5 s and then slows down uniformly to rest in another 5 s. In the following questions, use upwards as positive.

(i) Sketch a velocity–time graph for the journey from G to R.

On one occasion the lift stops for 5 s at R before returning to G.

(ii) Sketch a velocity–time graph for this journey from G to R and back.

(iii) Calculate the acceleration for each 5 s interval. Take care with the signs.

(iv) Sketch an acceleration–time graph for this journey.

6 A film of a dragster doing a 400 m run from a standing start yields the following positions at 1 second intervals.

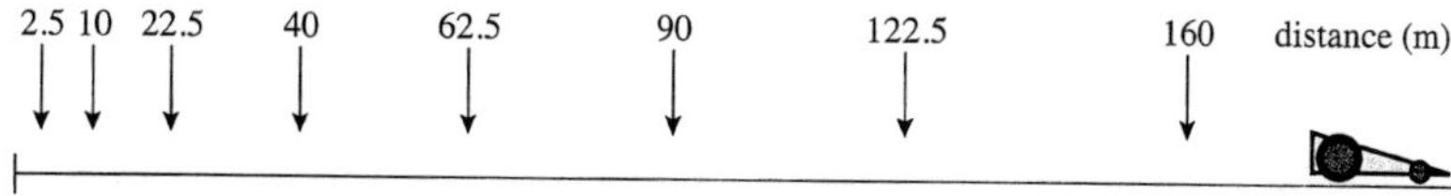

(i) Draw a displacement–time graph of its motion.

(ii) Use your graph to help you to sketch

(a) the velocity–time graph

(b) the acceleration–time graph.

Using areas to find distances and displacements

These graphs model the motion of a stone falling from rest.

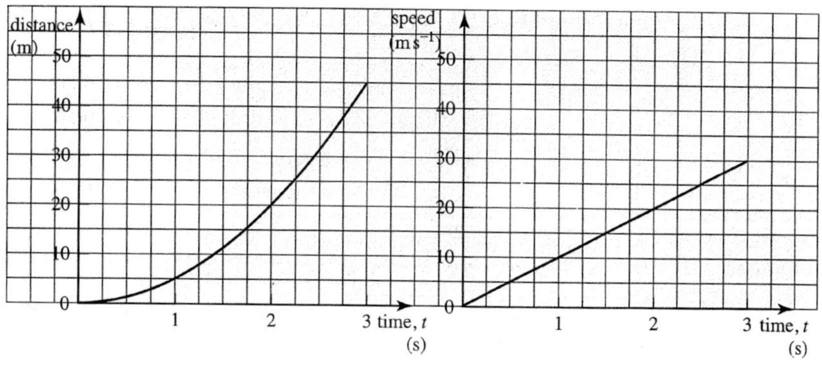

Figure 1.14 **Figure 1.15**

? Calculate the area between the speed–time graph and the time axis from

(i) $t = 0$ to 1 **(ii)** $t = 0$ to 2 **(iii)** $t = 0$ to 3.

Compare your answers with the distance that the stone has fallen, shown on the distance–time graph, at $t = 1$, 2 and 3. What conclusions do you reach?

- The area between a speed–time graph and the time axis represents the distance travelled.

There is further evidence for this if you consider the units on the graphs. Multiplying metres per second by seconds gives metres. A full justification relies on the calculus methods you will learn in Chapter 7.

Finding the area under speed–time graphs

Many of these graphs consist of straight-line sections. The area is easily found by splitting it up into triangles, rectangles or trapezia.

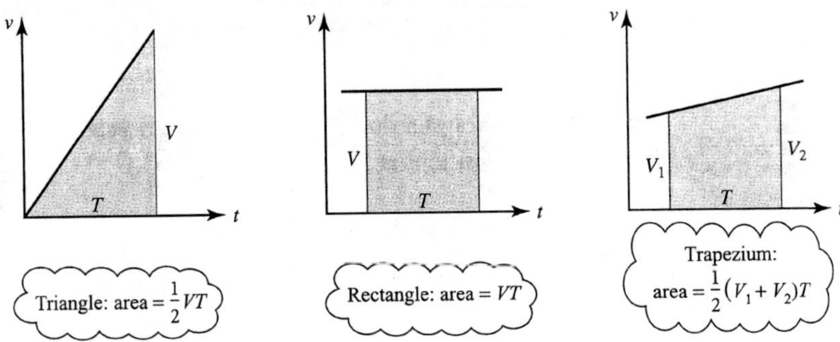

Figure 1.16

EXAMPLE 1.1

The graph shows Hinesh's journey from the time he turns on to the main road until he arrives home. How far does Hinesh cycle?

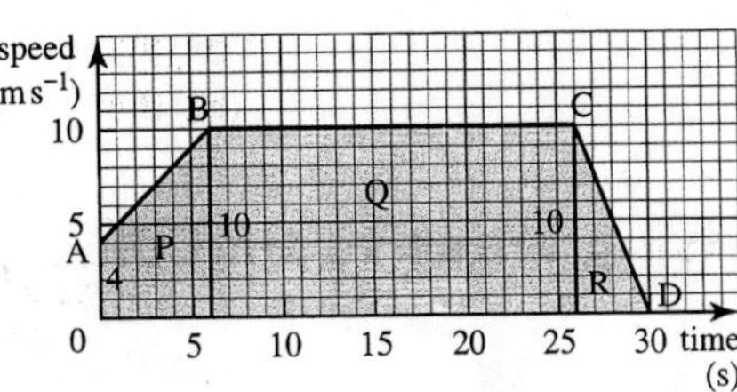

Figure 1.17

SOLUTION

Split the area under the speed–time graph into three regions.

P	trapezium:	area $= \frac{1}{2}(4 + 10) \times 6 =$	42 m
Q	rectangle:	area $= 10 \times 20$	= 200 m
R	triangle:	area $= \frac{1}{2} \times 10 \times 4$	= 20 m
		total area	= 262 m

Hinesh cycles 262 m.

? What is the meaning of the area between a velocity–time graph and the time axis?

The area between a velocity–time graph and the time axis

EXAMPLE 1.2 Sunil walks east for 6 s at $2\,\mathrm{m\,s^{-1}}$ then west for 2 s at $1\,\mathrm{m\,s^{-1}}$. Draw

(i) a diagram of the journey
(ii) the speed–time graph
(iii) the velocity–time graph.

Interpret the area under each graph.

SOLUTION

(i) Sunil's journey is illustrated below.

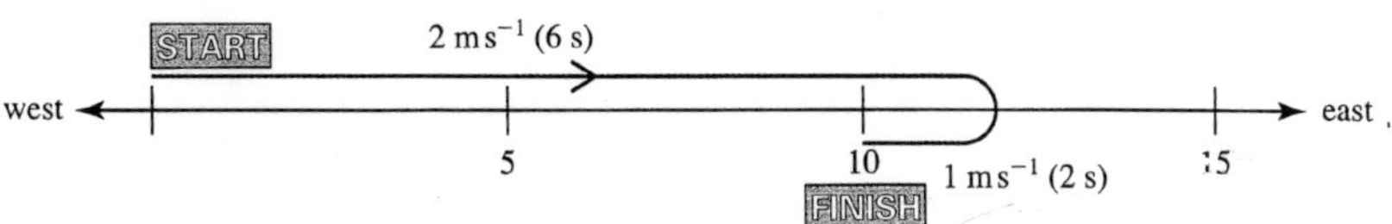

Figure 1.18

(ii) Speed–time graph

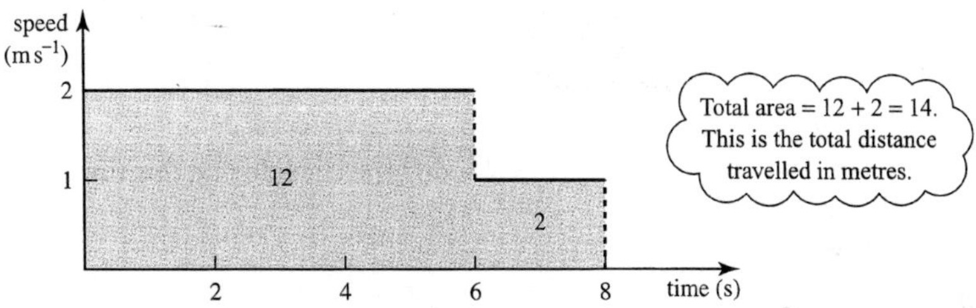

Figure 1.19

(iii) Velocity–time graph

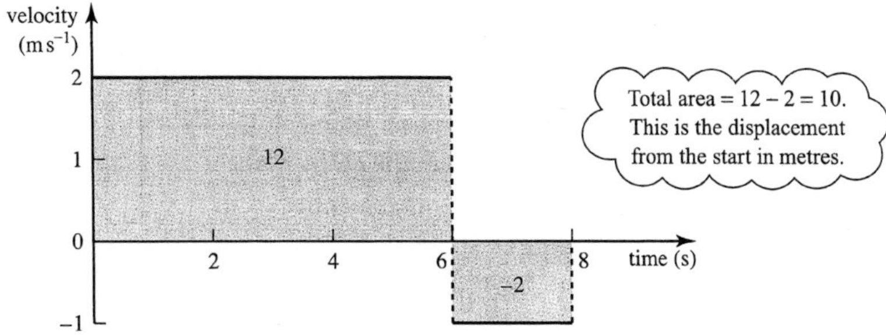

Figure 1.20

- The area between a velocity–time graph and the time axis represents the change in position, that is the displacement.

When the velocity is negative, the area is below the time axis and represents a displacement in the negative direction, west in this case.

Estimating areas

Sometimes the velocity–time graph does not consist of straight lines so you have to make the best estimate you can by counting the squares underneath it or by replacing the curve by a number of straight lines as for the trapezium rule (see *Pure Mathematics 2,* Chapter 5).

This speed–time graph shows the motion of a dog over a 60 s period.

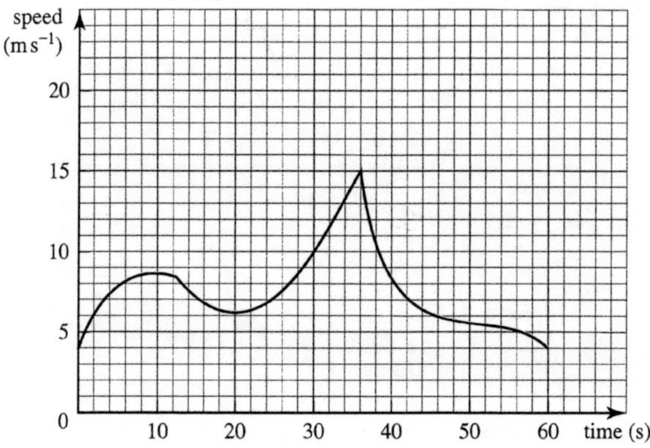

Figure 1.21

Estimate how far the dog travelled during this time.

EXAMPLE 1.3

On the London underground, Oxford Circus and Piccadilly Circus are 0.8 km apart. A train accelerates uniformly to a maximum speed when leaving Oxford Circus and maintains this speed for 90 s before decelerating uniformly to stop at Piccadilly Circus. The whole journey takes 2 minutes. Find the maximum speed.

SOLUTION

The sketch of the speed–time graph of the journey shows the given information, with suitable units. The maximum speed is $v\,\mathrm{m\,s^{-1}}$.

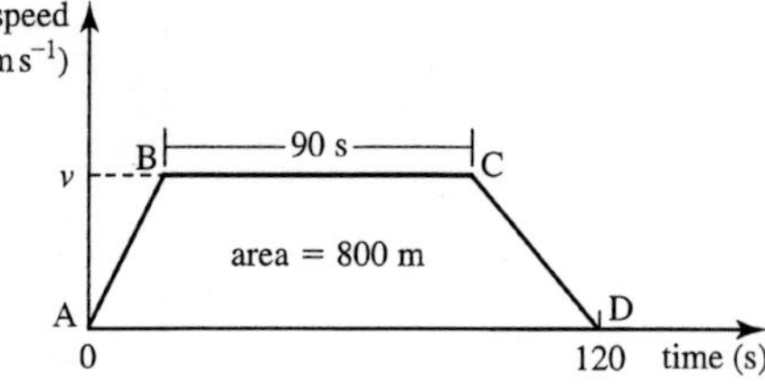

Figure 1.22

$$\text{The area is } \tfrac{1}{2}(120 + 90) \times v = 800$$

$$v = \frac{800}{105}$$

$$= 7.619$$

The maximum speed of the train is $7.6\,\mathrm{m\,s^{-1}}$ (to 2 s.f.).

Does it matter how long the train takes to speed up and slow down?

EXERCISE 1D

1 The graphs show the speeds of two cars travelling along a street.

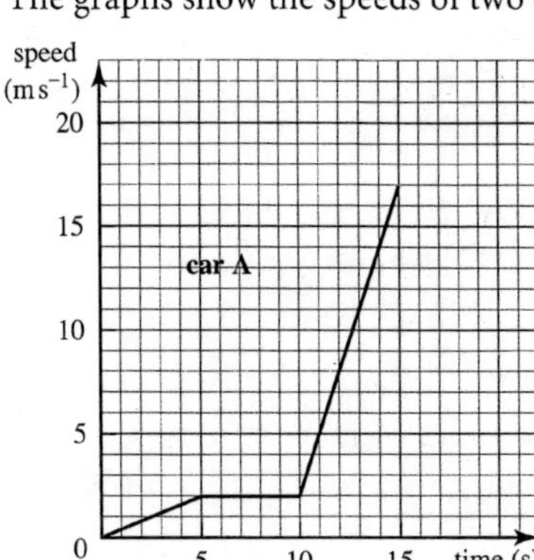

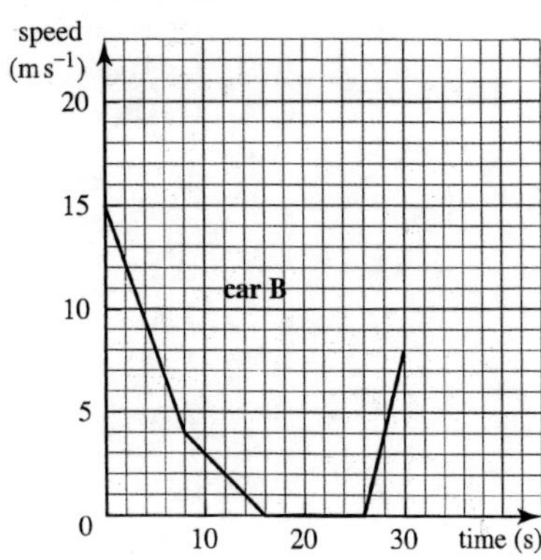

For each car find

(i) the acceleration for each part of its motion

(ii) the total distance it travels in the given time

(iii) its average speed.

2 The graph shows the speed of a lorry when it enters a very busy road.

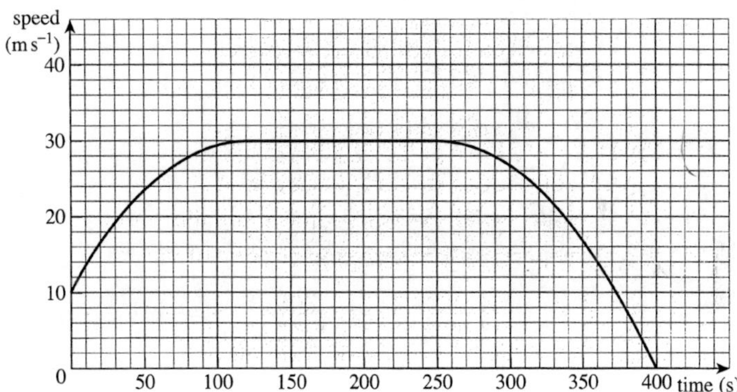

(i) Describe the journey over this time.

(ii) Use a ruler to make a tangent to the graph and hence estimate the acceleration at the beginning and end of the period.

(iii) Estimate the distance travelled and the average speed.

3 A train leaves a station where it has been at rest and picks up speed at a constant rate for 60 s. It then remains at a constant speed of $17\,m\,s^{-1}$ for 60 s before it begins to slow down uniformly as it approaches a set of signals. After 45 s it is travelling at $10\,m\,s^{-1}$ and the signal changes. The train again increases speed uniformly for 75 s until it reaches a speed of $20\,m\,s^{-1}$. A second set of signals then orders the train to stop, which it does after slowing down uniformly for 30 s.

(i) Draw a speed–time graph for the train.

(ii) Use your graph to find the distance that it has travelled from the station.

4 When a parachutist jumps from a helicopter hovering above an airfield her speed increases at a constant rate to $28\,\text{m}\,\text{s}^{-1}$ in the first 3 s of her fall. It then decreases uniformly to $8\,\text{m}\,\text{s}^{-1}$ in a further 6 s, remaining constant until she reaches the ground.

(i) Sketch a speed–time graph for the parachutist.

(ii) Find the height of the plane when the parachutist jumps out if the complete jump takes 1 minute.

5 A car is moving at $20\,\text{m}\,\text{s}^{-1}$ when it begins to increase speed. Every 10 s it gains $5\,\text{m}\,\text{s}^{-1}$ until it reaches its maximum speed of $50\,\text{m}\,\text{s}^{-1}$ which it retains.

(i) Draw the speed–time graph of the car.

(ii) When does the car reach its maximum speed of $50\,\text{m}\,\text{s}^{-1}$?

(iii) Find the distance travelled by the car after 150 s.

(iv) Write down expressions for the speed of the car t seconds after it begins to speed up.

6 A train takes 10 minutes to travel between two stations. The train accelerates at a rate of $0.5\,\text{m}\,\text{s}^{-2}$ for 30 s. It then travels at a constant speed and is finally brought to rest in 15 s with a constant deceleration.

(i) Sketch a velocity–time graph for the journey.

(ii) Find the steady speed, the rate of deceleration and the distance between the two stations.

7 A train was scheduled to travel at $50\,\text{m}\,\text{s}^{-1}$ for 15 minutes on part of its journey. The velocity–time graph illustrates the actual progress of the train which was forced to stop because of signals.

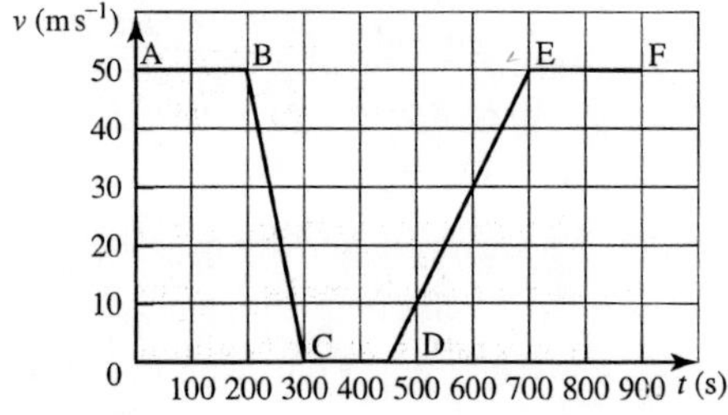

(i) Without carrying out any calculations, describe what was happening to the train in each of the stages BC, CD and DE.

(ii) Find the deceleration of the train while it was slowing down and the distance travelled during this stage.

(iii) Find the acceleration of the train when it starts off again and the distance travelled during this stage.

(iv) Calculate by how long the stop will have delayed the train.

(v) Sketch the distance–time graph for the journey between A and F, marking the points A, B, C, D, E and F. [MEI]

8 A car is travelling at $36\,\text{km}\,\text{h}^{-1}$ when the driver has to perform an emergency stop. During the time the driver takes to appreciate the situation and apply the brakes the car has travelled 7 m ('thinking distance'). It then pulls up with constant deceleration in a further 8 m ('braking distance') giving a total stopping distance of 15 m.

(i) Find the initial speed of the car in metres per second and the time that the driver takes to react.

(ii) Sketch the velocity–time graph for the car.

(iii) Calculate the deceleration once the car starts braking.

(iv) What is the stopping distance for a car travelling at $60\,\text{km}\,\text{h}^{-1}$ if the reaction time and the deceleration are the same as before?

9 The diagram shows the displacement–time graph for a car's journey. The graph consists of two curved parts AB and CD, and a straight line BC. The line BC is a tangent to the curve AB at B and a tangent to the curve CD at C. The gradient of the curves at $t = 0$ and $t = 600$ is zero, and the acceleration of the car is constant for $0 < t < 80$ and for $560 < t < 600$. The displacement of the car is 400 m when $t = 80$.

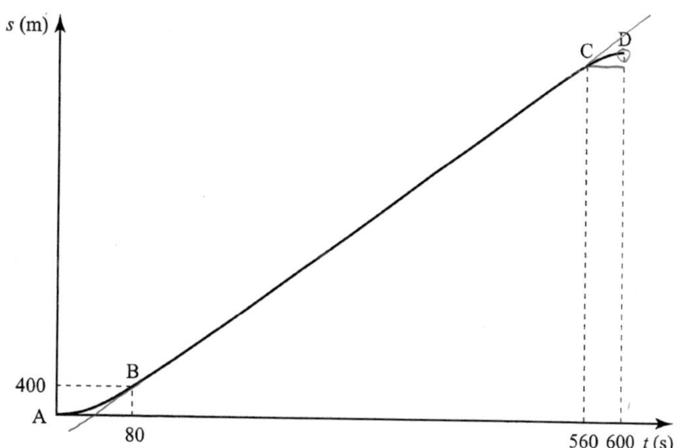

(i) Sketch the velocity–time graph for the journey.

(ii) Find the velocity at $t = 80$.

(iii) Find the total distance for the journey.

(iv) Find the acceleration of the car for $0 < t < 80$.

[Cambridge AS & A Level Mathematics 9709, Paper 4 Q5 November 2005]

10 A train travels from A to B, a distance of 20 000 m, taking 1000 s. The journey has three stages. In the first stage the train starts from rest at A and accelerates uniformly until its speed is $V \text{ m s}^{-1}$. In the second stage the train travels at constant speed $V \text{ m s}^{-1}$ for 600 s. During the third stage of the journey the train decelerates uniformly, coming to rest at B.

(i) Sketch the velocity–time graph for the train's journey.

(ii) Find the value of V.

(iii) Given that the acceleration of the train during the first stage of the journey is 0.15 m s^{-2}, find the distance travelled by the train during the third stage of the journey.

[Cambridge AS & A Level Mathematics 9709, Paper 4 Q6 November 2008]

11 The diagram shows the velocity–time graph for the motion of a machine's cutting tool. The graph consists of five straight line segments. The tool moves forward for 8 s while cutting and then takes 3 s to return to its starting position.

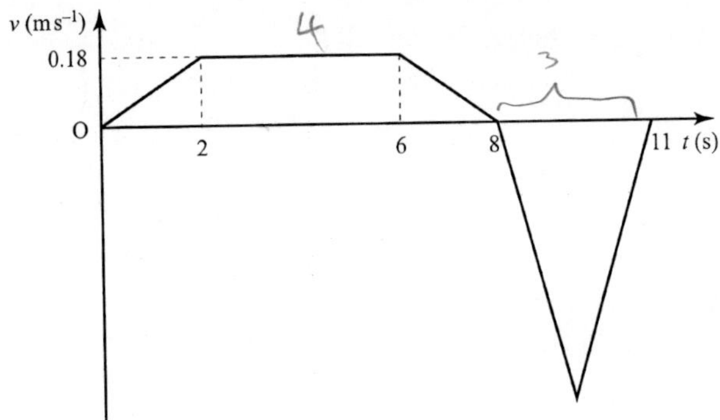

Find

(i) the acceleration of the tool during the first 2 s of the motion,

(ii) the distance the tool moves forward while cutting,

(iii) the greatest speed of the tool during the return to its starting position.

[Cambridge AS & A Level Mathematics 9709, Paper 41 Q2 June 2010]

INVESTIGATION

Train journey

If you look out of a train window in many countries you will see distance markers beside the track (in the UK they are every quarter of a mile). Take a train journey and record the time as you go past each marker. Use your figures to draw distance–time, speed–time and acceleration–time graphs. What can you conclude about the greatest acceleration, deceleration and speed of the train?

KEY POINTS

1 Vectors (with magnitude and direction) | **Scalars** (magnitude only)

Vectors (with magnitude and direction)	**Scalars** (magnitude only)
Displacement	Distance
Position – displacement from a fixed origin	
Velocity – rate of change of position	Speed – magnitude of velocity
Acceleration – rate of change of velocity	
	Time

- *Vertical* is towards the centre of the earth; *horizontal* is perpendicular to vertical.

2 Diagrams

- Motion along a line can be illustrated vertically or horizontally (as shown).

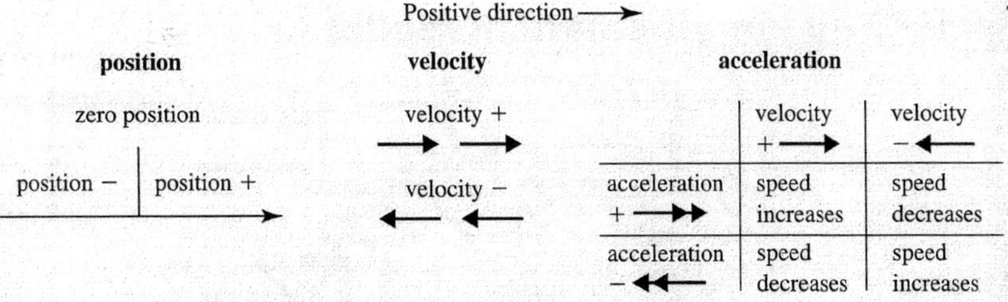

- $\text{Average speed} = \dfrac{\text{total distance travelled}}{\text{total time taken}}$
- $\text{Average velocity} = \dfrac{\text{displacement}}{\text{time taken}}$
- $\text{Average acceleration} = \dfrac{\text{change in velocity}}{\text{time taken}}$

3 Graphs

- *Position–time*
- *Velocity–time*

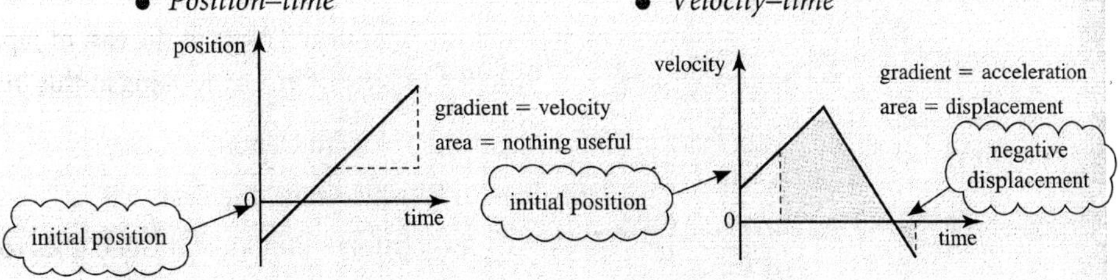

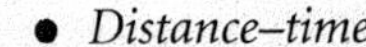

- *Distance–time*
- *Speed–time*

distance

gradient = speed

area = nothing useful

0

time

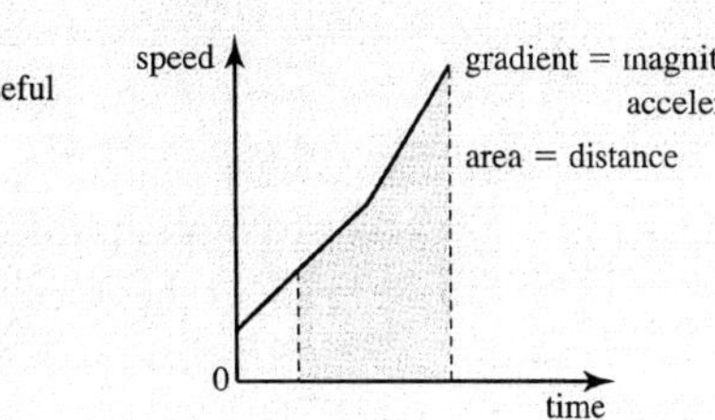

2 The constant acceleration formulae

The poetry of motion! The real way to travel! The only way to travel! Here today – in next week tomorrow! Villages skipped, towns and cities jumped – always somebody else's horizon! O bliss! O poop-poop! O my!

Kenneth Grahame (The Wind in the Willows)

Setting up a mathematical model

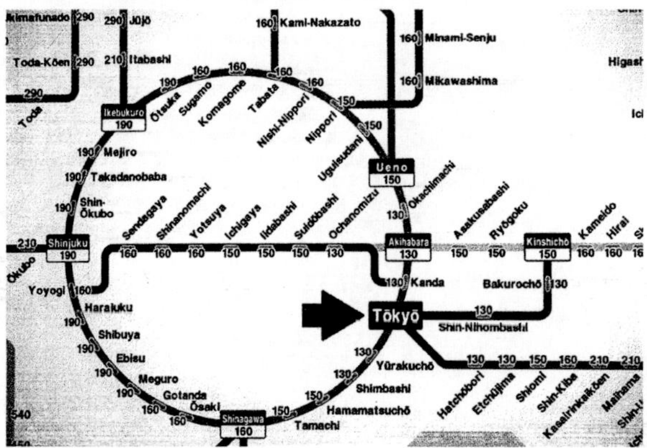

Figure 2.1

Figure 2.1 shows a map of railway lines near Tokyo in the east of Japan. Which of the following statements can you be sure of just by looking at this map?

(i) Aldhabara is on the line from Tokyo to Ueno.

(ii) The line from Aldhabara to Kinshico runs due East.

(iii) The line through Tokyo, Shinagawa, Shinjuku and Ueno goes round a perfect circle.

(iv) Shinjuku is a railway junction.

This is a *diagrammatic model* of the railway system which gives essential though by no means all the information you need for planning train journeys. You can be sure about the places a line passes through but distances and directions are only approximate and if you compare this map with an ordinary map you will see that statements **(ii)** and **(iii)** are false.

Making simplifying assumptions

When setting up a model, you first need to decide what is essential. For example, what would you take into account and what would you ignore when considering the motion of a car travelling from San Francisco to Los Angeles?

You will need to know the distance and the time taken for parts of the journey, but you might decide to ignore the dimensions of the car and the motion of the wheels. You would then be using the idea of a *particle* to model the car. *A particle has no dimensions.*

You might also decide to ignore the bends in the road and its width, and so treat it as a *straight line with only one dimension.* A length along the line would represent a length along the road in the same way as a piece of thread following a road on a map might be straightened out to measure its length.

You might decide to split the journey up into parts and assume that the speed is constant over these parts.

The process of making decisions like these is called *making simplifying assumptions* and is the first stage of setting up a *mathematical model* of the situation.

Defining the variables and setting up the equations

The next step in setting up a mathematical model is to *define the variables* with suitable units. These will depend on the problem you are trying to solve. Suppose you want to know where you ought to be at certain times in order to maintain a good average speed between San Francisco and Los Angeles. You might define your variables as follows:

- the total time since the car left San Francisco is t hours
- the distance from San Francisco at time t is x km
- the average speed up to time t is v km h^{-1}.

Then, at Kettleman City $t = t_1$ and $x = x_1$; etc.

You can then *set up equations* and go through the mathematics required to solve the problem. Remember to check that your answer is sensible. If it isn't, you might have made a mistake in your arithmetic or your simplifying assumptions might need reconsideration.

The theories of mechanics that you will learn about in this course, and indeed any other studies in which mathematics is applied, are based on mathematical models of the real world. When necessary, these models can become more complex as your knowledge increases.

 The simplest form of the San Francisco to Los Angeles model assumes that the speed remains constant over sections of the journey. Is this reasonable?

For a much shorter journey, you might need to take into account changes in the speed of the car. This chapter develops the mathematics required when an object can be modelled as a *particle moving in a straight line with constant acceleration.* In most real situations this is only the case for part of the motion – you wouldn't expect a car to continue accelerating at the same rate for very long – but it is a very useful model to use as a first approximation over a short time.

The constant acceleration formulae

The velocity–time graph shows part of the motion of a car on a fairground ride as it picks up speed. The graph is a straight line so the velocity increases at a constant rate and the car has a constant acceleration which is equal to the gradient of the graph.

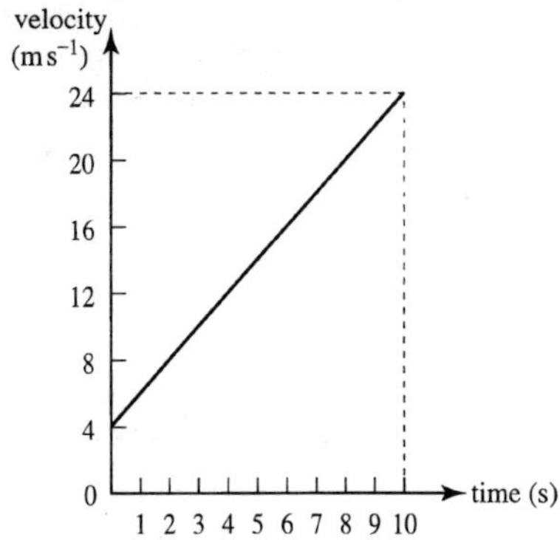

Figure 2.2

The velocity increases from $4\,\mathrm{m\,s^{-1}}$ to $24\,\mathrm{m\,s^{-1}}$ in 10 s so its acceleration is

$$\frac{24-4}{10} = 2\mathrm{ms^{-2}}.$$

In general, when the initial velocity is $u\,\mathrm{m\,s^{-1}}$ and the velocity a time t s later is $v\,\mathrm{m\,s^{-1}}$, as in figure 2.3 (on the next page), the increase in velocity is $(v-u)\,\mathrm{m\,s^{-1}}$ and the constant acceleration $a\,\mathrm{m\,s^{-2}}$ is given by

$$\frac{v-u}{t} = a$$

So $\quad v - u = at$

$$v = u + at. \qquad ①$$

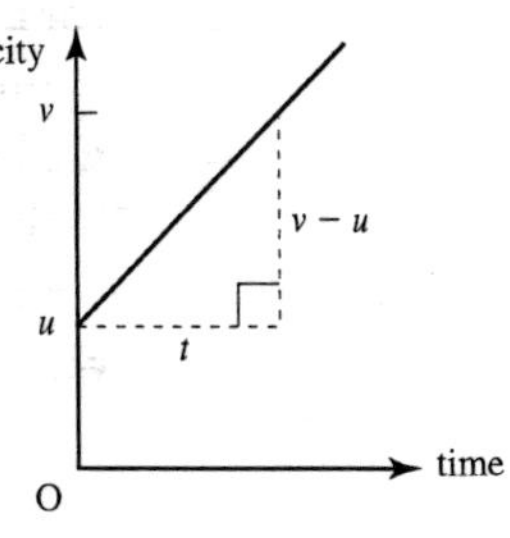

Figure 2.3

The area under the graph represents the distance travelled. For the fairground car, that is represented by a trapezium of area

$$\frac{(4+24)}{2} \times 10 = 140\,\text{m}.$$

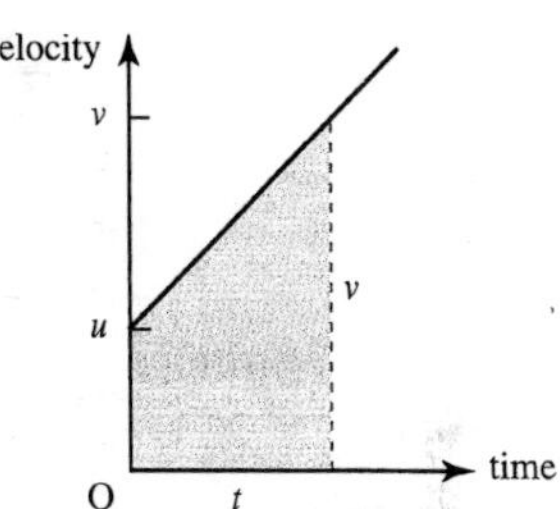

Figure 2.4

In the general situation, the area represents the displacement s metres and is

$$s = \frac{(u+v)}{2} \times t \qquad ②$$

❓ The two equations, ① and ②, can be used as formulae for solving problems when the acceleration is constant. Check that they work for the fairground ride.

There are other useful formulae as well. For example, you might want to find the displacement, s, without involving v in your calculations. This can be done by looking at the area under the velocity–time graph in a different way, using the rectangle R and the triangle T (see figure 2.5).

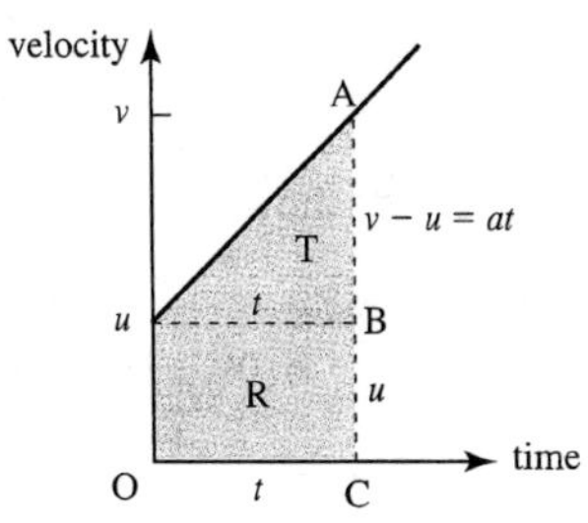

Figure 2.5

$$AC = v \text{ and } BC = u$$

so $\quad AB = v - u$

$$= at \qquad \text{from equation ①}$$

$$\text{total area} = \text{area of R} + \text{area of T}$$

so $\quad s = ut + \frac{1}{2} \times t \times at$

Giving $\quad s = ut + \frac{1}{2}at^2 \qquad ③$

To find a formula which does not involve t, you need to eliminate t. One way to do this is first to rewrite equations ① and ② as

$$v - u = at \quad \text{and} \quad v + u = \frac{2s}{t}$$

and then multiplying them gives

$$(v-u)(v+u) = at \times \frac{2s}{t}$$

$$v^2 - u^2 = 2as$$

$$v^2 = u^2 + 2as \qquad ④$$

You might have seen the equations ① to ④ before. They are sometimes called the *suvat* equations or formulae and they can be used whenever an object can be assumed to be moving with *constant acceleration.*

When solving problems it is important to remember the requirement for constant acceleration and also to remember to specify positive and negative directions clearly.

EXAMPLE 2.1

A bus leaving a bus stop accelerates at $0.8\,\text{m}\,\text{s}^{-2}$ for 5 s and then travels at a constant speed for 2 minutes before slowing down uniformly at $0.4\,\text{m}\,\text{s}^{-2}$ to come to rest at the next bus stop. Calculate

(i) the constant speed

(ii) the distance travelled while the bus is accelerating

(iii) the total distance travelled.

SOLUTION

(i) The diagram shows the information for the first part of the motion.

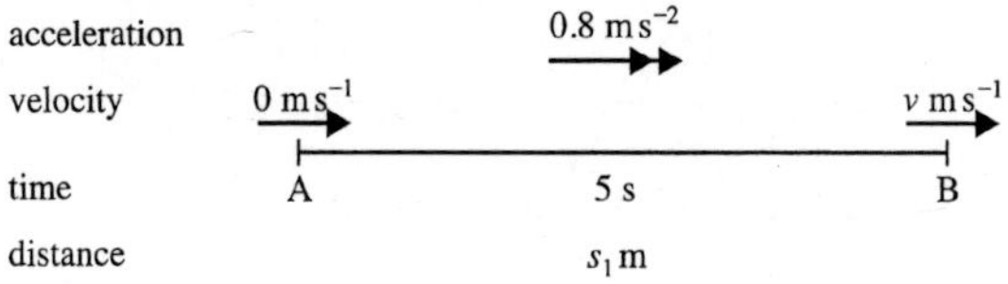

Figure 2.6

Let the constant speed be $v\,\mathrm{m\,s^{-1}}$.

$u = 0,\ a = 0.8,\ t = 5$, so use $v = u + at$

$v = 0 + 0.8 \times 5$
$= 4$

Want v
know $u = 0,\ t = 5,\ a = 0.8$
$v^2 = u^2 + 2as$ ✗
$v = u + at$ ✓

The constant speed is $4\,\mathrm{m\,s^{-1}}$.

Use the suffix '$_1$' because there are three distances to be found in this question.

(ii) Let the distance travelled be s_1 m.

$u = 0,\ a = 0.8,\ t = 5$, so use $s = ut + \frac{1}{2}at^2$

$s_1 = 0 + \frac{1}{2} \times 0.8 \times 5^2$
$= 10$

Want s
know $u = 0,\ t = 5,\ a = 0.8$
$s = \frac{1}{2}(u + v)t$ ✗
$s = ut + \frac{1}{2}at^2$ ✓

The bus accelerates over 10 m.

(iii) The diagram gives all the information for the rest of the journey.

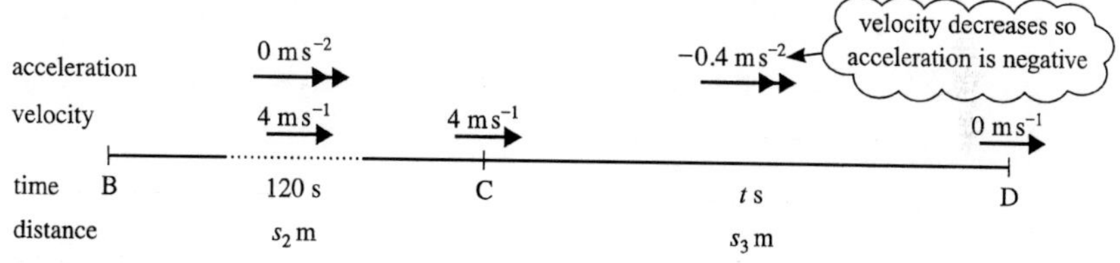

Figure 2.7

Between B and C the velocity is constant so the distance travelled is $4 \times 120 = 480$ m.

Let the distance between C and D be s_3 m.

$u = 4,\ a = -0.4,\ v = 0$, so use $v^2 = u^2 + 2as$

$$0 = 16 + 2(-0.4)s_3$$
$$0.8s_3 = 16$$
$$s_3 = 20$$

Want s
know $u = 4,\ a = -0.4,\ v = 0$
$v = u + at$ ✗
$s = ut + \frac{1}{2}at^2$ ✗
$s = \frac{1}{2}(u + v)t$ ✗
$v^2 = u^2 + 2as$ ✓

Distance taken to slow down = 20 m

The total distance travelled is
$(10 + 480 + 20)$ m = 510 m.

Units in the *suvat* formulae

Constant acceleration usually takes place over short periods of time so it is best to use $\mathrm{m\,s^{-2}}$ for this. When you don't need to use a value for the acceleration you can, if you wish, use the *suvat* formulae with other units provided they are consistent. This is shown in the next example.

EXAMPLE 2.2

When leaving a town, a car accelerates from $30\,\mathrm{km\,h^{-1}}$ to $60\,\mathrm{km\,h^{-1}}$ in 5 s. Assuming the acceleration is constant, find the distance travelled in this time.

SOLUTION

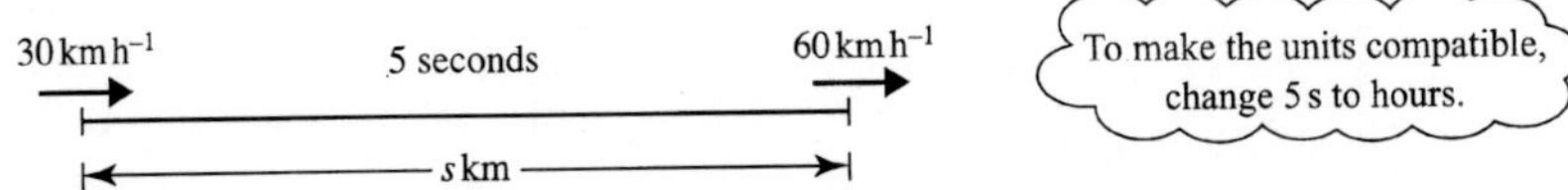

Figure 2.8

Let the distance travelled be s km. You want s and are given $u = 30$, $v = 60$ and $t = 5 \div 3600$ so you need a formula involving u, v, t and s.

$$s = \frac{(u+v)}{2} \times t$$

$$s = \frac{(30+60)}{2} \times \frac{5}{3600}$$

$$= \frac{1}{16}$$

The distance travelled is $\frac{1}{16}$ km or 62.5m.

⚠ In Examples 2.1 and 2.2, the bus and the car are always travelling in the positive direction so it is safe to use s for distance. Remember that s is not the same as the distance travelled if the direction changes during the motion.

The acceleration due to gravity

When a model ignoring air resistance is used, all objects falling freely under gravity fall with the same constant acceleration, $g\,\mathrm{m\,s^{-2}}$. This varies over the surface of the earth. In this book it is assumed that all the situations occur in a place where it is $10\,\mathrm{m\,s^{-2}}$. The value $9.8\,\mathrm{m\,s^{-2}}$ is also used. Most answers are given correct to three significant figures so that you can check your working.

EXAMPLE 2.3 A coin is dropped from rest at the top of a building of height 12 m and travels in a straight line with constant acceleration $10\,\mathrm{m\,s^{-2}}$.

Find the time it takes to reach the ground and the speed of impact.

SOLUTION

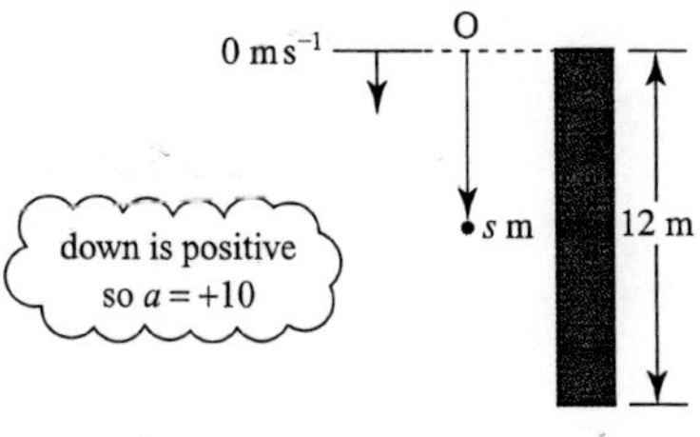

Figure 2.9

Suppose the time taken to reach the ground is t seconds. Using S.I. units, $u = 0$, $a = 10$ and $s = 12$ when the coin hits the ground, so you need to use a formula involving u, a, s and t.

$$s = ut + \frac{1}{2}at^2$$
$$12 = 0 + \frac{1}{2} \times 10 \times t^2$$
$$t^2 = 2.4$$
$$t = 1.55 \text{ (to 3 s.f.)}$$

To find the velocity, v, a formula involving s, u, a and v is required.

$$v^2 = u^2 + 2as$$
$$v^2 = 0 + 2 \times 10 \times 12$$
$$v^2 = 240$$
$$v = 15.5 \text{ (to 3 s.f.)}$$

The coin takes 1.55 s to hit the ground and has speed $15.5\,\mathrm{m\,s^{-1}}$ on impact.

Summary

The formulae for motion with *constant acceleration* are

① $v = u + at$ ② $s = \dfrac{(u+v)}{2} \times t$

③ $s = ut + \frac{1}{2}at^2$ ④ $v^2 = u^2 + 2as$

? Derive formula ③ algebraically by substituting for v from formula ① into formula ②.

If you look at these formulae you will see that each omits one variable. But there are five variables and only four formulae; there isn't one without u. A formula omitting u is

$$s = vt - \frac{1}{2}at^2 \quad ⑤$$

? How can you derive this by referring to a graph or using substitution?

When using these formulae make sure that the units you use are consistent. For example, when the time is t seconds and the distance s metres, any speed involved is in $\mathrm{m\,s^{-1}}$.

EXERCISE 2A

1 **(i)** Find v when $u = 10$, $a = 6$ and $t = 2$.

(ii) Find s when $v = 20$, $u = 4$ and $t = 10$.

(iii) Find s when $v = 10$, $a = 2$ and $t = 10$.

(iv) Find a when $v = 2$, $u = 12$, $s = 7$.

2 Decide which equation to use in each of these situations.

(i) Given u, s, a; find v.
(ii) Given a, u, t; find v.
(iii) Given u, a, t; find s.
(iv) Given u, v, s; find t.
(v) Given u, s, v; find a.
(vi) Given u, s, t; find a.
(vii) Given u, a, v; find s.
(viii) Given a, s, t; find v.

3 Assuming no air resistance, a ball has an acceleration of $10\,\mathrm{m\,s^{-2}}$ when it is dropped from a window (so its initial speed, when $t = 0$, is zero). Calculate

(i) its speed after 1 s and after 10 s

(ii) how far it has fallen after 1 s and after 10 s

(iii) how long it takes to fall 20 m.

Which of these answers are likely to need adjusting to take account of air resistance? Would you expect your answer to be an over- or underestimate?

4 A car starting from rest at traffic lights reaches a speed of $90\,\mathrm{km\,h^{-1}}$ in 12 s. Find the acceleration of the car (in $\mathrm{m\,s^{-2}}$) and the distance travelled. Write down any assumptions that you have made.

5 A top sprinter accelerates from rest to $9\,\mathrm{m\,s^{-1}}$ in 2 s. Calculate his acceleration, assumed constant, during this period and the distance travelled.

6 A van skids to a halt from an initial speed of $24\,\mathrm{m\,s^{-1}}$ covering a distance of 36 m. Find the acceleration of the van (assumed constant) and the time it takes to stop.

7 An object moves along a straight line with acceleration $-8\,\mathrm{m\,s^{-2}}$. It starts its motion at the origin with velocity $16\,\mathrm{m\,s^{-1}}$.

(i) Write down equations for its position and velocity at time t s.

(ii) Find the smallest non-zero time when

(a) the velocity is zero

(b) the object is at the origin.

(iii) Sketch the position–time, velocity–time and speed–time graphs for $0 \leqslant t \leqslant 4$.

Further examples

The next two examples illustrate ways of dealing with more complex problems. In Example 2.4, none of the possible formulae has only one unknown and there are also two situations, so simultaneous equations are used.

EXAMPLE 2.4

James practises using the stopwatch facility on his new watch by measuring the time between lamp posts on a car journey. As the car speeds up, two consecutive times are 1.2 s and 1 s. Later he finds out that the lamp posts are 30 m apart.

(i) Calculate the acceleration of the car (assumed constant) and its speed at the first lamp post.

(ii) Assuming the same acceleration, find the time the car took to travel the 30 m before the first lamp post.

SOLUTION

(i) The diagram shows all the information assuming the acceleration is $a\,\mathrm{m\,s^{-2}}$ and the velocity at A is $u\,\mathrm{m\,s^{-2}}$.

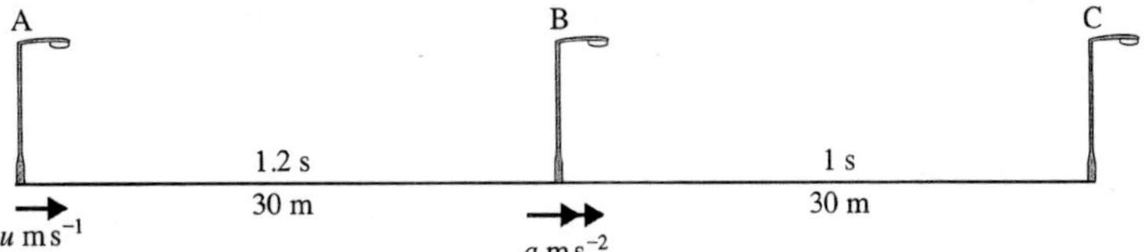

Figure 2.10

For AB, $s = 30$ and $t = 1.2$. You are using u and you want a so you use

$$s = ut + \tfrac{1}{2}at^2$$
$$30 = 1.2u + \tfrac{1}{2}a \times 1.2^2$$
$$30 = 1.2u + 0.72a \qquad ①$$

To use the same equation for the part BC you would need the velocity at B and this brings in another unknown. It is much better to go back to the beginning and consider the whole of AC with $s = 60$ and $t = 2.2$. Then again

using $s = ut + \tfrac{1}{2}at^2$

$$60 = 2.2u + \tfrac{1}{2}a \times 2.2^2$$
$$60 = 2.2u + 2.42a \qquad ②$$

These two simultaneous equations in two unknowns can be solved more easily if they are simplified. First make the coefficients of u integers.

	① × 10 ÷ 12	$25 = u + 0.6a$	③
	② × 5	$300 = 11u + 12.1a$	④
then	③ × 11	$275 = 11u + 6.6a$	⑤

Subtracting gives

$$25 = 0 + 5.5a$$
$$a = 4.545$$

Now substitute 4.545 for a in ③ to find

$$u = 25 - 0.6 \times 4.545 = 22.273$$

The acceleration of the car is 4.55 m s^{-2} and the initial speed is 22.3 m s^{-1} (correct to 3 s.f.).

(ii)

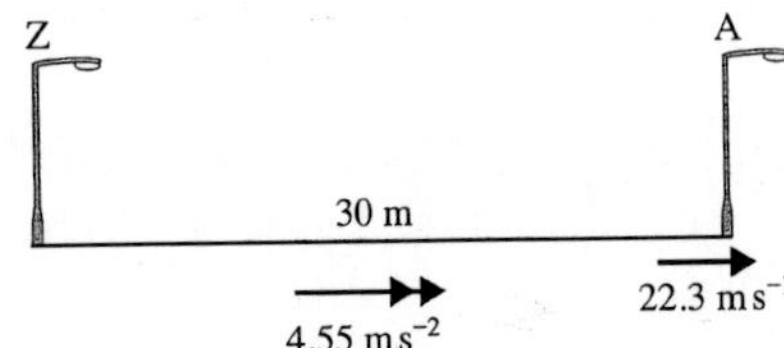

Figure 2.11

For this part, you know that $s = 30$, $v = 22.3$ and $a = 4.55$ and you want t so you use the fifth formula.

$$s = vt - \tfrac{1}{2}at^2$$

$$30 = 22.3 \times t - \tfrac{1}{2} \times 4.55 \times t^2$$

$$\Rightarrow \quad 2.275t^2 - 22.3t + 30 = 0$$

Solving this using the quadratic formula gives $t = 1.61$ and $t = 8.19$.

The most sensible answer to this particular problem is 1.61 s.

? Calculate u when $t = 8.19$, $v = 22.3$ and $a = 4.55$. Is $t = 8.19$ a possible answer?

Using a non-zero initial displacement

What, in the constant acceleration formulae, are v and s when $t = 0$?

Putting $t = 0$ in the *suvat* formulae gives the *initial values*, u for the velocity and $s = 0$ for the position.

Sometimes, however, it is convenient to use an origin which gives a non-zero value for s when $t = 0$. For example, when you model the motion of an eraser thrown vertically upwards you might decide to find its height above the ground rather than above the point from which it was thrown.

What is the effect on the various *suvat* formulae if the initial position is s_0 rather than zero?

If the height of the eraser above the ground is s at time t and s_0 when $t = 0$, the displacement over time t is $s - s_0$. You then need to replace formula ③ with

$$s - s_0 = ut + \tfrac{1}{2}at^2$$

The next example avoids this in the first part but it is very useful in part **(ii)**.

EXAMPLE 2.5

A juggler throws a ball up in the air with initial speed $5\,\text{m s}^{-1}$ from a height of 1.2 m. It has a constant acceleration of $10\,\text{m s}^{-1}$ vertically downwards duc to gravity.

(i) Find the maximum height of the ball above the ground and the time it takes to reach it.

At the instant that the ball reaches its maximum height, the juggler throws up another ball with the same speed and from the same height.

(ii) Where and when will the balls pass each other?

SOLUTION

(i) In this example it is very important to draw a diagram and to be clear about the position of the origin. When O is 1.2 m above the ground and s is the height in metres above O after t s, the diagram looks like figure 2.12.

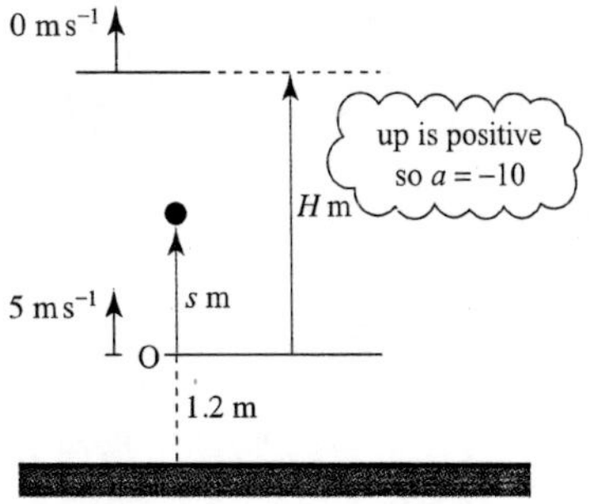

Figure 2.12

At the point of maximum height, let $s = H$ and $t = t_1$.

Use the suffix because there are two times to be found in this question.

The ball stops instantaneously before falling so at the top $v = 0$.

A formula involving u, v, a and s is required.

$$v^2 = u^2 + 2as$$
$$0 = 5^2 + 2 \times (-10) \times H$$
$$H = 1.25$$

The acceleration given is constant, $a = -10$; $u = +5$; $v = 0$ and $s = H$.

The maximum height of the ball above the ground is 1.25 + 1.2 = 2.45 m.

To find t_1, given $v = 0$, $a = -10$ and $u = +5$ requires a formula in v, u, a and t.

$$v = u + at$$
$$0 = 5 + (-10)t_1$$
$$t_1 = 0.5$$

The ball takes half a second to reach its maximum height.

(ii) Now consider the motion from the instant the first ball reaches the top of its path and the second is thrown up.

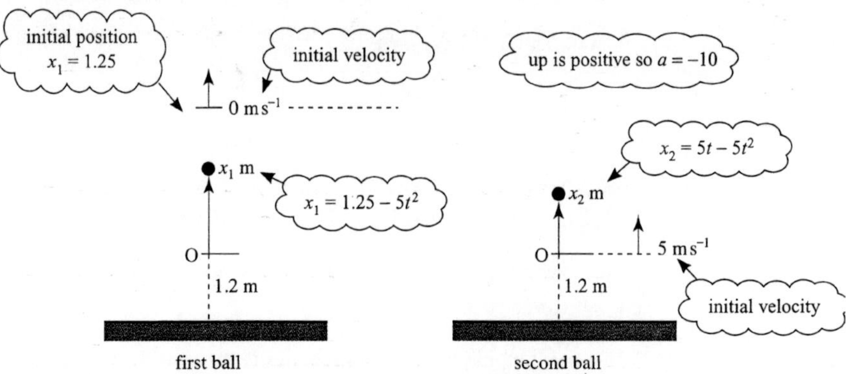

Figure 2.13

Suppose that the balls have displacements *above* the origin of x_1 m and x_2 m, as shown in the diagram, at a general time t s after the second ball is thrown up. The initial position of the second ball is zero, but the initial position of the first ball is +1.25 m.

For each ball you know u and a. You want to involve t and s so you use

$$s - s_0 = ut + \tfrac{1}{2}at^2$$

i.e. $$s = s_0 + ut + \tfrac{1}{2}at^2$$

For the first ball:

$$x_1 = 1.25 + 0 \times t + \tfrac{1}{2} \times (-10) \times t^2$$

This makes $x_1 = 1.25$ when $t = 0$.

$$x_1 = 1.25 - 5t^2 \quad ①$$

x_1 decreases as t increases.

For the second ball:

$$x_2 = 0 + 5 \times t + \tfrac{1}{2} \times (-10) \times t^2$$
$$x_2 = 5t - 5t^2 \quad ②$$

Suppose the balls pass after a time t s. This is when they are at the same height, so equate x_1 and x_2 from equations ① and ②.

$$1.25 - 5t^2 = 5t - 5t^2$$
$$1.25 = 5t$$
$$t = 0.25$$

Then substituting $t = 0.25$ in ① and ② gives

$$x_1 = 1.25 - 5 \times 0.25^2 = 0.9375$$

and

$$x_2 = 5 \times 0.25 - 5 \times 0.25^2 = 0.9375$$

These are the same, as expected.

The balls pass after 0.25 seconds at a height of 1.2 m + 0.94 m = 2.14 m above the ground (correct to the nearest centimetre).

? Try solving part **(ii)** of this example by supposing that the first ball falls x m and the second rises $(1.25 - x)$m in t seconds.

Note

The balls pass after half the time to reach the top, but *not* half-way up.

? Why don't they travel half the distance in half the time?

EXERCISE 2B

Use $g = 10\,\text{m s}^{-2}$ in this exercise.

1 A car is travelling along a straight road. It accelerates uniformly from rest to a speed of $15\,\text{m s}^{-1}$ and maintains this speed for 10 minutes. It then decelerates uniformly to rest. If the acceleration and deceleration are $5\,\text{m s}^{-2}$ and $8\,\text{m s}^{-2}$ respectively, find the total journey time and the total distance travelled during the journey.

2 A skier pushes off at the top of a slope with an initial speed of $2\,\text{m s}^{-1}$. She gains speed at a constant rate throughout her run. After 10 s she is moving at $6\,\text{m s}^{-1}$.

(i) Find an expression for her speed t seconds after she pushes off.

(ii) Find an expression for the distance she has travelled at time t seconds.

(iii) The length of the ski slope is 400 m. What is her speed at the bottom of the slope?

3 Towards the end of a half-marathon Sabina is 100 m from the finish line and is running at a constant speed of $5\,\text{m s}^{-1}$. Daniel, who is 140 m from the finish and is running at $4\,\text{m s}^{-1}$, decides to accelerate to try to beat Sabina. If he accelerates uniformly at $0.25\,\text{m s}^{-2}$ does he succeed?

4 Rupal throws a ball upwards at $8\,\text{m s}^{-1}$ from a window which is 4 m above ground level.

(i) Write down an equation for the height h m of the ball above the ground after t s (while it is still in the air).

(ii) Use your answer to part **(i)** to find the time the ball hits the ground.

(iii) How fast is the ball moving just before it hits the ground?

(iv) In what way would you expect your answers to parts **(ii)** and **(iii)** to change if you were able to take air resistance into account?

5 Nathan hits a tennis ball straight up into the air from a height of 1.25 m above the ground. The ball hits the ground after 2.5 seconds. Find

(i) the speed Nathan hits the ball
(ii) the greatest height above the ground reached by the ball
(iii) the speed the ball hits the ground
(iv) how high the ball bounces if it loses 0.2 of its speed on hitting the ground.
(v) Is your answer to part **(i)** likely to be an over- or underestimate given that you have ignored air resistance?

6 A ball is dropped from a building of height 30 m and at the same instant a stone is thrown vertically upwards from the ground so that it hits the ball. In modelling the motion of the ball and stone it is assumed that each object moves in a straight line with a constant downward acceleration of magnitude $10\,\mathrm{m\,s^{-2}}$. The stone is thrown with initial speed of $15\,\mathrm{m\,s^{-1}}$ and is h_s metres above the ground t seconds later.

(i) Draw a diagram of the ball and stone before they collide, marking their positions.
(ii) Write down an expression for h_s at time t.
(iii) Write down an expression for the height h_b of the ball at time t.
(iv) When do the ball and stone collide?
(v) How high above the ground do the ball and stone collide?

7 When Kim rows her boat, the two oars are both in the water for 3 s and then both out of the water for 2 s. This 5 s cycle is then repeated. When the oars are in the water the boat accelerates at a constant $1.8\,\mathrm{m\,s^{-2}}$ and when they are not in the water it decelerates at a constant $2.2\,\mathrm{m\,s^{-2}}$.

(i) Find the change in speed that takes place in each 3 s period of acceleration.
(ii) Find the change in speed that takes place in each 2 s period of deceleration.
(iii) Calculate the change in the boat's speed for each 5 s cycle.
(iv) A race takes Kim 45 s to complete. If she starts from rest what is her speed as she crosses the finishing line?
(v) Discuss whether this is a realistic speed for a rowing boat.

8 A ball is dropped from a tall building and falls with acceleration of magnitude $10\,\mathrm{m\,s^{-2}}$. The distance between floors in the block is constant. The ball takes 0.5 s to fall from the 14th to the 13th floor and 0.3 s to fall from the 13th floor to the 12th. What is the distance between floors?

9 Two clay pigeons are launched vertically upwards from exactly the same spot at 1 s intervals. Each clay pigeon has initial speed $30\,\mathrm{m\,s^{-1}}$ and acceleration $10\,\mathrm{m\,s^{-2}}$ downwards. How high above the ground do they collide?

10 A train accelerates along a straight, horizontal section of track. The driver notes that he reaches a bridge 120 m from the station in 8 s and that he crosses the bridge, which is 31.5 m long, in a further 2 s.

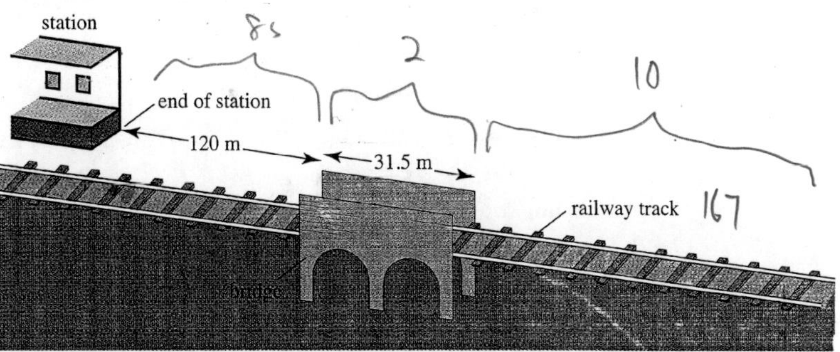

The motion of the train is modelled by assuming constant acceleration. Take the speed of the train when leaving the station to be $u\,\text{m s}^{-1}$ and the acceleration to have the value $a\,\text{m s}^{-2}$.

(i) By considering the part of the journey from the station to the bridge, show that $u + 4a = 15$.

(ii) Find a second equation involving u and a.

(iii) Solve the two equations for u and a to show that a is 0.15 and find the value of u.

(iv) If the driver also notes that he travels 167 m in the 10 s after he crosses the bridge, have you any evidence to reject the modelling assumption that the acceleration is constant?

[MEI]

11 The diagram shows the velocity–time graph for a lift moving between floors in a building. The graph consists of straight line segments. In the first stage the lift travels downwards from the ground floor for 5 s, coming to rest at the basement after travelling 10 m.

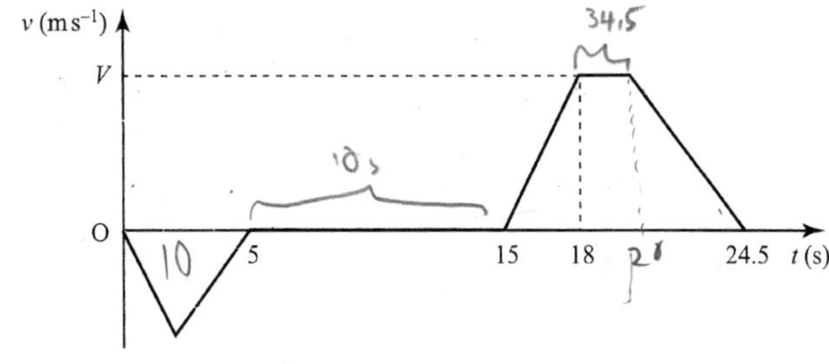

(i) Find the greatest speed reached during this stage.

The second stage consists of a 10 s wait at the basement. In the third stage, the lift travels upwards until it comes to rest at a floor 34.5 m above the basement, arriving 24.5 s after the start of the first stage. The lift accelerates at $2\,\mathrm{m\,s^{-2}}$ for the first 3 s of the third stage, reaching a speed of $V\,\mathrm{m\,s^{-1}}$. Find

(ii) the value of V,

(iii) the time during the third stage for which the lift is moving at constant speed,

(iv) the deceleration of the lift in the final part of the third stage.

[Cambridge AS and A Level Mathematics 9709, Paper 4 Q6 June 2005]

12 A particle is projected vertically upwards from a point O with initial speed $12.5\,\mathrm{m\,s^{-1}}$. At the same instant another particle is released from rest at a point 10 m vertically above O. Find the height above O at which the particles meet.

[Cambridge AS and A Level Mathematics 9709, Paper 4 Q2 November 2007]

INVESTIGATION

The situation described below involves mathematical modelling. You will need to take these steps to help you.

(i) Make a list of the assumptions you need to make to simplify the situation to the point where you can apply mathematics to it.

(ii) Make a list of the quantities involved.

(iii) Find out any information you require such as safe stopping distances or a value for the acceleration and deceleration of a car on a housing estate.

(iv) Assign suitable letters for your unknown quantities. (Don't vary too many things at once.)

(v) Set up your equations and solve them. You might find it useful to work out several values and draw a suitable graph.

(vi) Decide whether your results make sense, preferably by checking them against some real data.

(vii) If you think your results need adjusting, decide whether any of your initial assumptions should be changed and, if so, in what way.

Speed bumps

The residents of a housing estate are worried about the danger from cars being driven at high speed. They request that speed bumps be installed.

How far apart should the bumps be placed to ensure that drivers do not exceed a speed of $40\,\mathrm{km\,h^{-1}}$? Some of the things to consider are the maximum sensible velocity over each bump and the time taken to speed up and slow down.

KEY POINTS

1 The *suvat* formulae

- The formulae for motion with constant acceleration are

① $v = u + at$

② $s = \frac{(u+v)}{2} \times t$

③ $s = ut + \frac{1}{2}at^2$

④ $v^2 = u^2 + 2as$

⑤ $s = vt - \frac{1}{2}at^2$

- a is the constant acceleration; s is the displacement from the starting position at time t; v is the velocity at time t; u is the velocity when $t = 0$.

  If $s = s_0$ when $t = 0$, replace s in each formula with $(s - s_0)$.

2 Vertical motion under gravity

- The acceleration due to gravity ($g\,\mathrm{m\,s^{-2}}$) is vertically downwards and is often taken to be $10\,\mathrm{m\,s^{-2}}$. The value $9.8\,\mathrm{m\,s^{-2}}$ is also used.
- Always draw a diagram and decide in advance where your origin is and which way is positive.
- Make sure that your units are compatible.

3 Using a mathematical model

- Make simplifying assumptions by deciding what is most relevant.
  For example: a car is a *particle* with no dimensions
  a road is a *straight line* with one dimension
  acceleration is constant.
- Define variables and set up equations.
- Solve the equations.
- Check that the answer is sensible. If not, think again.

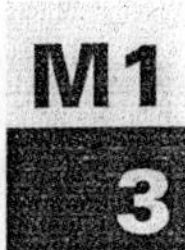

3 Forces and Newton's laws of motion

**Nature and Nature's Laws lay hid in Night.
God said, Let Newton be! and All was Light.**

Alexander Pope

Force diagrams

The picture shows crates of supplies being dropped into a remote area by parachute. What forces are acting on a crate of supplies and the parachute?

One force which acts on every object near the earth's surface is its own *weight.* This is the force of gravity pulling it towards the centre of the earth. The weight of the crate acts on the crate and the weight of the parachute acts on the parachute.

The parachute is designed to make use of *air resistance.* A resistance force is present whenever a solid object moves through a liquid or gas. It acts in the opposite direction to the motion and depends on the speed of the object. The crate also experiences air resistance, but to a lesser extent than the parachute.

Other forces are the *tensions* in the guy lines attaching the crate to the parachute. These pull upwards on the crate and downwards on the parachute.

All these forces can be shown most clearly if you draw *force diagrams* for the crate and the parachute.

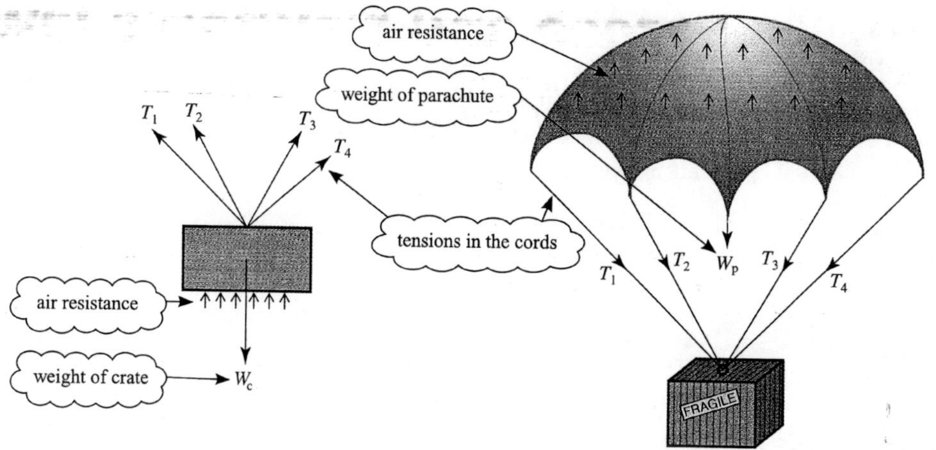

Figure 3.1 *Forces acting on the crate* **Figure 3.2** *Forces acting on the parachute*

Force diagrams are essential for the understanding of most mechanical situations. A force is a vector: it has a magnitude, or size, and a direction. It also has a *line of action*. This line often passes through a point of particular interest. Any force diagram should show clearly

- the direction of the force
- the magnitude of the force
- the line of action.

In figures 3.1 and 3.2 each force is shown by an arrow along its line of action. The air resistance has been depicted by a lot of separate arrows but this is not very satisfactory. It is much better if the combined effect can be shown by one arrow. When you have learned more about vectors, you will see how the tensions in the guy lines can also be combined into one force if you wish. The forces on the crate and parachute can then be simplified.

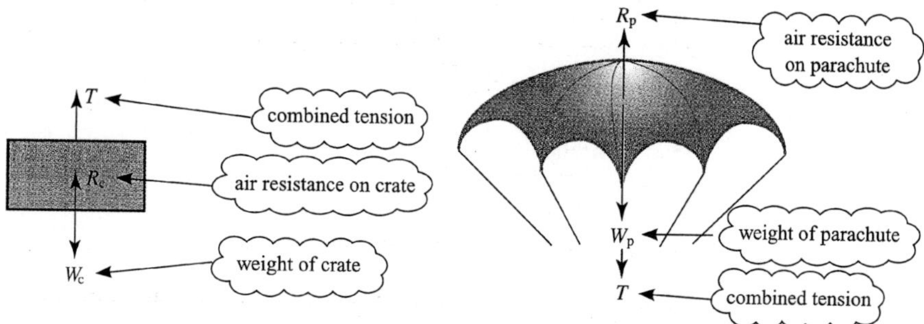

Figure 3.3 *Forces acting on the crate* **Figure 3.4** *Forces acting on the parachute*

Centre of mass and the particle model

When you combine forces you are finding their *resultant*. The weights of the crate and parachute are also found by combining forces; they are the resultant of the weights of all their separate parts. Each weight acts through a point called the *centre of mass* or centre of gravity.

Think about balancing a pen on your finger. The diagrams show the forces acting on the pen.

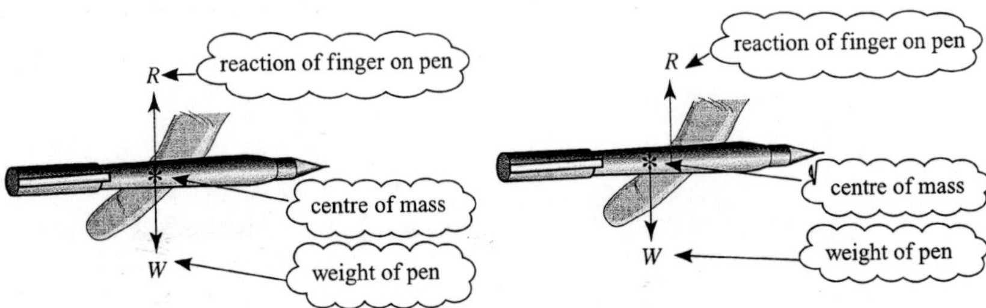

Figure 3.5 **Figure 3.6**

So long as you place your finger under the centre of mass of the pen, as in figure 3.5, it will balance. There is a force called a *reaction* between your finger and the pen which balances the weight of the pen. The forces on the pen are then said to be *in equilibrium*. If you place your finger under another point, as in figure 3.6, the pen will fall. The pen can only be in equilibrium if the two forces have the same line of action.

If you balance the pen on two fingers, there is a reaction between each finger and the pen at the point where it touches the pen. These reactions can be combined into one resultant vertical reaction acting through the centre of mass.

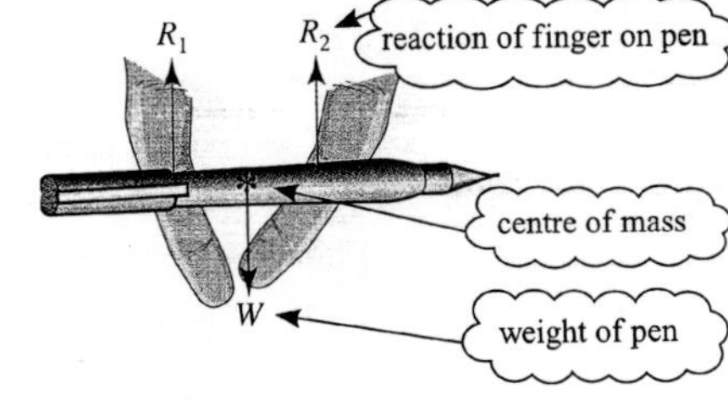

Figure 3.7

The behaviour of objects which are liable to rotate under the action of forces is covered in *Mechanics 2* Chapter 11. In *Mechanics 1* you will only deal with situations where the resultant of the forces does not cause rotation. An object can then be modelled as a particle, that is a point mass, situated at its centre of mass.

Newton's third law of motion

Sir Isaac Newton (1642–1727) is famous for his work on gravity and the mechanics you learn in this course is often called Newtonian Mechanics because it is based entirely on Newton's three laws of motion. These laws provide us with an extremely powerful model of how objects, ranging in size from specks of dust to planets and stars, behave when they are influenced by forces.

We start with Newton's *third law* which says that

- When one object exerts a force on another there is always a reaction of the same kind which is equal, and opposite in direction, to the acting force.

You might have noticed that the combined tensions acting on the parachute and the crate in figures 3.3 and 3.4 are both marked with the same letter, *T*. The crate applies a force on the parachute through the supporting guy lines and the parachute applies an equal and opposite force on the crate. When you apply a force to a chair by sitting on it, it responds with an equal and opposite force on you. Figure 3.8 shows the forces acting when someone sits on a chair.

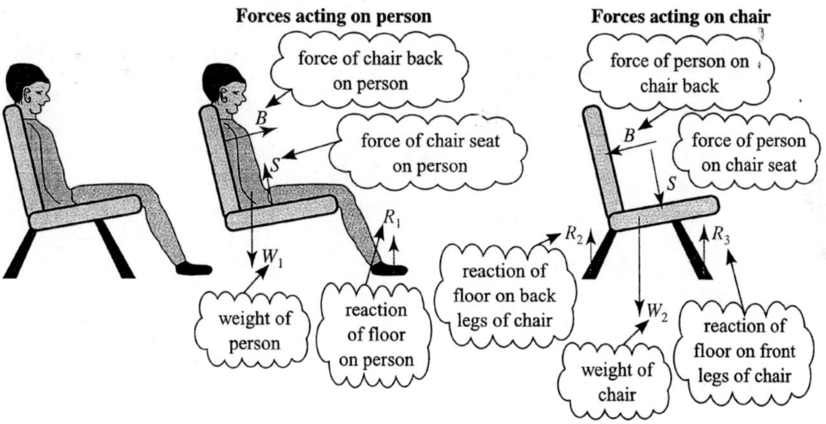

Figure 3.8

The reactions of the floor on the chair and on your feet act where there is contact with the floor. You can use R_1, R_2 and R_3 to show that they have different magnitudes. There are equal and opposite forces acting on the floor, but the forces on the floor are not being considered and so do not appear here.

? Why is the weight of the person *not* shown on the force diagram for the chair?

Gravitational forces obey Newton's third law just as other forces between bodies. According to Newton's universal law of gravitation, the earth pulls us towards its centre and we pull the earth in the opposite direction. However, in this book we are only concerned with the gravitational force on us and not the force we exert on the earth.

All the forces you meet in mechanics apart from the gravitational force are the result of physical contact. This might be between two solids or between a solid and a liquid or gas.

Friction and normal reaction

When you push your hand along a table, the table reacts in two ways.

- Firstly there are forces which stop your hand going through the table. Such forces are always present when there is any contact between your hand and the table. They are at right angles to the surface of the table and their resultant is called the *normal reaction* between your hand and the table.
- There is also another force which tends to prevent your hand from sliding. This is the *friction* and it acts in a direction which opposes the sliding.

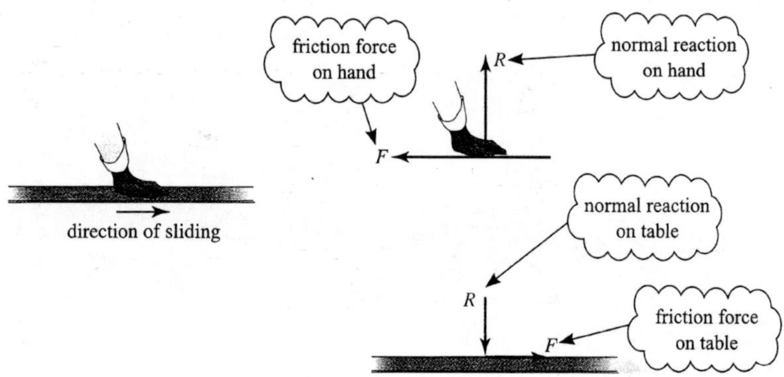

Figure 3.9

Figure 3.9 shows the reaction forces acting on your hand and on the table. By Newton's third law they are equal and opposite to each other. The frictional force is due to tiny bumps on the two surfaces (see electronmicrograph below). When you hold your hands together you will feel the normal reaction between them. When you slide them against each other you will feel the friction.

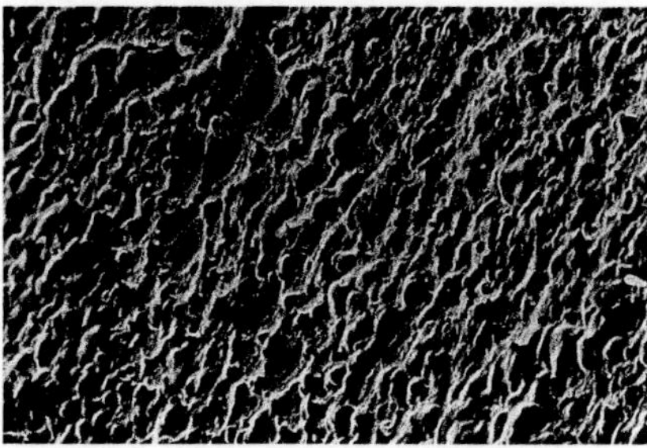

Etched glass magnified to high resolution, showing the tiny bumps.

When the friction between two surfaces is negligible, at least one of the surfaces is said to be *smooth*. This is a modelling assumption which you will meet frequently in this book. Oil can make surfaces smooth and ice is often modelled as a smooth surface.

- When the contact between two surfaces is smooth, the only forces between them are normal reactions which act at right angles to any possible sliding.

? What direction is the reaction between the sweeper's broom and the smooth ice?

EXAMPLE 3.1

A TV set is standing on a small table. Draw a diagram to show the forces acting on the TV and on the table as seen from the front.

SOLUTION

The diagram shows the forces acting on the TV and on the table. They are all vertical because the weights are vertical and there are no horizontal forces acting.

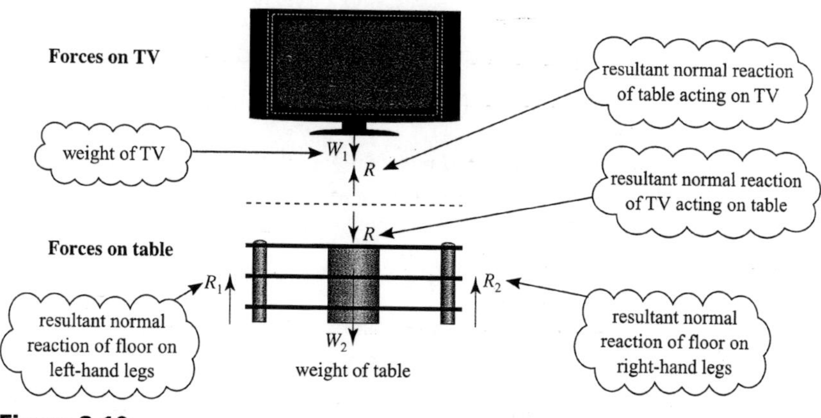

Figure 3.10

EXAMPLE 3.2

Draw diagrams to show the forces acting on a tennis ball which is hit downwards across the court

(i) at the instant it is hit by the racket

(ii) as it crosses the net

(iii) at the instant it lands on the other side.

SOLUTION

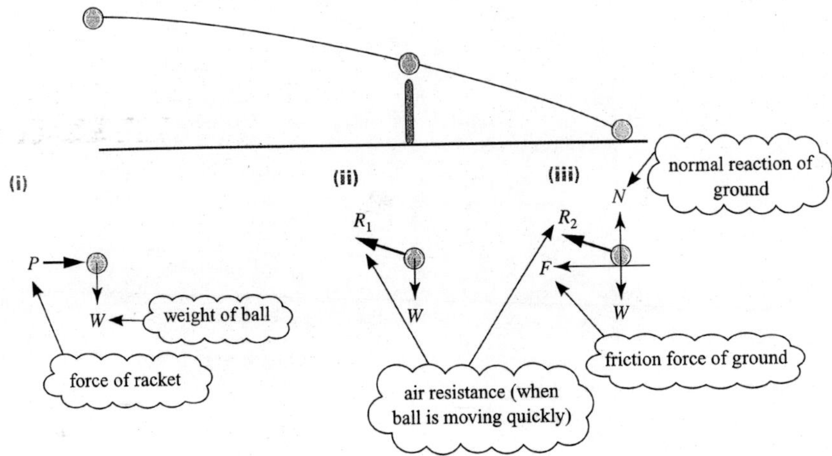

Figure 3.11

EXERCISE 3A

In this exercise draw clear diagrams to show the forces acting on the objects named in italics. Clarity is more important than realism when drawing these diagrams.

1 A *gymnast* hanging at rest on a bar.

2 A *light bulb* hanging from a ceiling.

3 A *book* lying at rest on a table.

4 A *book* at rest on a table but being pushed by a small horizontal force.

5 *Two books* lying on a table, one on top of the other.

6 A *horizontal plank* being used to bridge a stream.

7 A *snooker ball* on a table which can be assumed to be smooth

(i) as it lies at rest on the table

(ii) at the instant it is hit by the cue.

8 An *ice hockey puck*

(i) at the instant it is hit when standing on smooth ice

(ii) at the instant it is hit when standing on rough ice.

9 A *cricket ball* which follows the path shown on the right.
Draw diagrams for each of the three positions A, B and C (include air resistance).

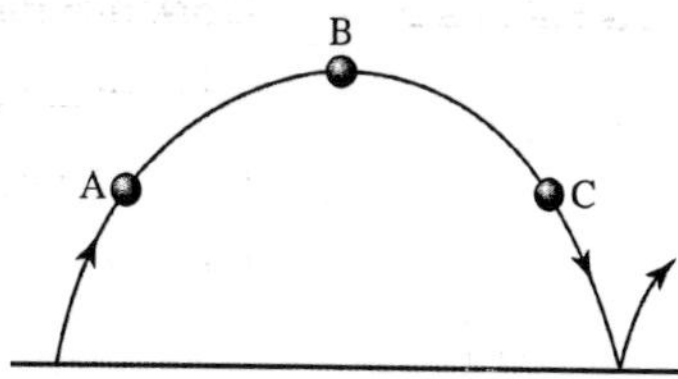

10 **(i)** *Two balls* colliding in mid-air.
(ii) *Two balls* colliding on a snooker table.

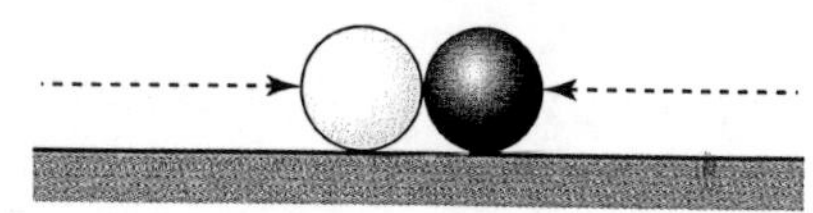

11 A *paving stone* leaning against a wall.

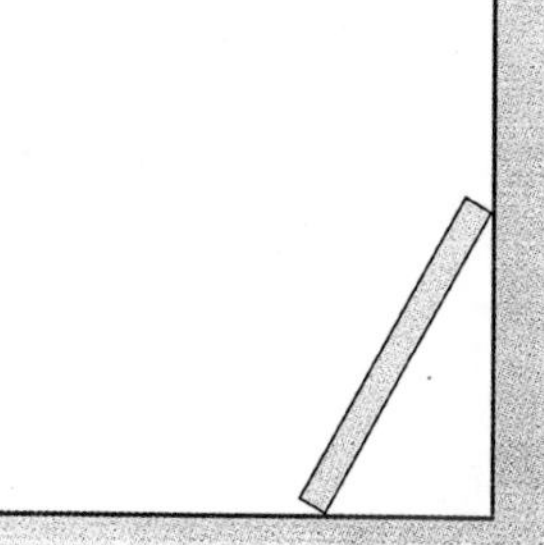

12 A *cylinder* at rest on smooth surfaces.

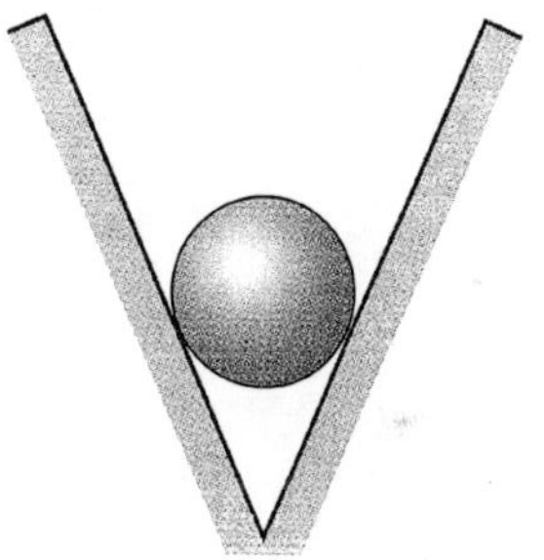

Force and motion

? How are the rails and handles provided in buses and trains used by standing passengers?

Newton's first law

Newton's *first law* can be stated as follows.

- Every particle continues in a state of rest or uniform motion in a straight line unless acted on by a resultant external force.

Newton's first law provides a reason for the handles on trains and buses. When you are on a train which is stationary or moving at constant speed in a straight line you can easily stand without support. But when the velocity of the train changes, a force is required to change your velocity to match. This happens when the train slows down or speeds up. It also happens when the train goes round a bend even if the speed does not change. The velocity changes because the direction changes.

? Why is Josh's car in the pond?

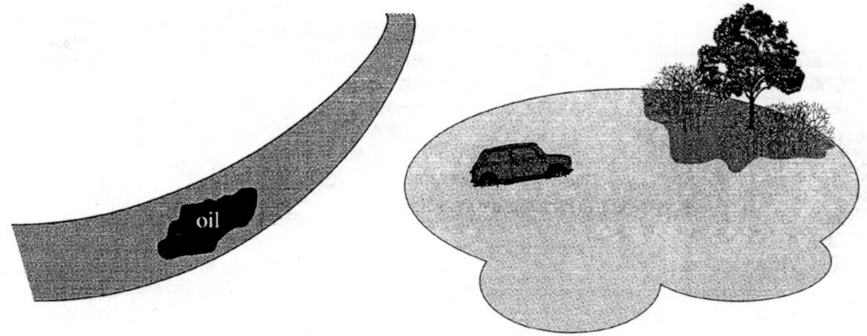

Figure 3.12

EXAMPLE 3.3

A coin is balanced on your finger and then you move it upwards.

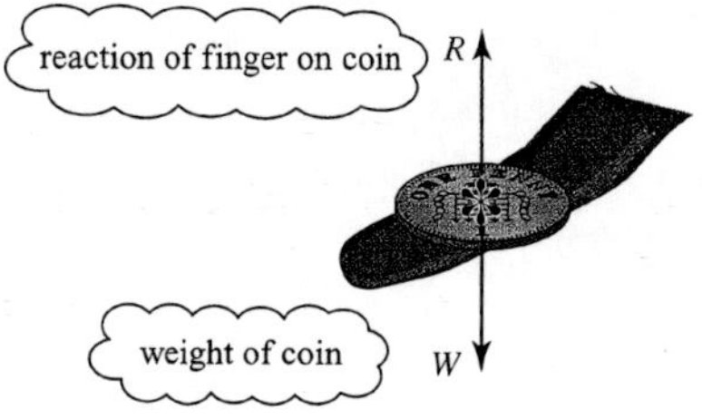

Figure 3.13

By considering Newton's first law, what can you say about W and R in each of these situations?

(i) The coin is stationary.
(ii) The coin is moving upwards with a constant velocity.
(iii) The speed of the coin is increasing as it moves upwards.
(iv) The speed of the coin is decreasing as it moves upwards.

SOLUTION

(i) When the coin is stationary the velocity does not change. The forces are in equilibrium and $R = W$.

(ii) When the coin is moving upwards with a constant velocity the velocity does not change. The forces are in equilibrium and $R = W$.

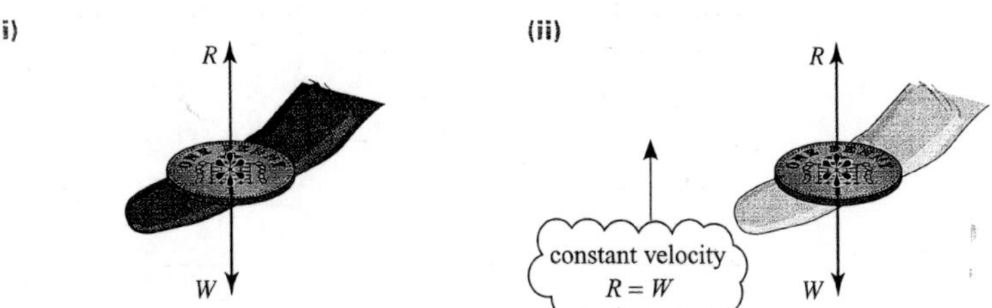

Figure 3.14 **Figure 3.15**

(iii) When the speed of the coin is increasing as it moves upwards there must be a net upward force to make the velocity increase in the upward direction so $R > W$. The net force is $R - W$.

(iv) When the speed of the coin is decreasing as it moves upwards there must be a net downward force to make the velocity decrease and slow the coin down as it moves upwards. In this case $W > R$ and the net force is $W - R$.

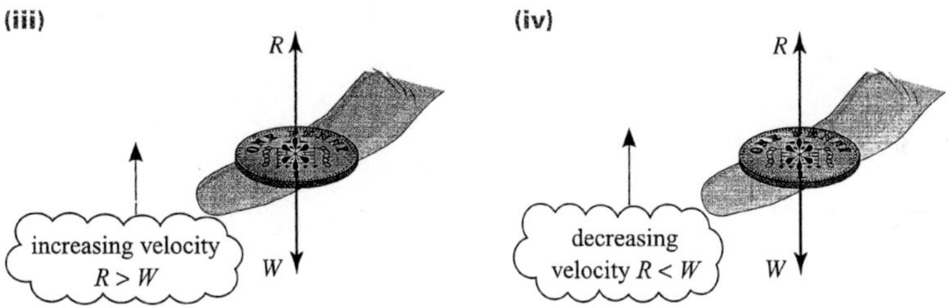

Figure 3.16 **Figure 3.17**

EXERCISE 3B

1 A book is resting on an otherwise empty table.

(i) Draw diagrams showing the forces acting on

(a) the book

(b) the table as seen from the side.

(ii) Write down equations connecting the forces acting on the book and on the table.

2 You balance a coin on your finger and move it up and down. The reaction of your finger on the coin is R and its weight is W. Decide in each case whether R is greater than, less than or equal to W and describe the net force.

(i) The coin is moving downwards with a constant velocity.

(ii) The speed of the coin is increasing as it moves downwards.

(iii) The speed of the coin is decreasing as it moves downwards.

3 In each of the following situations say whether the forces acting on the object are in equilibrium by deciding whether its motion is changing.

(i) A car that has been stationary, as it moves away from a set of traffic lights.

(ii) A motorbike as it travels at a steady $60\,\text{km}\,\text{h}^{-1}$ along a straight road.

(iii) A parachutist descending at a constant rate.

(iv) A box in the back of a lorry as the lorry picks up speed along a straight, level motorway.

(v) An ice hockey puck sliding across a smooth ice rink.

(vi) A book resting on a table.

(vii) A plane flying at a constant speed in a straight line, but losing height at a constant rate.

(viii) A car going round a corner at constant speed.

4 Explain each of the following in terms of Newton's laws.

(i) Seat belts should be worn in cars.

(ii) Head rests are necessary in a car to prevent neck injuries when there is a collision from the rear.

Driving forces and resistances to the motion of vehicles

In problems about such things as cycles, cars and trains, all the forces acting along the line of motion will usually be reduced to two or three: the *driving force* forwards, the *resistance* to motion (air resistance, etc.) and possibly a *braking force* backwards.

Resistances due to air or water always act in a direction opposite to the velocity of a vehicle or boat and are usually more significant for fast-moving objects.

Tension and thrust

The lines joining the crate of supplies to the parachute described at the beginning of this chapter are in tension. They pull upwards on the crate and downwards on the parachute. You are familiar with tensions in ropes and strings, but rigid objects can also be in tension.

When you hold the ends of a pencil, one with each hand, and pull your hands apart, you are pulling on the pencil. What is the pencil doing to each of your hands? Draw the forces acting on your hands and on the pencil.

Now draw the forces acting on your hands and on the pencil when you push the pencil inwards.

Your first diagram might look like figure 3.18. The pencil is in tension so there is an inward *tension* force on each hand.

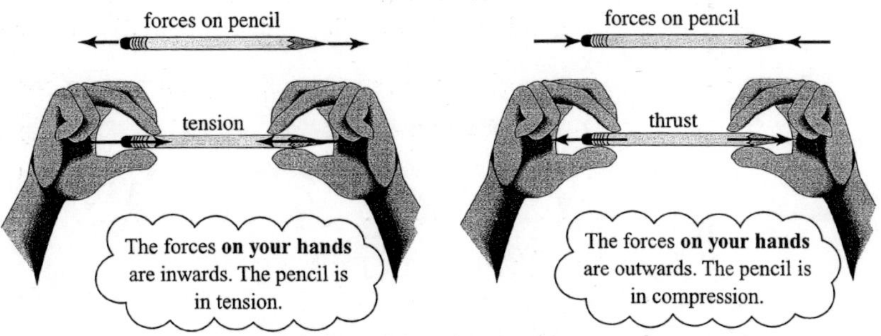

Figure 3.18 **Figure 3.19**

When you push the pencil inwards the forces on your hands are outwards as in figure 3.19. The pencil is said to be *in compression* and the outward force on each hand is called a *thrust.*

If each hand applies a force of 2 units on the pencil, the tension or thrust acting on each hand is also 2 units because each hand is in equilibrium.

Which of the above diagrams is still possible if the pencil is replaced by a piece of string?

Resultant forces and equilibrium

You have already met the idea that a single force can have the same effect as several forces acting together. Imagine that several people are pushing a car. A single rope pulled by another car can have the same effect. The force of the rope is equivalent to the resultant of the forces of the people pushing the car. When there is no resultant force, the forces are in equilibrium and there is no change in motion.

EXAMPLE 3.4

A car is using a tow bar to pull a trailer along a straight, level road. There are resisting forces R acting on the car and S acting on the trailer. The driving force of the car is D and its braking force is B.

Draw diagrams showing the horizontal forces acting on the car and the trailer

(i) when the car is moving at constant speed
(ii) when the speed of the car is increasing
(iii) when the car brakes and slows down rapidly.

In each case write down the resultant force acting on the car and on the trailer.

SOLUTION

(i) When the car moves at constant speed, the forces are as shown in figure 3.20 (overleaf). The tow bar is in tension and the effect is a forward force on the trailer and an equal and opposite backward force on the car.

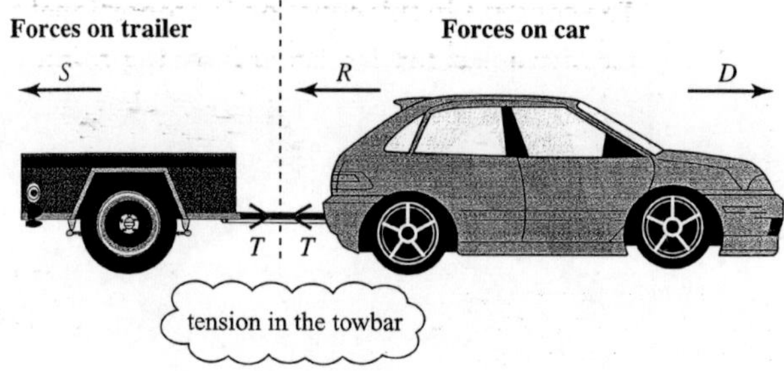

Figure 3.20 *Car travelling at constant speed*

There is no resultant force on either the car or the trailer when the speed is constant; the forces on each are in equilibrium.
For the trailer: $T - S = 0$
For the car: $D - R - T = 0$

(ii) When the car speeds up, the same diagram will do, but now the magnitudes of the forces are different. There is a resultant *forward* force on both the car and the trailer.
For the trailer: resultant $= T - S$
For the car: resultant $= D - R - T$

(iii) When the car brakes a resultant *backward* force is required to slow down the trailer. When the resistance S is not sufficiently large to do this, a thrust in the tow bar comes into play as shown in the figure 3.21.

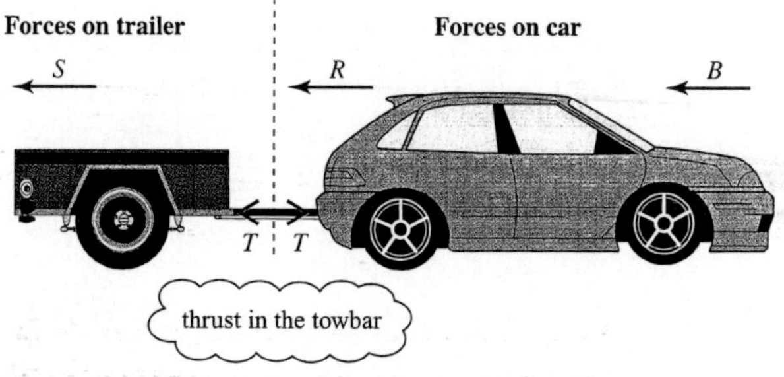

Figure 3.21 *Car braking*

For the trailer: resultant $= T + S$
For the car: resultant $= B + R - \mathrm{T}$

Newton's second law

Newton's *second law* gives us more information about the relationship between the magnitude of the resultant force and the change in motion. Newton said that

- The change in motion is proportional to the force.

For objects with constant mass, this can be interpreted as *the force is proportional to the acceleration.*

Resultant force = a constant × acceleration ①

The constant in this equation is proportional to the mass of the object: a more massive object needs a larger force to produce the same acceleration. For example, you and your friends would be able to give a car a greater acceleration than you would be able to give a lorry.

Newton's second law is so important that a special unit of force, the *newton* (N), has been defined so that the constant in equation ① is actually equal to the mass. A force of 1 newton will give a mass of 1 kilogram an acceleration of $1\,\mathrm{m\,s^{-2}}$. The equation then becomes:

$$\text{Resultant force} = \text{mass} \times \text{acceleration} \qquad ②$$

This is written: $F = ma$

The resultant force and the acceleration are always in the same direction.

Relating mass and weight

The *mass* of an object is related to the amount of matter in the object. It is a *scalar*. The *weight* of an object is a force. It has magnitude and direction and so is a vector.

The mass of an astronaut on the moon is the same as his mass on the earth but his weight is only about one-sixth of his weight on the earth. This is why he can bounce around more easily on the moon. The gravitational force on the moon is less because the mass of the moon is less than that of the earth.

When Buzz Aldrin made the first landing on the moon in 1969 with Neil Armstrong, one of the first things he did was to drop a feather and a hammer to demonstrate that they fell at the same rate. Their accelerations due to the gravitational force of the moon were equal, even though they had very different masses. The same is true on earth. If other forces were negligible all objects would fall with an acceleration g.

When the weight is the only force acting on an object, Newton's second law means that

$$\text{Weight in newtons} = \text{mass in kg} \times g \text{ in m s}^{-2}.$$

Using standard letters:

$$W = mg$$

Even when there are other forces acting, the weight can still be written as mg. A good way to visualise a force of 1 N is to think of the weight of an apple. 1 kg of apples weighs approximately (1×10) N = 10 N. There are about 10 small to medium-sized apples in 1 kg, so each apple weighs about 1 N.

Note

Anyone who says 1 kg of apples *weighs* 1 kg is not strictly correct. The terms weight and mass are often confused in everyday language but it is very important for your study of mechanics that you should understand the difference.

EXAMPLE 3.5

What is the weight of

(i) a baby of mass 3 kg
(ii) a golf ball of mass 46 g?

SOLUTION

(i) The baby's weight is $3 \times 10 = 30$ N.
(ii) Mass of golf ball $= 46$ g

$$= 0.046 \text{ kg}$$

$$\text{Weight} = 0.046 \times 10 = 0.46 \text{ N}.$$

EXERCISE 3C

Data: On the earth $g = 10\,\text{m s}^{-2}$. *On the moon* $g = 1.6\,\text{m s}^{-2}$.
1000 newtons (N) = 1 kilonewton (kN).

1 Calculate the magnitude of the force of gravity on the following objects on the earth.

(i) A suitcase of mass 15 kg.
(ii) A car of mass 1.2 tonnes. (1 tonne = 1000 kg)
(iii) A letter of mass 50 g.

2 Find the mass of each of these objects on the earth.

(i) A girl of weight 600 N.
(ii) A lorry of weight 11 kN.

3 A person has mass 65 kg. Calculate the force of gravity

(i) of the earth on the person
(ii) of the person on the earth.

4 What reaction force would an astronaut of mass 70 kg experience while standing on the moon?

5 Two balls of the same shape and size but with masses 1 kg and 3 kg are dropped from the same height.

(i) Which hits the ground first?
(ii) If they were dropped on the moon what difference would there be?

6 **(i)** Estimate your mass in kilograms.
(ii) Calculate your weight when you are on the earth's surface.
(iii) What would your weight be if you were on the moon?
(iv) When people say that a baby weighs 4 kg, what do they mean?

? Most weighing machines have springs or some other means to measure force even though they are calibrated to show mass. Would something appear to weigh the same on the moon if you used one of these machines? What could you use to find the mass of an object irrespective of where you measure it?

Pulleys

In the remainder of this chapter weight will be represented by *mg*. You will learn to apply Newton's second law more generally in the next chapter.

A pulley can be used to change the direction of a force; for example it is much easier to pull down on a rope than to lift a heavy weight. When a pulley is well designed it takes a relatively small force to make it turn and such a pulley is modelled as being *smooth and light.* Whatever the direction of the string passing over this pulley, its tension is the same on both sides.

Figure 3.22 shows the forces acting when a pulley is used to lift a heavy parcel.

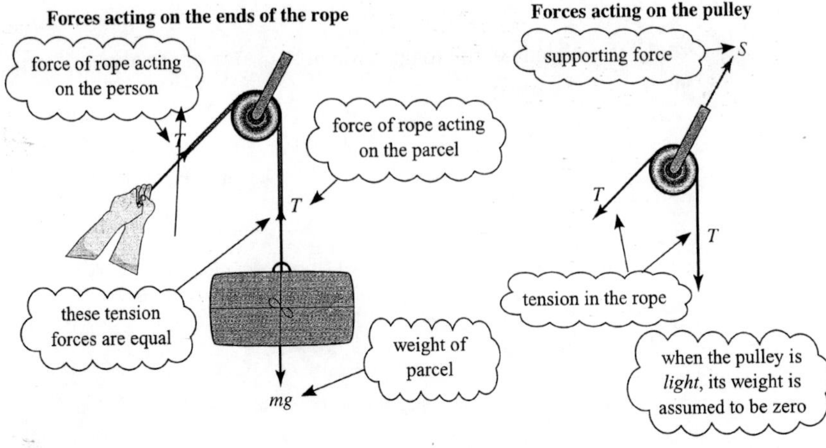

Figure 3.22

Note

The rope is in tension. It is not possible for a rope to exert a thrust force.

EXAMPLE 3.6

In this diagram the pulley is smooth and light and the 2 kg block, A, is on a rough surface.

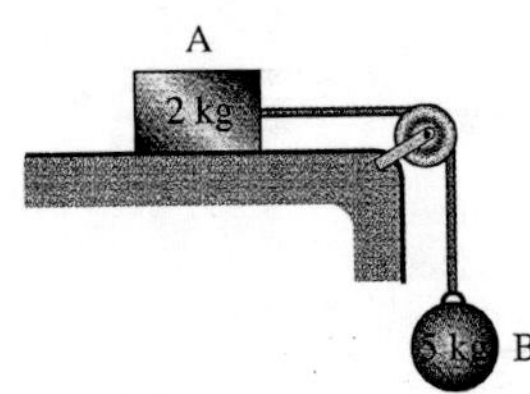

Figure 3.23

(i) Draw diagrams to show the forces acting on each of A and B.

(ii) If the block A does not slip, find the tension in the string and calculate the magnitude of the friction force on the block.

(iii) Write down the resultant force acting on each of A and B if the block slips and accelerates.

SOLUTION

(i)

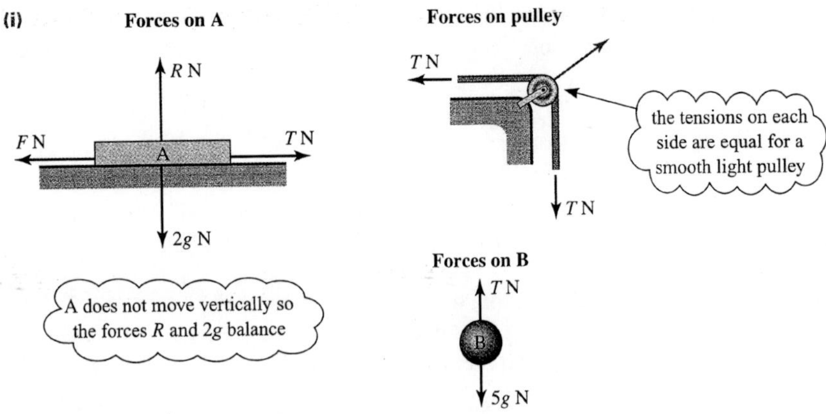

Figure 3.24

Note

The masses of 2 kg and 5 kg are not shown in the force diagram. The weights $2g$ N and $5g$ N are more appropriate.

(ii) When the block does not slip, the forces on B are in equilibrium so

$$5g - T = 0$$
$$T = 5g$$

The tension throughout the string is $5g$ N.
For A, the resultant horizontal force is zero so

$$T - F = 0$$
$$F = T = 5g$$

The friction force is $5g$ N towards the left.

(iii) When the block slips, the forces are not in equilibrium and T and F have different magnitudes.

The resultant horizontal force on A is $(T - F)$ N towards the right.
The resultant force on B is $(5g - T)$ N vertically downwards.

EXERCISE 3D

In this exercise you are asked to draw force diagrams using the various types of force you have met in this chapter. Remember that all the forces you need, other than weight, occur when objects are in contact or joined together in some way. Where motion is involved, indicate its direction clearly.

1 Draw labelled diagrams showing the forces acting on the objects in italics.

(i) *A car* towing a caravan.

(ii) *A caravan* being towed by a car.

(iii) *A person* pushing a supermarket trolley.

(iv) *A suitcase* on a horizontal moving pavement (as at an airport)

(a) immediately after it has been put down

(b) when it is moving at the same speed as the pavement.

(v) *A sledge* being pulled uphill.

2 Ten boxes each of mass 5 kg are stacked on top of each other on the floor.

(i) What forces act on the top box?

(ii) What forces act on the bottom box?

3 The diagrams show a box of mass m under different systems of forces.

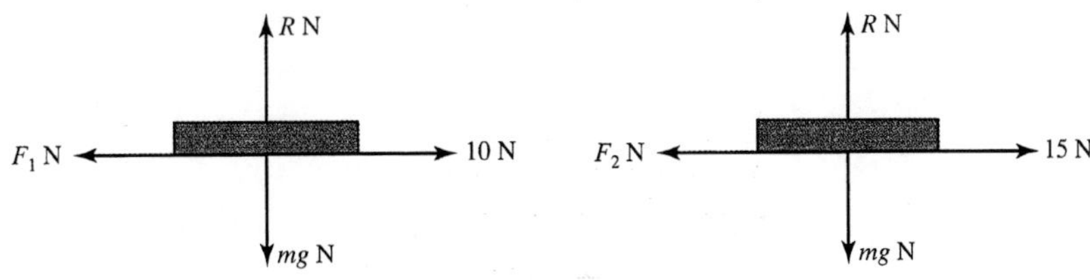

(i) In the first case the box is at rest. State the value of F_1.

(ii) In the second case the box is slipping. Write down the resultant horizontal force acting on it.

4 In this diagram the pulleys are smooth and light, the strings are light, and the table is rough.

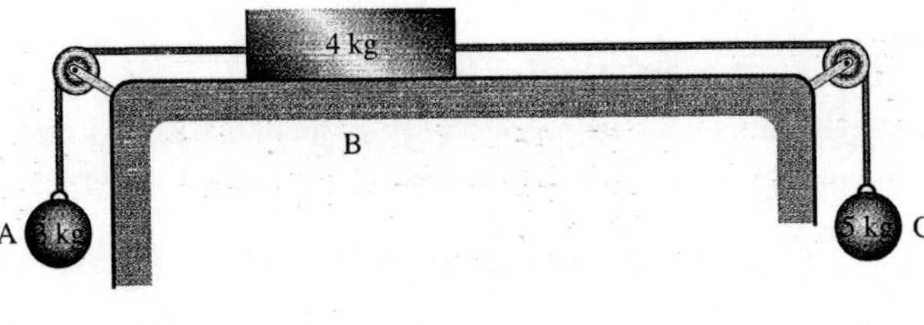

(i) What is the direction of the friction force on the block B?

(ii) Draw clear diagrams to show the forces on each of A, B and C.

(iii) By considering the equilibrium of A and C, calculate the tensions in the strings when there is no slipping.

(iv) Calculate the magnitude of the friction when there is no slipping.

Now suppose that there is insufficient friction to stop the block from slipping.

(v) Write down the resultant force acting on each of A, B and C.

5 A man who weighs 720 N is doing some repairs to a shed. In each of these situations draw diagrams showing

(a) the forces the man exerts on the shed

(b) all the forces acting on the man (ignore any tools he might be using).

In each case, compare the reaction between the man and the floor with his weight of 720 N.

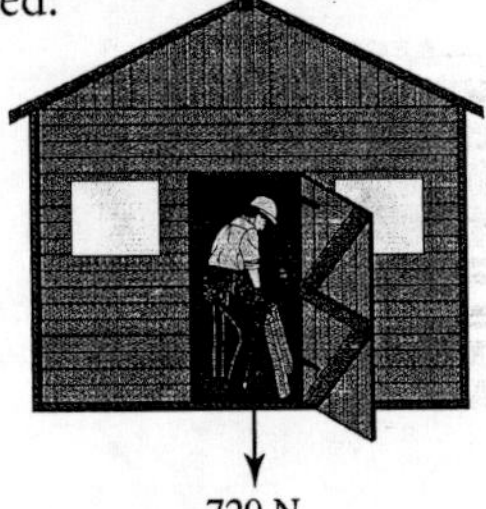

(i) He is pushing upwards on the ceiling with force U N.

(ii) He is pulling downwards on the ceiling with force D N.

(iii) He is pulling upwards on a nail in the floor with force F N.

(iv) He is pushing downwards on the floor with force T N.

6 The diagram shows a train, consisting of an engine of mass 50 000 kg pulling two trucks, A and B, each of mass 10 000 kg. The force of resistance on the engine is 2000 N and that on each of the trucks 200 N. The train is travelling at constant speed.

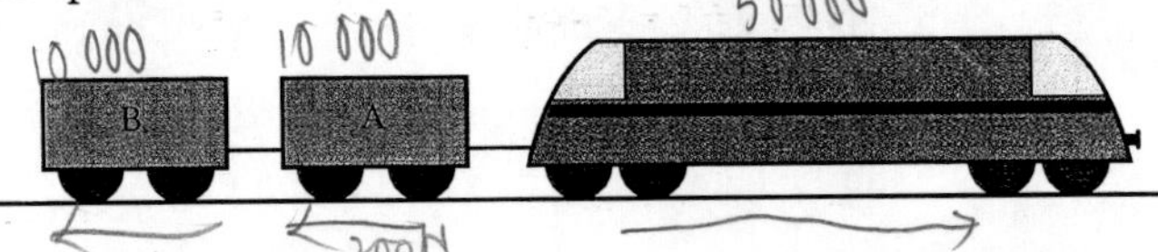

(i) Draw a diagram showing the horizontal forces on the train as a whole. Hence, by considering the equilibrium of the train as a whole, find the driving force provided by the engine.

The coupling connecting truck A to the engine exerts a force T_1 N on the engine and the coupling connecting truck B to truck A exerts a force T_2 N on truck B.

(ii) Draw diagrams showing the horizontal forces on the engine and on truck B.

(iii) By considering the equilibrium of the engine alone, find T_1.

(iv) By considering the equilibrium of truck B alone, find T_2.

(v) Show that the forces on truck A are also in equilibrium.

Historical note

Isaac Newton was born in Lincolnshire in 1642. He was not an outstanding scholar either as a schoolboy or as a university student, yet later in life he made remarkable contributions in dynamics, optics, astronomy, chemistry, music theory and theology. He became Member of Parliament for Cambridge University and later Warden of the Royal Mint. His tomb in Westminster Abbey reads 'Let mortals rejoice that there existed such and so great an Ornament to the Human Race'.

Reviewing a mathematical model: air resistance

In mechanics you express the real world as mathematical models. The process of modelling involves the cycle shown in Figure 3.25 and this is used in the example that follows.

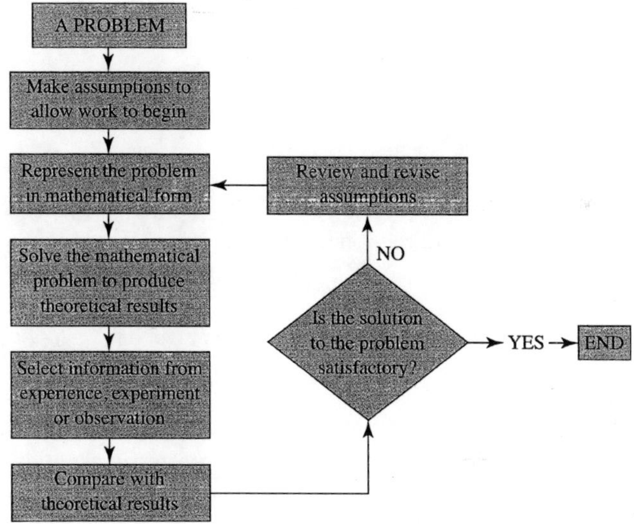

Figure 3.25

? Why does a leaf or a feather or a piece of paper fall more slowly than other objects?

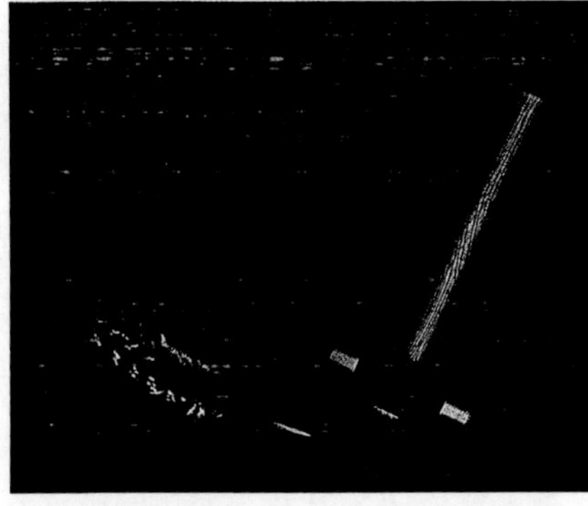

Model 1: The model you have used so far for falling objects has assumed no air resistance and this is clearly unrealistic in many circumstances. There are several possible models for air resistance but it is usually better when modelling to try simple models first. Having rejected the first model you could try a second one as follows.

Model 2: Air resistance is constant and the same for all objects.

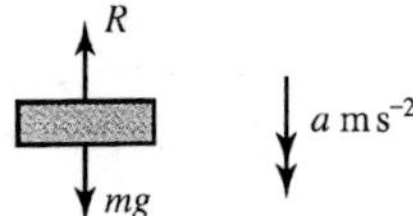

Figure 3.26

Assume an object of mass m falls vertically through the air.

The equation of motion is $mg - R = ma$

$$a = g - \frac{R}{m}$$

The model predicts that a heavy object will have a greater acceleration than a lighter one because $\frac{R}{m}$ is smaller for larger m.

This seems to agree with our experience of dropping a piece of paper and a book, for example. The heavier book has a greater acceleration.

? However, think again about air resistance. Is there a property of the object other than its mass which might affect its motion as it falls? How do people and animals maximise or minimise the force of the air?

Try dropping two identical sheets of paper from a horizontal position, but fold one of them. The folded one lands first even though they have the same mass.

This contradicts the prediction of model 2. A large surface at right angles to the motion seems to increase the resistance.

Model 3: Air resistance is proportional to the area perpendicular to the motion.

Assume the air resistance is kA where k is constant and A is the area of the surface perpendicular to the motion.

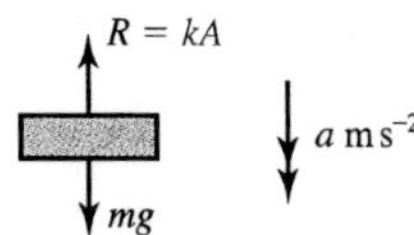

Figure 3.27

The equation of motion is now $mg - kA = ma$

$$a = g - \frac{kA}{m}$$

According to this model, the acceleration depends on the ratio of the area to the mass.

EXPERIMENT

Testing the new model

For this experiment you will need some rigid corrugated card such as that used for packing or in grocery boxes (cereal box card is too thin), scissors and tape.

Cut out ten equal squares of side 8 cm. Stick two together by binding the edges with tape to make them smooth. Then stick three and four together in the same way so that you have four blocks A to D of different thickness as shown in the diagram.

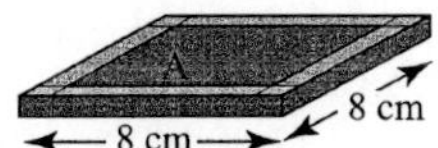

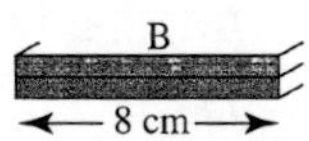

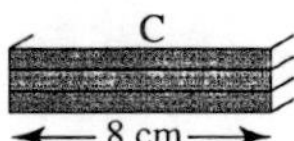

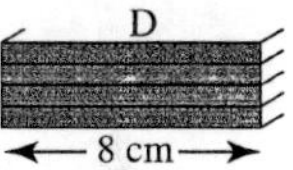

Figure 3.28

Cut out ten larger squares with 12 cm sides. Stick them together in the same way to make four blocks E to H.

Observe what happens when you hold one or two blocks horizontally at a height of about 2 m and let them fall. You do not need to measure anything in this experiment, unless you want to record the area and mass of each block, but write down your observations in an orderly fashion.

1 Drop each one separately. Could its acceleration be constant?

2 Compare A with B and C with D. Make sure you drop each pair from the same height and at the same instant of time. Do they take the same time to fall? Predict what will happen with other combinations and test your predictions.

3 Experiment in a similar way with E to H.

4 Now compare A with E, B with F, C with G and D with H. Compare also the two blocks whose dimensions are all in the same ratio, i.e. B and G.

? Do your results suggest that model 3 might be better than model 2?

If you want to be more certain, the next step would be to make accurate measurements. Nevertheless, this model explains why small animals can be relatively unscathed after falling through heights which would cause serious injury to human beings.

? All the above models ignore one important aspect of air resistance. What is that?

KEY POINTS

1 Newton's laws of motion

I Every object continues in a state of rest or uniform motion in a straight line unless it is acted on by a resultant external force.

II Resultant force = mass × acceleration or $F = ma$.

III When one object exerts a force on another there is always a reaction which is equal, and opposite in direction, to the acting force.

- *Force* is a vector; *mass* is a scalar.
- The *weight* of an object is the force of gravity pulling it towards the centre of the earth. Weight = mg vertically downwards.

2 S.I. units

- length: metre (m)
- time: second (s)
- velocity: m s^{-1}
- acceleration: m s^{-2}
- mass: kilogram (kg)

3 Force

1 newton (N) is the force required to give a mass of 1 kg an acceleration of 1 m s^{-2}.

A force of 1000 newtons (N) = 1 kilonewton (kN).

4 Types of force

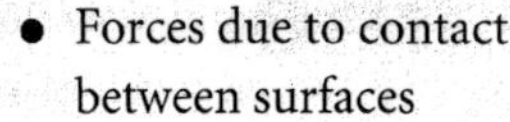

- Forces due to contact between surfaces
- Forces in a joining rod or string

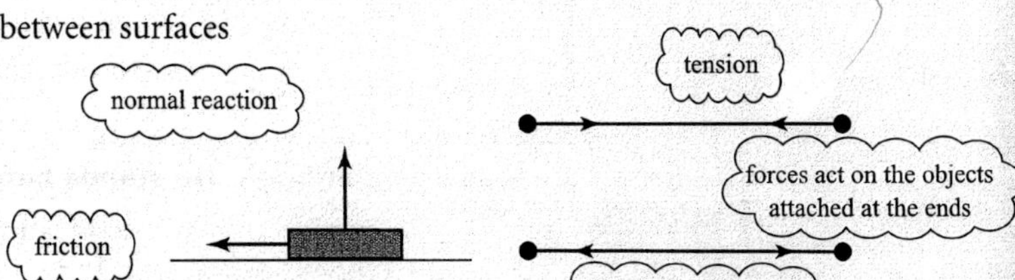

- A smooth light pulley

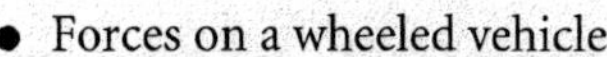

- Forces on a wheeled vehicle

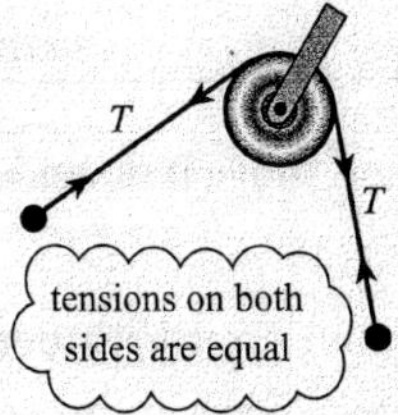

5 Commonly used modelling terms

- inextensible — does not vary in length
- light — negligible mass
- negligible — small enough to ignore
- particle — negligible dimensions
- smooth — negligible friction
- uniform — the same throughout

6 Reviewing a model

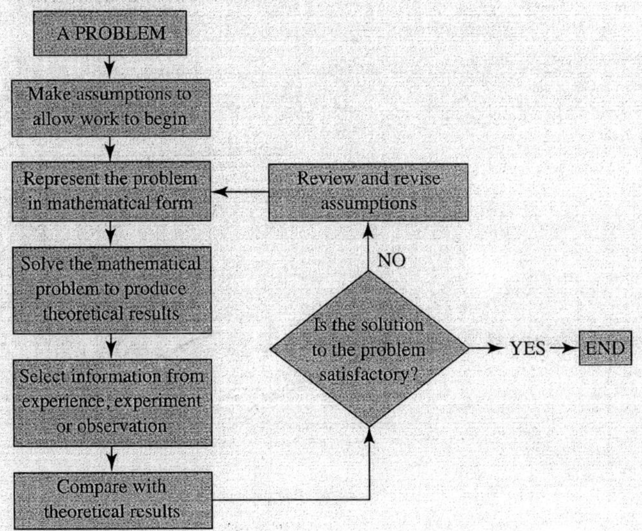

Applying Newton's second law along a line

Nature to him was an open book. He stands before us, strong, certain and alone.

Einstein on Newton

Newton's second law

Attach a weight to a spring balance and move it up and down. What happens to the pointer on the balance?

What would you observe if you stood on some bathroom scales in a moving lift?

Hold a heavy book on your hand and move it up and down.

What force do you feel on your hand?

Equation of motion

Suppose you make the book accelerate upwards at $a\,\mathrm{m\,s^{-2}}$. Figure 4.1 shows the forces acting on the book and the acceleration.

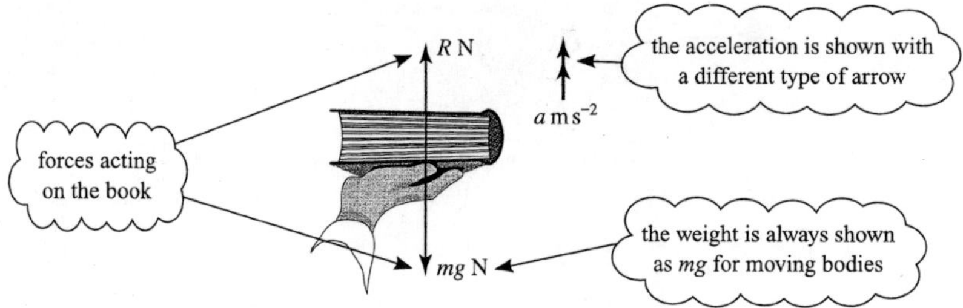

Figure 4.1

By Newton's first law, a resultant force is required to produce an acceleration. In this case the resultant upward force is $R - mg$ newtons.

You were introduced to Newton's second law in Chapter 3. When the forces are in newtons, the mass in kilograms and the acceleration in metres per second squared, this law is:

Resultant force = mass × a ← (where force and acceleration are in the same direction)

So for the book: $R - mg = ma$ ①

When Newton's second law is applied, the resulting equation is called *the equation of motion.*

When you give a book of mass 0.8 kg an acceleration of 0.5 m s^{-2} equation ① becomes

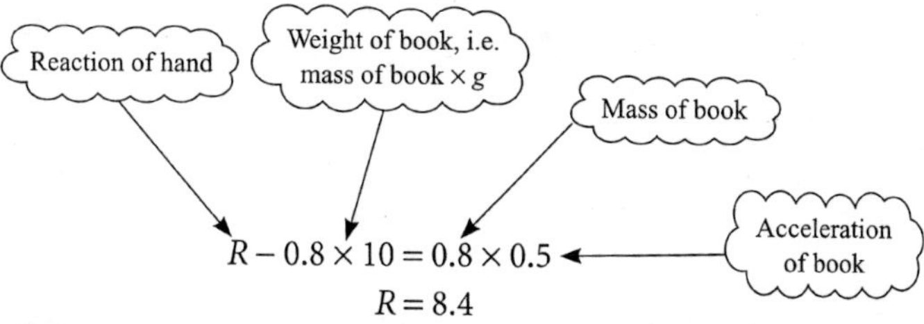

$$R - 0.8 \times 10 = 0.8 \times 0.5$$
$$R = 8.4$$

When the book is accelerating upwards the reaction force of your hand on the book is 8.4 N. This is equal and opposite to the force experienced by you so the book feels heavier than its actual weight, mg, which is $0.8 \times 10 = 8$ N.

EXERCISE 4A

1 Calculate the resultant force in newtons required to produce the following accelerations.

(i) A car of mass 400 kg has acceleration 2 m s^{-2}.

(ii) A blue whale of mass 177 tonnes has acceleration $\frac{1}{2}$ m s^{-2}.

(iii) A pygmy mouse of mass 7.5 g has acceleration 3 m s^{-2}.

(iv) A freight train of mass 42 000 tonnes brakes with deceleration 0.02 m s^{-2}.

(v) A bacterium of mass 2×10^{-16} g has acceleration 0.4 m s^{-2}.

(vi) A woman of mass 56 kg falling off a high building has acceleration 9.8 m s^{-2}.

(vii) A jumping flea of mass 0.05 mg accelerates at 1750 m s^{-2} during take-off.

(viii) A galaxy of mass 10^{42} kg has acceleration 10^{-12} m s^{-2}.

2 A resultant force of 100 N is applied to a body. Calculate the mass of the body when its acceleration is

(i) 0.5 m s^{-2} **(ii)** 2 m s^{-2}

(iii) 0.01 m s^{-2} **(iv)** 10g.

3 What is the reaction between a book of mass 0.8 kg and your hand when it is

(i) accelerating downwards at 0.3 m s^{-2}?

(ii) moving upwards at constant speed?

EXAMPLE 4.1

A lift and its passengers have a total mass of 400 kg. Find the tension in the cable supporting the lift when

(i) the lift is at rest
(ii) the lift is moving at constant speed
(iii) the lift is accelerating upwards at $0.8\,\mathrm{m\,s^{-2}}$
(iv) the lift is accelerating downwards at $0.6\,\mathrm{m\,s^{-2}}$.

SOLUTION

Before starting the calculations you must define a direction as positive. In this example the upward direction is chosen to be positive.

(i) At rest
As the lift is at rest the forces must be in equilibrium. The equation of motion is

$$T - mg = 0$$
$$T - 400 \times 10 = 0$$
$$T = 4000$$

The tension in the cable is 4000 N.

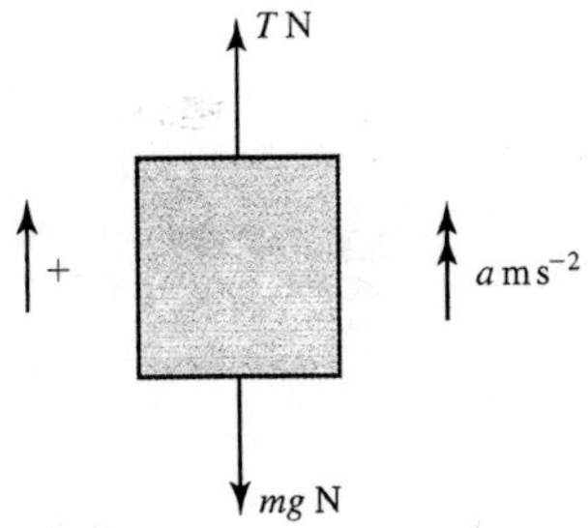

Figure 4.2

(ii) Moving at constant speed
Again, the forces on the lift must be in equilibrium because it is moving at a constant speed, so the tension is 4000 N.

(iii) Accelerating upwards
The resultant upward force on the lift is $T - mg$ so the equation of motion is

$$T - mg = ma$$

which in this case gives

$$T - 400 \times 10 = 400 \times 0.8$$
$$T - 4000 = 320$$
$$T = 4320$$

The tension in the cable is 4320 N.

(iv) Accelerating downwards
The equation of motion is

$$T - mg = ma$$

In this case, a is negative so

$$T - 400 \times 10 = 400 \times (-0.6)$$
$$T - 4000 = -240$$
$$T = 3760$$

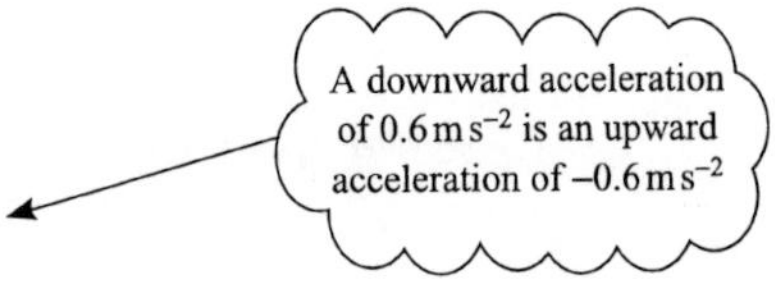

A downward acceleration of $0.6\,\mathrm{m\,s^{-2}}$ is an upward acceleration of $-0.6\,\mathrm{m\,s^{-2}}$

The tension in the cable is 3760 N.

❓ How is it possible for the tension to be 3760 N upwards but the lift to accelerate downwards?

EXAMPLE 4.2

This example shows how the *suvat* formulae for motion with constant acceleration, which you met in Chapter 2, can be used with Newton's second law.

A supertanker of mass 500 000 tonnes is travelling at a speed of $10\,\mathrm{m\,s^{-1}}$ when its engines fail. It then takes half an hour for the supertanker to stop.

(i) Find the force of resistance, assuming it to be constant, acting on the supertanker.

When the engines have been repaired it takes the supertanker 10 minutes to return to its full speed of $10\,\mathrm{m\,s^{-1}}$.

(ii) Find the driving force produced by the engines, assuming this also to be constant.

SOLUTION

Use the direction of motion as positive.

You have to be very careful with signs: the resultant force and acceleration are both positive towards the right.

(i) First find the acceleration of the supertanker, which is constant for constant forces. Figure 4.3 shows the velocities and acceleration.

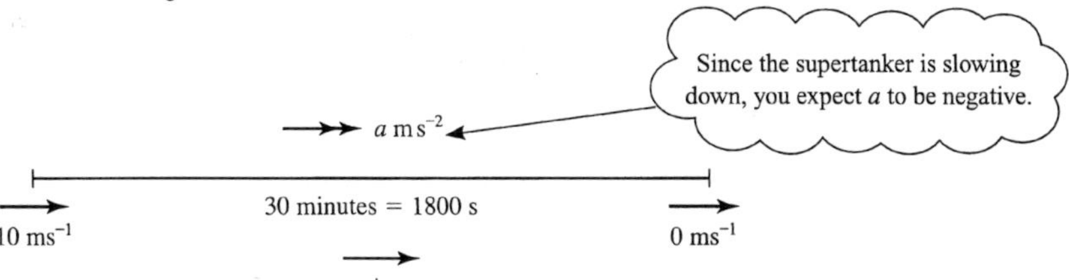

Figure 4.3

You know $u = 10$, $v = 0$, $t = 1800$ and you want a, so use $v = u + at$.

$$0 = 10 + 1800a$$
$$a = -\frac{1}{180}$$

The acceleration is negative because the supertanker is slowing down.

Now we can use Newton's second law (Newton II) to write down the equation of motion. Figure 4.4 shows the horizontal forces and the acceleration.

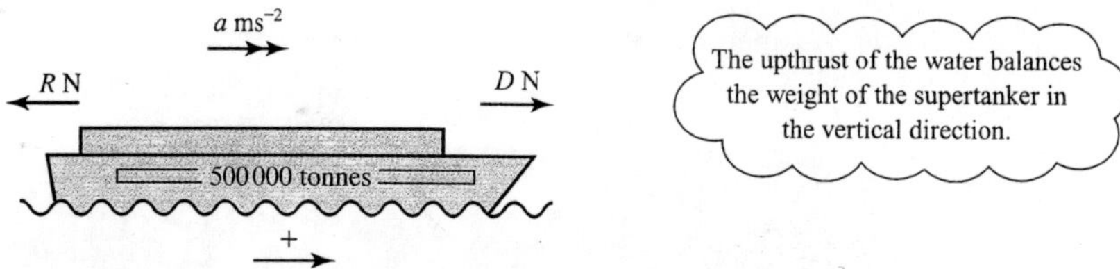

Figure 4.4

The resultant forward force is D – R newtons. When there is no driving force $D = 0$ so Newton II gives

the mass must be in kg

$$0 - R = 500\,000\,000 \times a$$

so when $a = -\frac{1}{180}$, $\quad -R = 500\,000\,000 \times \left(-\frac{1}{180}\right)$

The resistance to motion is 2.78×10^6 N or 2780 kN (correct to 3 s.f.).

(ii) Now $u = 0$, $v = 10$ and $t = 600$, and you want a, so use $v = u + at$ again.

$$10 = 0 + a \times 600$$
$$a = \frac{1}{60}$$

Using Newton's second law again

$$D - R = 500\,000\,000 \times a$$
$$D - 2.78 \times 10^6 = 500\,000\,000 \times \frac{1}{60}$$
$$D = 2.78 \times 10^6 + 8.33 \times 10^6$$

The driving force is 11.11×10^6 N or 11 100 kN (correct to 3 s.f.).

Tackling mechanics problems

When you tackle mechanics problems such as these you will find them easier if you:

- always draw a clear diagram
- clearly indicate the positive direction
- label each object (A, B, etc. or whatever is appropriate)
- show all the forces acting on each object
- make it clear which object you are referring to when writing an equation of motion.

EXERCISE 4B

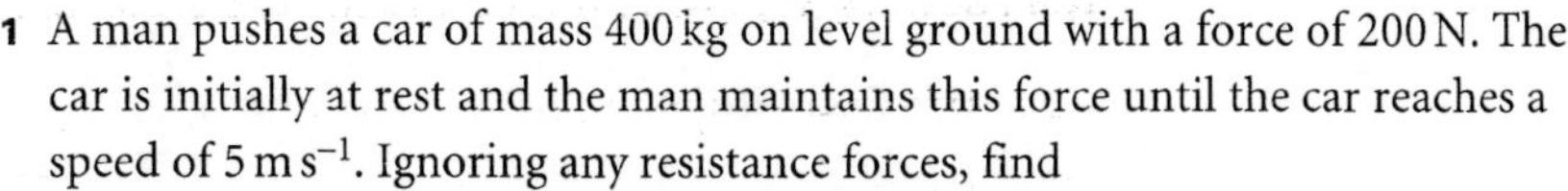

1 A man pushes a car of mass 400 kg on level ground with a force of 200 N. The car is initially at rest and the man maintains this force until the car reaches a speed of $5\,\text{m}\,\text{s}^{-1}$. Ignoring any resistance forces, find

(i) the acceleration of the car

(ii) the distance the car travels while the man is pushing.

2 The engine of a car of mass 1.2 tonnes can produce a driving force of 2000 N. Ignoring any resistance forces, find

(i) the car's resulting acceleration

(ii) the time taken for the car to go from rest to $27\,\text{m}\,\text{s}^{-1}$ (about 60 mph).

3 A top sprinter of mass 65 kg starting from rest reaches a speed of $10\,\text{m}\,\text{s}^{-1}$ in 2 s.

(i) Calculate the force required to produce this acceleration, assuming it is uniform.

(ii) Compare this to the force exerted by a weight lifter holding a mass of 180 kg above the ground.

4 An ice skater of mass 65 kg is initially moving with speed $2\,\text{m}\,\text{s}^{-1}$ and glides to a halt over a distance of 10 m. Assuming that the force of resistance is constant, find

(i) the size of the resistance force

(ii) the distance he would travel gliding to rest from an initial speed of $6\,\text{m}\,\text{s}^{-1}$

(iii) the force he would need to apply to maintain a steady speed of $10\,\text{m}\,\text{s}^{-1}$.

5 A helicopter of mass 1000 kg is taking off vertically.

(i) Draw a labelled diagram showing the forces on the helicopter as it lifts off and the direction of its acceleration.

(ii) Its initial upward acceleration is $1.5\,\text{m}\,\text{s}^{-2}$. Calculate the upward force its rotors exert. Ignore the effects of air resistance.

6 Pat and Nicholas are controlling the movement of a canal barge by means of long ropes attached to each end. The tension in the ropes may be assumed to be horizontal and parallel to the line and direction of motion of the barge, as shown in the diagrams.

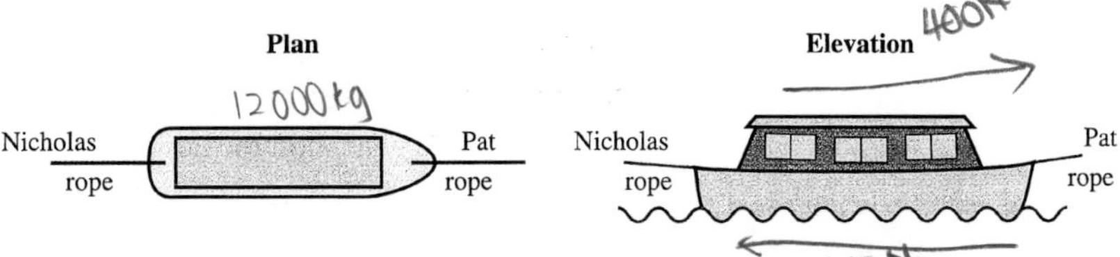

The mass of the barge is 12 tonnes and the total resistance to forward motion may be taken to be 250 N at all times. Initially Pat pulls the barge forwards from rest with a force of 400 N and Nicholas leaves his rope slack.

(i) Write down the equation of motion for the barge and hence calculate its acceleration.

Pat continues to pull with the same force until the barge has moved 10 m.

(ii) What is the speed of the barge at this time and for what length of time did Pat pull?

Pat now lets her rope go slack and Nicholas brings the barge to rest by pulling with a constant force of 150 N.

(iii) Calculate

(a) how long it takes the barge to come to rest

(b) the total distance travelled by the barge from when it first moved

(c) the total time taken for the motion. [MEI]

7 A spaceship of mass 5000 kg is stationary in deep space. It fires its engines, producing a forward thrust of 2000 N for 2.5 minutes, and then turns them off.

(i) What is the speed of the spaceship at the end of the 2.5 minute period?

(ii) Describe the subsequent motion of the spaceship.

The spaceship then enters a cloud of interstellar dust which brings it to a halt after a further distance of 7200 km.

(iii) What is the force of resistance (assumed constant) on the spaceship from the interstellar dust cloud?

The spaceship is travelling in convoy with another spaceship which is the same in all respects except that it is carrying an extra 500 kg of equipment. The second spaceship carries out exactly the same procedure as the first one.

(iv) Which spaceship travels further into the dust cloud?

8 A crane is used to lift a hopper full of cement to a height of 20 m on a building site. The hopper has mass 200 kg and the cement 500 kg. Initially the hopper accelerates upwards at $0.05\,\mathrm{m\,s^{-2}}$, then it travels at constant speed for some time before decelerating at $0.1\,\mathrm{m\,s^{-2}}$ until it is at rest. The hopper is then emptied.

(i) Find the tension in the crane's cable during each of the three phases of the motion and after emptying.

The cable's maximum safe load is 10 000 N.

(ii) What is the greatest mass of cement that can safely be transported in the same manner?

The cable is in fact faulty and on a later occasion breaks without the hopper leaving the ground. On that occasion the hopper is loaded with 720 kg of cement.

(iii) What can you say about the strength of the cable?

9 The police estimate that for good road conditions the frictional force, F, on a skidding vehicle of mass m is given by $F = 0.8\,mg$. A car of mass 450 kg skids to a halt narrowly missing a child. The police measure the skid marks and find they are 12.0 m long.

(i) Calculate the deceleration of the car when it was skidding to a halt.

The child's mother says the car was travelling well over the speed limit of $50\,\text{km h}^{-1}$ but the driver of the car says she was travelling at $48\,\text{km h}^{-1}$ and the child ran out in front of her.

(ii) Calculate the speed of the car when it started to skid.

Who was telling the truth?

Newton's second law applied to connected objects

This section is about using Newton's second law for more than one object. It is important to be very clear which forces act on which object in these cases.

A stationary helicopter is raising two people of masses 90 kg and 70 kg as shown in the diagram.

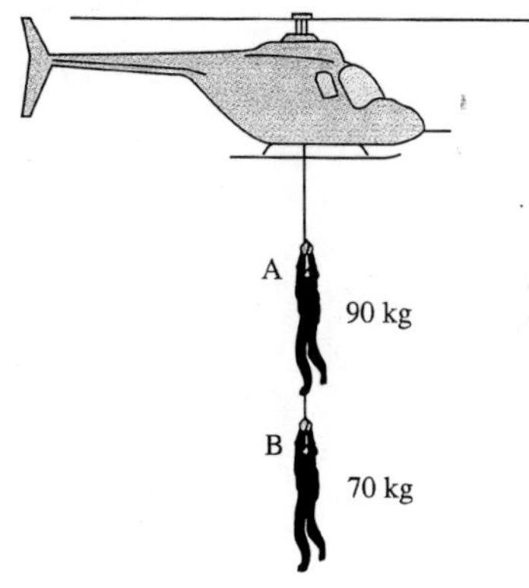

Figure 4.5

? Imagine that you are each person in turn. Your eyes are shut so you cannot see the helicopter or the other person. What forces act on you?

Remember that all the forces acting, apart from your weight, are due to contact between you and something else.

Which forces acting on A and B are equal in magnitude? What can you say about their accelerations?

EXAMPLE 4.3

(i) Draw a diagram to show the forces acting on the two people being raised by the helicopter in figure 4.5 and their acceleration.

(ii) Write down the equation of motion for each person.

(iii) When the force applied to the first person, A, by the helicopter is $180g$ N, calculate

(a) the acceleration of the two people being raised

(b) the tension in the ropes.

SOLUTION

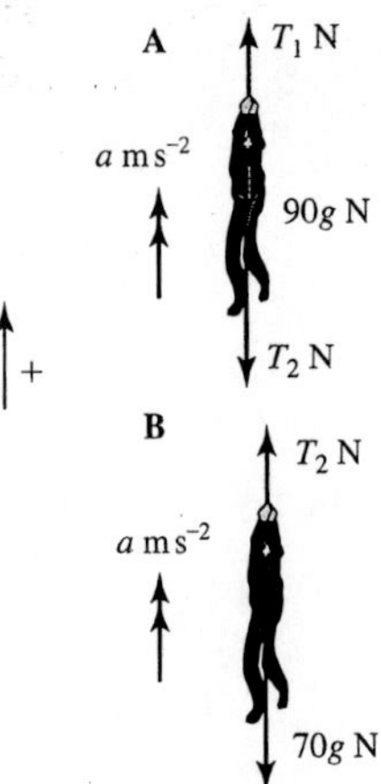

Figure 4.6

(i) Figure 4.6 shows the acceleration and forces acting on the two people.

(ii) When the helicopter applies a force T_1 N to A, the resultant upward forces are

A $(T_1 - 90g - T_2)$N

B $(T_2 - 70g)$ N

Their equations of motion are

A (↑) $T_1 - 90g - T_2 = 90a$ ①

B (↑) $T_2 - 70g = 70a$ ②

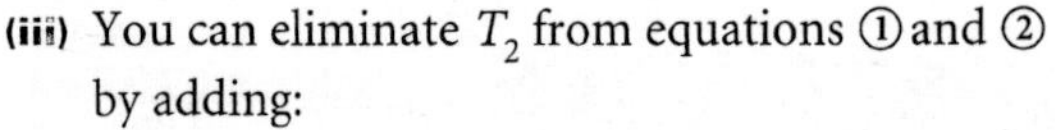

(iii) You can eliminate T_2 from equations ① and ② by adding:

$$T_1 - 90g - T_2 + T_2 - 70g = 90a + 70a$$

$$T_1 - 160g = 160a \quad ③$$

When the force, T_1, applied by the helicopter is $180g$

$$20g = 160a$$

$$a = 1.25$$

Substituting for a in equation ② gives $T_2 = 70 \times 1.25 + 70g$

$$= 787.5$$

The acceleration is $1.25\,\mathrm{m\,s^{-2}}$ and the tensions in the ropes are 1800 N and 787.5 N.

❓ The force pulling downwards on A is 787.5 N. This is *not* equal to B's weight (700 N). Why are they different?

Treating the system as a whole

When two objects are moving in the same direction with the same velocity at all times they can be treated as one. In Example 4.3 the two people can be treated as one object and then the equal and opposite forces T_2 cancel out. They are *internal forces* similar to the forces between your head and your body.

The resultant upward force on both people is $T_1 - 90g - 70g$ and the total mass is 160 kg so the equation of motion is:

$$T_1 - 90g - 70g = 160a$$ ← as equation ③ above

So you can find a directly

$$\text{when } T_1 = 180g$$
$$20g = 160a$$
$$a = 1.25$$

Treating the system as a whole finds a, but not the internal force T_2.

You need to consider the motion of B separately to obtain equation ②.

$$T_2 - 70g = 70a \quad ②$$
$$T_2 = 787.5$$ ← as before

Using this method, equation ① can be used to check your answers. Alternatively, you could use equation ① to find T_2 and equation ② to check your answers.

When several objects are joined there are always more equations possible than are necessary to solve a problem and they are not all independent. In the above example, only two of the equations were necessary to solve the problem. The trick is to choose the most relevant ones.

A note on mathematical modelling

Several modelling assumptions have been made in the solution to Example 4.3. It is assumed that:

- the only forces acting on the people are their weights and the tensions in the ropes (forces due to the wind or air turbulence are ignored)
- the motion is vertical and nobody swings from side to side
- the ropes do not stretch (i.e. they are inextensible) so the accelerations of the two people are equal
- the people are rigid bodies which do not change shape and can be treated as particles.

All these modelling assumptions make the problem simpler. In reality, if you were trying to solve such a problem you might work through it first using these assumptions. You would then go back and decide which ones needed to be modified to produce a more realistic solution.

In the next example one person is moving vertically and the other horizontally. You might find it easier to decide on which forces are acting if you imagine you are Alvin or Bernard and you can't see the other person.

EXAMPLE 4.4

Alvin is using a snowmobile to pull Bernard out of a crevasse. His rope passes over a smooth block of ice at the top of the crevasse as shown in figure 4.7 and Bernard hangs freely away from the side. Alvin and his snowmobile together have a mass of 300 kg and Bernard's mass is 75 kg. Ignore any resistance to motion.

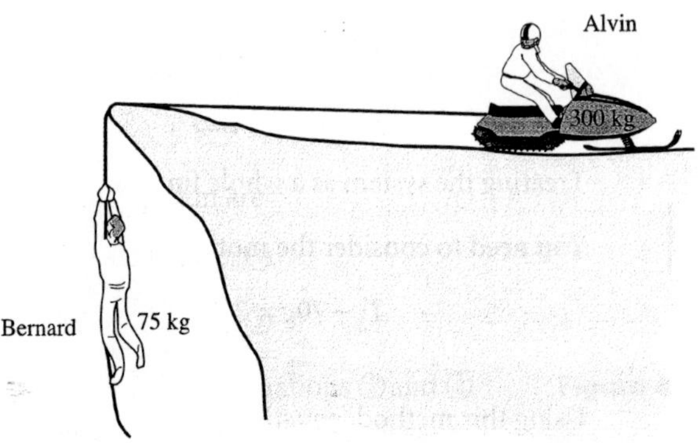

Figure 4.7

(i) Draw diagrams showing the forces on the snowmobile (including Alvin) and on Bernard.

(ii) Calculate the driving force required for the snowmobile to give Bernard an upward acceleration of $0.5\,\mathrm{m\,s^{-2}}$ and the tension in the rope for this acceleration.

(iii) How long will it take for Bernard's speed to reach $5\,\mathrm{m\,s^{-1}}$ starting from rest and how far will he have been raised in this time?

SOLUTION

(i) The diagram shows the essential features of the problem.

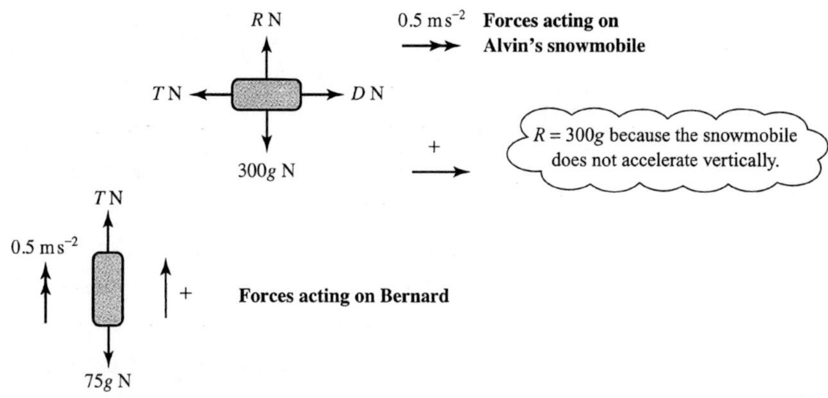

Figure 4.8

(ii) Alvin and Bernard have the same acceleration providing the rope does not stretch. The tension in the rope is T newtons and Alvin's driving force is D newtons.

The equations of motion are:

Alvin ($\rightarrow$)

$$D - T = 300 \times 0.5$$
$$D - T = 150 \quad ①$$

The force towards the right = mass × acceleration towards the right

Bernard ($\uparrow$)

$$T - 75g = 75 \times 0.5$$
$$T - 75g = 37.5 \quad ②$$
$$T = 37.5 + 75g$$
$$T = 787.5$$

The upward force = mass × upward acceleration

Substituting in equation ①

$$D - 787.5 = 150$$
$$D = 937.5$$

The driving force required is 937.5 N and the tension in the rope is 787.5 N.

(iii) When $u = 0$, $v = 5$, $a = 0.5$ and t is required

$$v = u + at$$
$$5 = 0 + 0.5 \times t$$
$$t = 10$$

The time taken is 10 seconds. Using $s = ut + \frac{1}{2}at^2$ to find s gives

$$s = 0 + \frac{1}{2}at^2$$
$$s = \frac{1}{2} \times 0.5 \times 100$$
$$s = 25$$

$v^2 = u^2 + 2as$ would also give s

The distance he has been raised is 25 m.

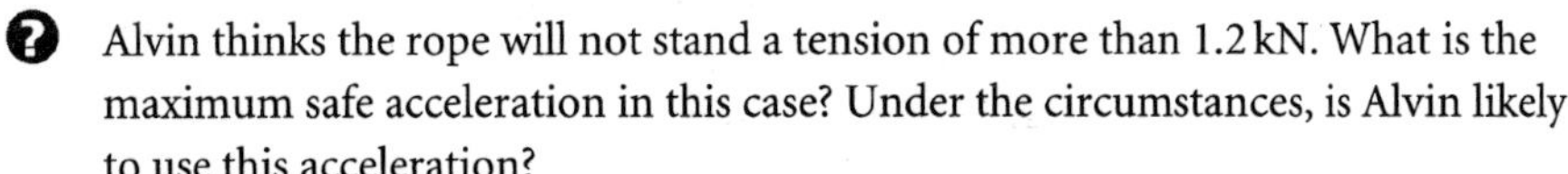

Alvin thinks the rope will not stand a tension of more than 1.2 kN. What is the maximum safe acceleration in this case? Under the circumstances, is Alvin likely to use this acceleration?

Make a list of the modelling assumptions made in this example and suggest what effect a change in each of these assumptions might have on the solution.

EXAMPLE 4.5

A woman of mass 60 kg is standing in a lift.

(i) Draw a diagram showing the forces acting on the woman.

Find the normal reaction of the floor of the lift on the woman in the following cases.

(ii) The lift is moving upwards at a constant speed of $3\,\text{m}\,\text{s}^{-1}$.
(iii) The lift is moving upwards with an acceleration of $2\,\text{m}\,\text{s}^{-2}$ upwards.
(iv) The lift is moving downwards with an acceleration of $2\,\text{m}\,\text{s}^{-2}$ downwards.
(v) The lift is moving downwards and slowing down with a deceleration of $2\,\text{m}\,\text{s}^{-2}$.

In order to calculate the maximum number of occupants that can safely be carried in the lift, the following assumptions are made:

The lift has mass 300 kg, all resistances to motion may be neglected, the mass of each occupant is 75 kg and the tension in the supporting cable should not exceed 12 000 N.

(vi) What is the greatest number of occupants that can be carried safely if the magnitude of the acceleration does not exceed $3\,\text{m}\,\text{s}^{-2}$? [MEI]

SOLUTION

(i) The diagram shows the forces acting on the woman and her acceleration.

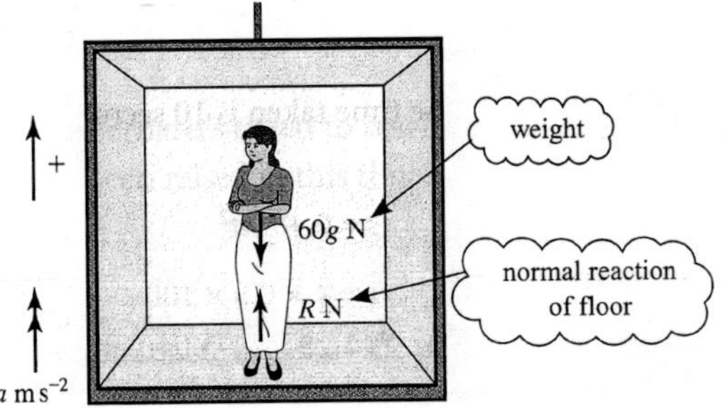

Figure 4.9

In general, when positive is upwards, her equation of motion is

$(\uparrow) \qquad R - 60g = 60a$

This equation contains all the mathematics in the situation. It can be used to solve parts (ii) to (iv).

(ii) When the speed is constant $a = 0$ so $R = 60g = 600$.

The normal reaction is 600 N.

(iii) When $a = 2$

$$\begin{aligned} R - 60g &= 60 \times 2 \\ R &= 120 + 600 \\ &= 720 \end{aligned}$$

The normal reaction is 720 N.

(iv) When the acceleration is downwards, $a = -2$ so

$$\begin{aligned} R - 60g &= 60 \times (-2) \\ R &= 480 \end{aligned}$$

The normal reaction is 480 N.

(v) When the lift is moving downwards and slowing down, the acceleration is negative downwards, so it is positive upwards, and $a = +2$. Then $R = 720$ as in part **(iii)**.

(vi) When there are n passengers in the lift, the combined mass of these and the lift is $(300 + 75n)$ kg and their weight is $(300 + 75n)g$ N.

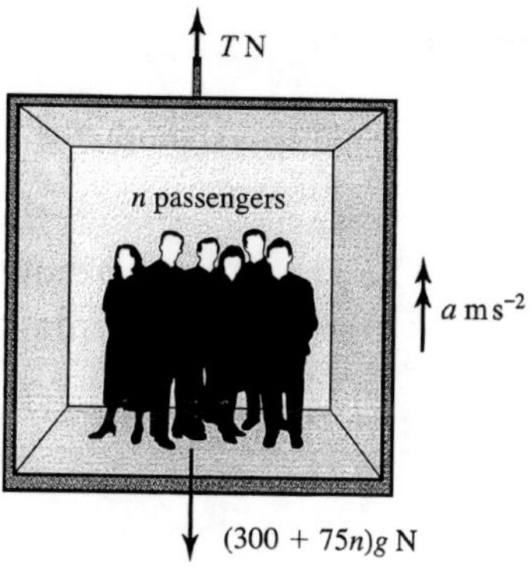

Figure 4.10

The equation of motion for the lift and passengers together is

$$T - (300 + 75n)g = (300 + 75n)a$$

So when $a = 3$ and $g = 10$,

$$\begin{aligned} T &= (300 + 75n) \times 3 + (300 + 75n) \times 10 \\ &= 13(300 + 75n) \end{aligned}$$

For a maximum tension of 12 000 N

$$\begin{aligned} 12\,000 &= 13(300 + 75n) \\ 12\,000 &= 3900 + 975n \\ 8100 &= 975n \\ n &= 8.31 \text{ (to 3 s.f.)} \end{aligned}$$

The lift cannot carry more than 8 passengers.

EXAMPLE 4.6

Two particles A and B, of masses 0.6 kg and 0.4 kg respectively, are connected by a light inextensible string which passes over a smooth fixed pulley. The particles hang freely, as shown in the diagram, and are released from rest.

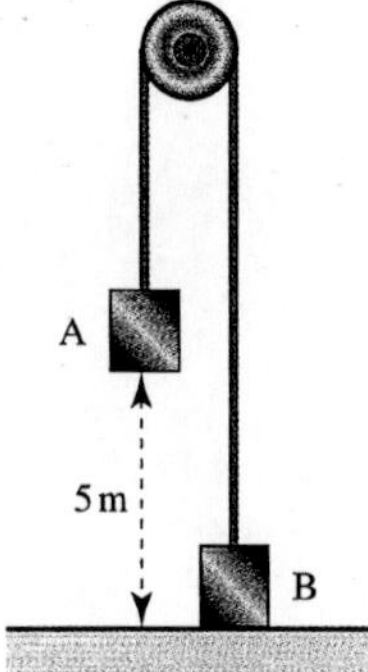

Figure 4.11

(i) Find the acceleration of the system and the tension in the string.

After 2 seconds the string is cut and in the subsequent motion both particles move freely under gravity.

(ii) Find the height of both particles at the instant that the string is cut.

SOLUTION

(i) Since the pulley is smooth, the tension, T N, is the same throughout the string.

When the particles start to move, particle A accelerates downwards and particle B accelerates upwards. Let their acceleration be $a\,\text{m}\,\text{s}^{-2}$.

Draw separate force diagrams for each particle.

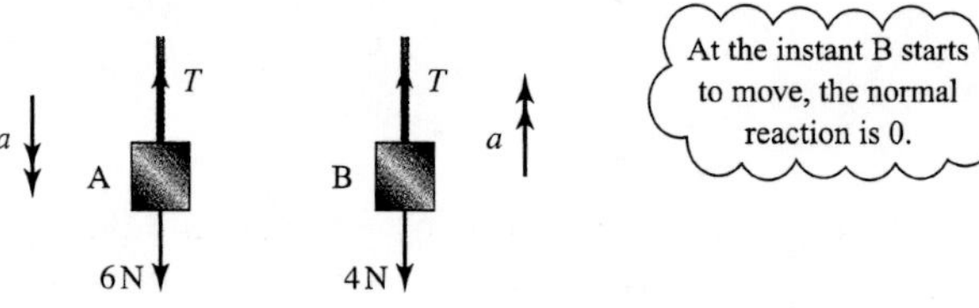

Applying $F = ma$ to each particle gives:

Particle A: $6 - T = 0.6a$ ①

Particle B: $T - 4 = 0.4a$ ②

Adding equations ① and ② gives:

$$2 = 1a$$

$$\Rightarrow \quad a = 2$$

Substituting $a = 2$ into equation ① or ② gives $T = 4.8$.

The acceleration is $2\ \text{m}\,\text{s}^{-2}$ and the tension is 4.8 N.

(ii) Let the particles' initial velocity be $u\,\mathrm{m\,s^{-1}}$ and the distance they have travelled $t\,\mathrm{s}$ after they are released be $s\,\mathrm{m}$.

$u = 0$ as the particles are initially at rest.

To find the height, s, use

$$s = ut + \frac{1}{2}at^2$$

with $a = 2$ and $t = 2$.

$$s = \frac{1}{2} \times 2 \times 2^2$$
$$s = 4$$

Particle A moves down 4 m and particle B moves up 4 m so that when the string is cut:

particle A is 5 m − 4 m = 1 m above the ground
particle B is 4 m above the ground.

EXERCISE 4C

Remember: Always make it clear which object each equation of motion refers to.

1 Masses A of 100 g and B of 200 g are attached to the ends of a light, inextensible string which hangs over a smooth pulley as shown in the diagram.

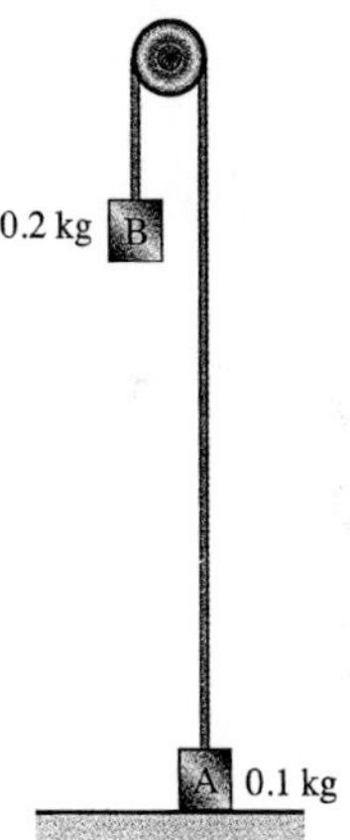

Initially B is held at rest 2 m above the ground and A rests on the ground with the string taut. Then B is let go.

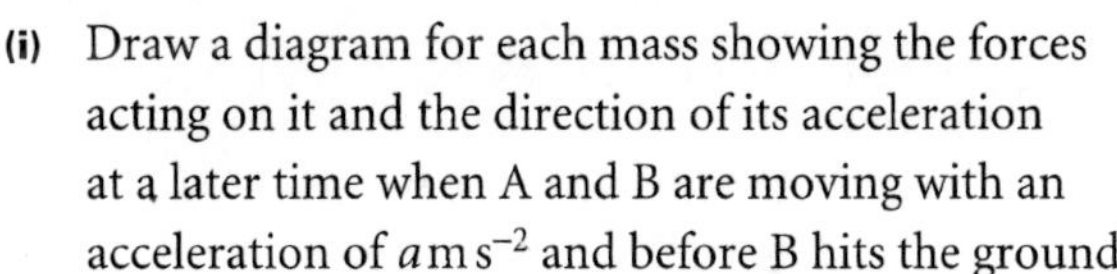

(i) Draw a diagram for each mass showing the forces acting on it and the direction of its acceleration at a later time when A and B are moving with an acceleration of $a\,\mathrm{m\,s^{-2}}$ and before B hits the ground.

(ii) Write down the equation of motion of each mass in the direction it moves using Newton's second law.

(iii) Use your equations to find a and the tension in the string.

(iv) Find the time taken for B to hit the ground.

2 The diagram shows a block of mass 5 kg lying on a smooth table. It is attached to blocks of mass 2 kg and 3 kg by strings which pass over smooth pulleys. The tensions in the strings are T_1 and T_2, as shown, and the blocks have acceleration $a\,\mathrm{m\,s^{-2}}$.

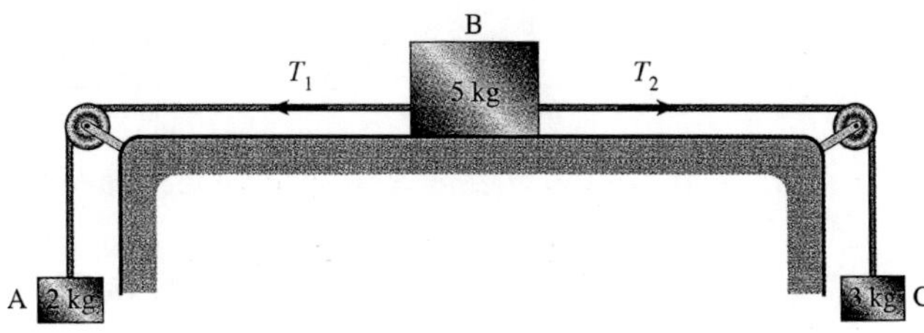

(i) Draw a diagram for each block showing all the forces acting on it and its acceleration.

(ii) Write down the equation of motion for each of the blocks.

(iii) Use your equations to find the values of a, T_1 and T_2.

In practice, the table is not truly smooth and a is found to be $0.5\,\text{m s}^{-2}$.

(iv) Repeat parts **(i)** and **(ii)** including a frictional force on B and use your new equations to find the frictional force that would produce this result.

3 A car of mass 800 kg is pulling a caravan of mass 1000 kg along a straight, horizontal road. The caravan is connected to the car by means of a light, rigid tow bar. The car is exerting a driving force of 1270 N. The resistances to the forward motion of the car and caravan are 400 N and 600 N respectively; you may assume that these resistances remain constant.

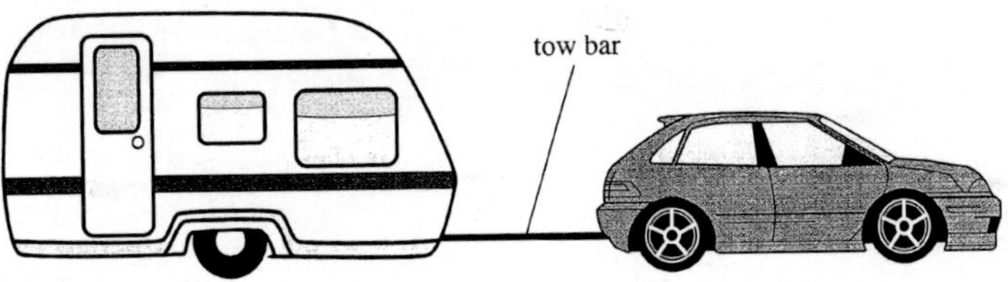

(i) Show that the acceleration of the car and caravan is $0.15\,\text{m s}^{-2}$.

(ii) Draw a diagram showing all the forces acting on the caravan along the line of its motion. Calculate the tension in the tow bar.

The driving force is removed but the car's brakes are not applied.

(iii) Determine whether the tow bar is now in tension or compression.

The car's brakes are then applied gradually. The brakes of the caravan come on automatically when the tow bar is subjected to a compression force of at least 50 N.

(iv) Show that the acceleration of the caravan just before its brakes come on automatically is $0.65\,\text{m s}^{-2}$ in the direction of its motion. Hence, calculate the braking force on the car necessary to make the caravan brakes come on.

[MEI]

4 The diagram shows a goods train consisting of an engine of mass 40 tonnes and two trucks of 20 tonnes each. The engine is producing a driving force of 5×10^4 N, causing the train to accelerate. The ground is level and resistance forces may be neglected.

(i) By considering the motion of the whole train, find its acceleration.

(ii) Draw a diagram to show the forces acting on the engine and use this to help you to find the tension in the first coupling.

(iii) Find the tension in the second coupling.

The brakes on the first truck are faulty and suddenly engage, causing a resistance of 10^4 N.

(iv) What effect does this have on the tension in the coupling to the last truck?

[MEI, adapted]

5 A short train consists of two locomotives, each of mass 20 tonnes, with a truck of mass 10 tonnes coupled between them, as shown in the diagram. The resistances to forward motion are 0.5 kN on the truck and 1 kN on each of the locomotives. The train is travelling along a straight, horizontal section of track.

Initially there is a driving force of 15 kN from the front locomotive only.

(i) Calculate the acceleration of the train.

(ii) Draw a diagram indicating the horizontal forces acting on each part of the train, including the forces in each of the couplings. Calculate the forces acting on the truck due to each coupling.

On another occasion each of the locomotives produces a driving force of 7.5 kN in the same direction and the resistances remain as before.

(iii) Find the acceleration of the train and the forces now acting on the truck due to each of the couplings. Compare your answer to this part with your answer to part **(ii)** and comment briefly.

[MEI]

6 The diagram shows a lift containing a single passenger.

(i) Make clear diagrams to show the forces acting on the passenger and the forces acting on the lift using the following letters:

the tension in the cable, T N
the reaction of the lift on the passenger, R_P N
the reaction of the passenger on the lift, R_L N
the weight of the passenger, mg N
the weight of the lift, Mg N.

The masses of the lift and the passenger are 450 kg and 50 kg respectively.

(ii) Calculate T, R_P and R_L when the lift is stationary.

The lift then accelerates upwards at $0.8\,\mathrm{m\,s^{-2}}$.

(iii) Find the new values of T, R_P and R_L.

7 A man of mass 70 kg is standing in a lift which has an upward acceleration $a\,\mathrm{m\,s^{-2}}$.

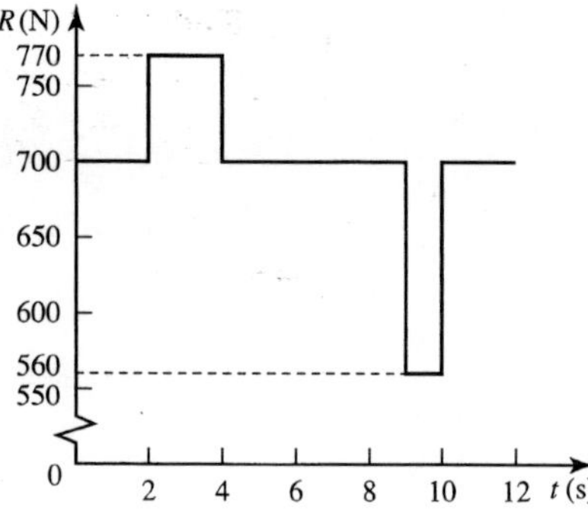

(i) Draw a diagram showing the man's weight, the force, R N, that the lift floor exerts on him and the direction of his acceleration.

(ii) Find the value of a when $R = 770$ N.

The graph shows the value of R from the time ($t = 0$) when the man steps into the lift to the time ($t = 12$) when he steps out.

(iii) Explain what is happening in each section of the journey.

(iv) Draw the corresponding speed–time graph.

(v) To what height does the man ascend?

8 A lift in a mine shaft takes exactly one minute to descend 500 m. It starts from rest, accelerates uniformly for 12.5 seconds to a constant speed which it maintains for some time and then decelerates uniformly to stop at the bottom of the shaft.

The mass of the lift is 5 tonnes and on the day in question it is carrying 12 miners whose average mass is 80 kg.

(i) Sketch the speed–time graph of the lift.

During the first stage of the motion the tension in the cable is 53 640 N.

(ii) Find the acceleration of the lift during this stage.

(iii) Find the length of time for which the lift is travelling at constant speed and find the final deceleration.

(iv) What is the maximum value of the tension in the cable?

(v) Just before the lift stops one miner experiences an upthrust of 1002 N from the floor of the lift. What is the mass of the miner?

9 Particles P and Q, of masses 0.6 kg and 0.2 kg respectively, are attached to the ends of a light inextensible string which passes over a smooth fixed peg. The particles are held at rest with the string taut. Both particles are at a height of 0.9 m above the ground (see diagram). The system is released and each of the particles moves vertically.

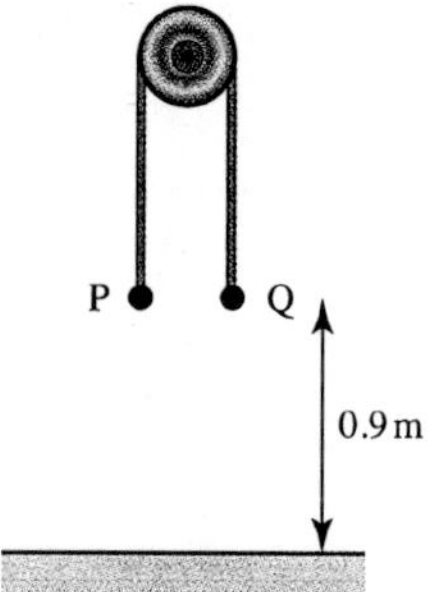

Find

(i) the acceleration of P and the tension in the string before P reaches the ground,

(ii) the time taken for P to reach the ground.

[Cambridge AS & A Level Mathematics 9709, Paper 4 Q4 June 2007]

10 Particles A and B are attached to the ends of a light inextensible string which passes over a smooth pulley. The system is held at rest with the string taut and its straight parts vertical. Both particles are at a height of 0.36 m above the floor (see diagram). The system is released and A begins to fall, reaching the floor after 0.6 s.

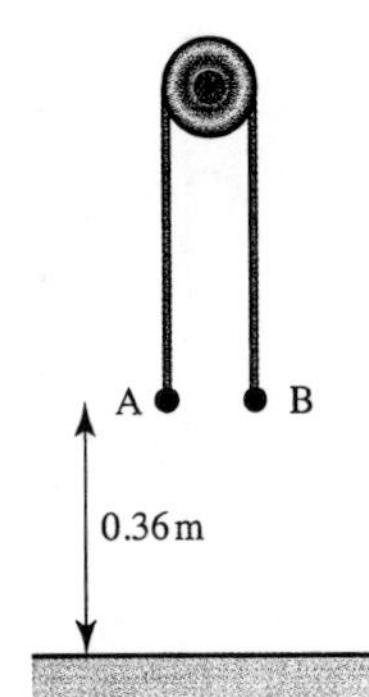

(i) Find the acceleration of A as it falls.

The mass of A is 0.45 kg. Find

(ii) the tension in the string while A is falling,

(iii) the mass of B,

(iv) the maximum height above the floor reached by B.

[Cambridge AS and A Level Mathematics 9709, Paper 4 Q6 June 2009]

11 Particles A and B, of masses 0.5 kg and m kg respectively, are attached to the ends of a light inextensible string which passes over a smooth fixed pulley. Particle B is held at rest on the horizontal floor and particle A hangs in equilibrium (see diagram). B is released and each particle starts to move vertically. A hits the floor 2 s after B is released. The speed of each particle when A hits the floor is $5\,\mathrm{m\,s^{-1}}$.

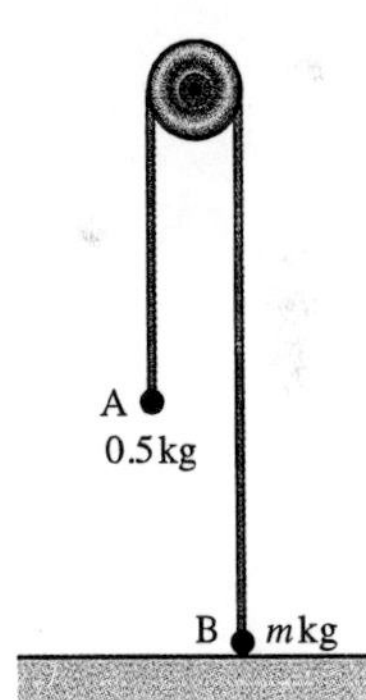

(i) For the motion while A is moving downwards, find

(a) the acceleration of A,

(b) the tension in the string.

(ii) Find the value of m.

[Cambridge AS and A Level Mathematics 9709, Paper 4 Q5 November 2008]

12 Particles P and Q, of masses 0.55 kg and 0.45 kg respectively, are attached to the ends of a light inextensible string which passes over a smooth fixed pulley. The particles are held at rest with the string taut and its straight parts vertical. Both particles are at a height of 5 m above the ground (see diagram). The system is released.

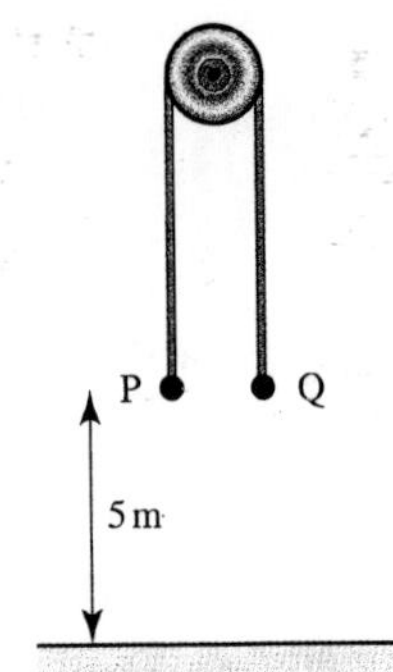

(i) Find the acceleration with which P starts to move.

The string breaks after 2 s and in the subsequent motion P and Q move vertically under gravity.

(ii) At the instant that the string breaks, find

(a) the height above the ground of P and of Q,

(b) the speed of the particles.

(iii) Show that Q reaches the ground 0.8 s later than P.

[Cambridge AS and A Level Mathematics 9709, Paper 41 Q6 November 2009]

KEY POINTS

1 The equation of motion

Newton's second law *gives the equation of motion* for an object.

The resultant force = mass × acceleration or $F = ma$

The acceleration is always in the same direction as the resultant force.

2 Connected objects

- Reaction forces between two objects (such as tension forces in joining rods or strings) are equal and opposite.
- When connected objects are moving along a line, the equations of motion can be obtained for each one separately or for a system containing more than one object. The number of independent equations is equal to the number of separate objects.

5 Vectors

But the principal failing occurred in the sailing
And the bellman, perplexed and distressed,
Said he had hoped, at least, when the wind blew due East
That the ship would not travel due West.

Lewis Carroll

Adding vectors

❓ If you walk 12 m east and then 5 m north, how far and in what direction will you be from your starting point?

A bird is caught in a wind blowing east at 12 m s^{-1} and flies so that its speed would be 5 m s^{-1} north in still air. What is its actual velocity?

A sledge is being pulled by two children with forces of 12 N east and 5 N north. What single force would have the same effect?

All these situations involve vectors. A *vector* has size (magnitude) and direction. By contrast a *scalar* quantity has only magnitude. There are many vector quantities; in this book you meet four of them: displacement, velocity, acceleration and force. When two or more dimensions are involved, the ideas underlying vectors are very important; however, in one dimension, along a straight line, you can use scalars to solve problems involving these quantities.

Although they involve quite different situations, the three problems above can be reduced to one by using the same vector techniques for finding magnitude and direction.

Displacement vectors

The instruction 'walk 12 m east and then 5 m north' can be modelled mathematically using a scale diagram, as in figure 5.1. The arrowed lines AB and BC are examples of vectors.

We write the vectors as $\overrightarrow{AB}$ and $\overrightarrow{BC}$. The arrow above the letters is very important as it indicates the direction of the vector. $\overrightarrow{AB}$ means from A to B. $\overrightarrow{AB}$ and $\overrightarrow{BC}$ are examples of *displacement vectors*. Their lengths represent the magnitude of the displacements.

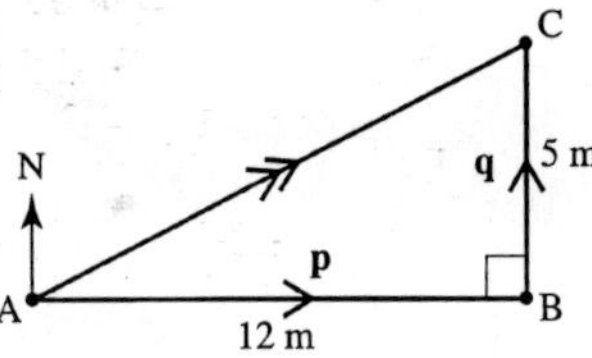

Figure 5.1

It is often more convenient to use a single letter to denote a vector. For example you might see the displacement vectors $\overrightarrow{AB}$ and $\overrightarrow{BC}$ written as **p** and **q** (i.e. in bold print). When writing these vectors yourself, you should underline your letters, e.g. $\underline{p}$ and $\underline{q}$.

The magnitudes of **p** and **q** are then shown as $|\mathbf{p}|$ and $|\mathbf{q}|$ or p and q (in italics).
These are scalar quantities.

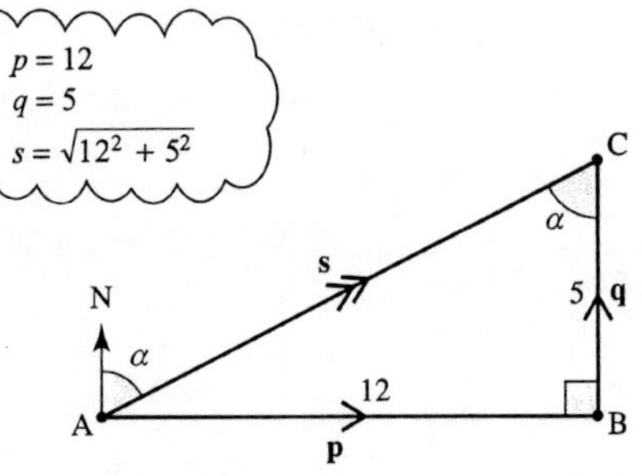

Figure 5.2

The combined effect of the two displacements $\overrightarrow{AB}$ (= **p**) and $\overrightarrow{BC}$ (= **q**) is $\overrightarrow{AC}$ and this is called the *resultant vector*. It is marked with two arrows to distinguish it from **p** and **q**. The process of combining vectors in this way is called *vector addition*. We write $\overrightarrow{AB} + \overrightarrow{BC} = \overrightarrow{AC}$ or $\mathbf{p} + \mathbf{q} = \mathbf{s}$.

You can calculate the resultant using Pythagoras' theorem and trigonometry.

In triangle ABC $\quad AC = \sqrt{12^2 + 5^2} = 13$

and $\quad \tan\alpha = \frac{12}{5}$

$$\alpha = 67° \text{ (to the nearest degree)}$$

The distance from the starting point is 13 m and the direction is 067°.

A special case of a displacement is a *position vector*. This is the displacement of a point from the origin.

Velocity and force

The other two problems that begin this chapter are illustrated in these diagrams.

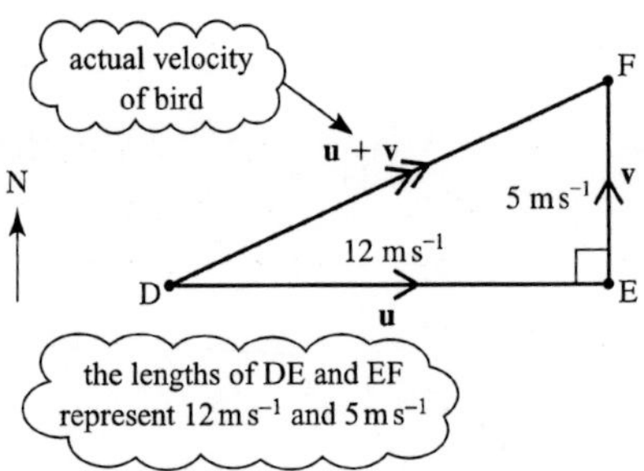

Figure 5.3

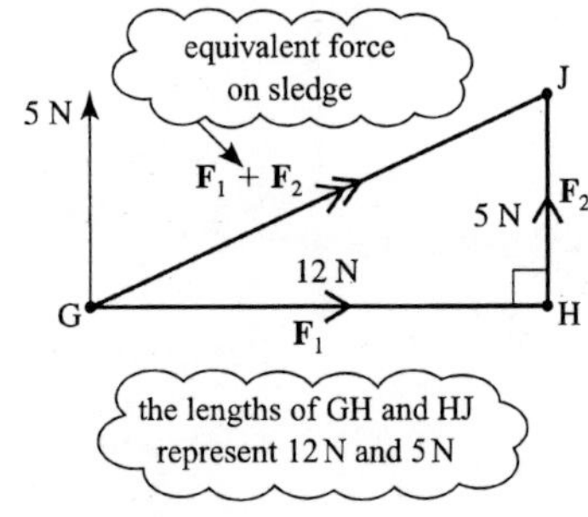

Figure 5.4

When DF represents the velocity (**u**) of the wind and EF represents the velocity (**v**) of the bird in still air, the vector $\overrightarrow{DF}$ represents the resultant velocity, **u** + **v**.

Why does the bird move in the direction DF? Think what happens in very small intervals of time.

In figure 5.4, the vector $\overrightarrow{GJ}$ represents the equivalent (resultant) force. You know that it acts at the same point on the sledge as the children's forces, but its magnitude and direction can be found using the triangle GHJ which is similar to the two triangles, ABC and DEF.

The same diagram does for all, you just have to supply the units. The bird travels at $13\,\text{m}\,\text{s}^{-1}$ in the direction of 067° and one child would have the same effect as the others by pulling with a force of 13 N in the direction 067°. In most of this chapter vectors are treated in the abstract. You can then apply what you learn to different real situations.

Components of a vector

It is often convenient to write one vector in terms of two others called *components*.

The vector **a** in the diagram can be split into two components in an infinite number of ways. All you need to do is to make **a** one side of a triangle. It is most sensible, however, to split vectors into components in convenient directions and these directions are usually perpendicular.

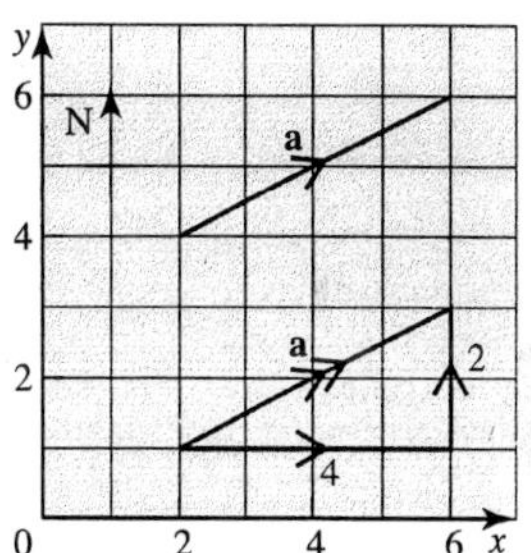

Figure 5.5

Using the given grid, **a** is 4 units east combined with 2 units north.

You can write **a** in figure 5.5 as $\begin{pmatrix}4\\2\end{pmatrix}$. This is called a *column vector*. The *unit vector* $\begin{pmatrix}1\\0\end{pmatrix}$ is in the direction east and the unit vector $\begin{pmatrix}0\\1\end{pmatrix}$ is in the directions north.

Alternatively **a** can be written as $4\mathbf{i} + 2\mathbf{j}$ but this notation is not used in this book.

Note

You have already used components in your work and so have met the idea of vectors. For example, the total reaction between two surfaces is often split into two components. One (friction) is opposite to the direction of possible sliding and the other (normal reaction) is perpendicular to it.

EXAMPLE 5.1

Four forces **a**, **b**, **c** and **d** are shown in the diagram. The units are in newtons.

(i) Write them in component form.

(ii) Draw a diagram to show 2**c** and −**d** and write them in component form.

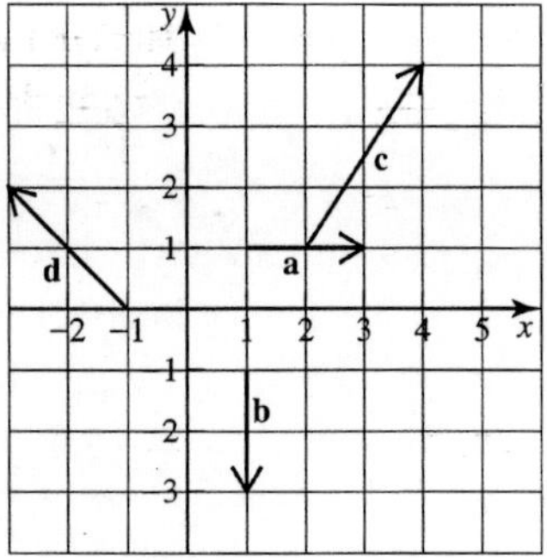

Figure 5.6

(i) $\mathbf{a} = \begin{pmatrix} 2 \\ 0 \end{pmatrix}$

$\mathbf{b} = \begin{pmatrix} 0 \\ -2 \end{pmatrix}$

$\mathbf{c} = \begin{pmatrix} 2 \\ 3 \end{pmatrix}$

$\mathbf{d} = \begin{pmatrix} -2 \\ 2 \end{pmatrix}$

(ii) $2\mathbf{c} = 2\begin{pmatrix} 2 \\ 3 \end{pmatrix}$

$= \begin{pmatrix} 4 \\ 6 \end{pmatrix}$

$-\mathbf{d} = -\begin{pmatrix} -2 \\ 2 \end{pmatrix}$

$= \begin{pmatrix} 2 \\ -2 \end{pmatrix}$

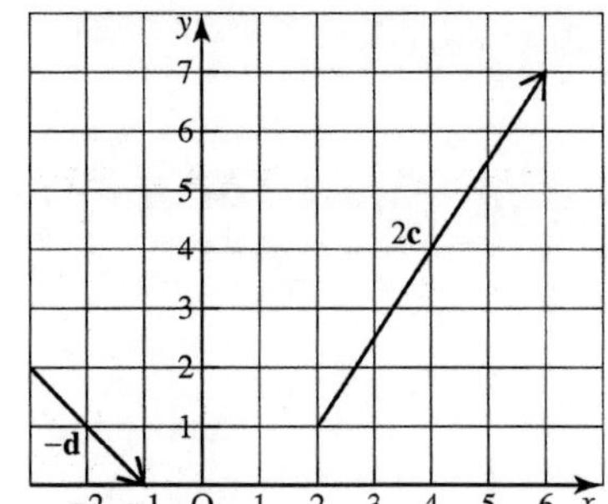

Figure 5.7

Equal vectors and parallel vectors

When two vectors, **p** and **q**, are *equal* then they must be equal in both magnitude and direction. If they are written in component form their components must be equal.

So if $\mathbf{p} = \begin{pmatrix} a_1 \\ b_1 \end{pmatrix}$

and $\mathbf{q} = \begin{pmatrix} a_2 \\ b_2 \end{pmatrix}$

then $a_1 = a_2$ and $b_1 = b_2$.

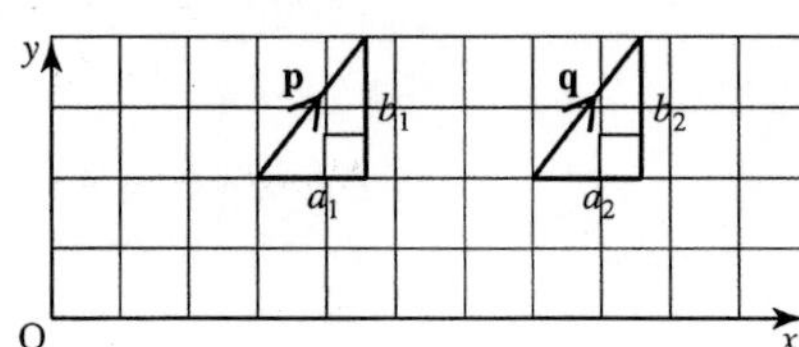

Figure 5.8

Thus in two dimensions, the statement **p** = **q** is the equivalent of two equations (and in three dimensions, three equations).

If **p** and **q** are *parallel but not equal*, they make the same angle with the x axis.

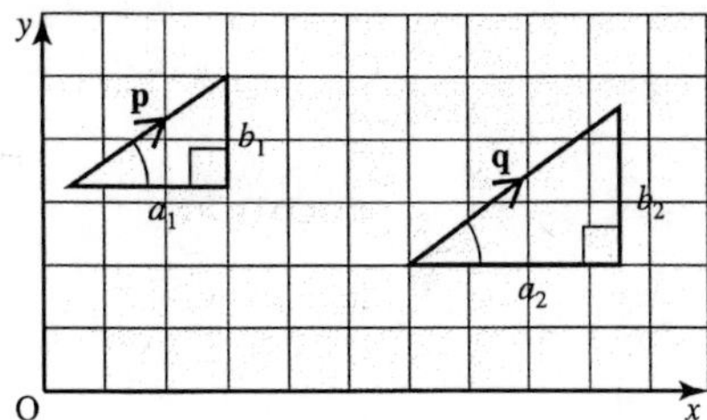

Figure 5.9

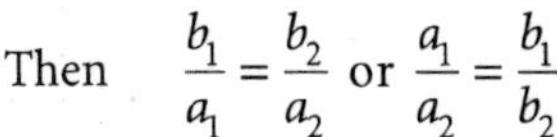

Then $\frac{b_1}{a_1} = \frac{b_2}{a_2}$ or $\frac{a_1}{a_2} = \frac{b_1}{b_2}$

If $\begin{pmatrix} 4 \\ 3 \end{pmatrix}$ is parallel to $\begin{pmatrix} -8 \\ y \end{pmatrix}$ what is y?

You will often meet parallel vectors when using Newton's second law, as in the following example.

EXAMPLE 5.2

A force of $\begin{pmatrix} 6 \\ 8 \end{pmatrix}$ N acts on an object of mass 2 kg. Find the object's acceleration as a column vector.

SOLUTION

Using Force = Mass × Acceleration

$$\begin{pmatrix} 6 \\ 8 \end{pmatrix} = 2 \times \text{Acceleration}$$

So the acceleration is $\begin{pmatrix} 3 \\ 4 \end{pmatrix}$ m s^{-2}.

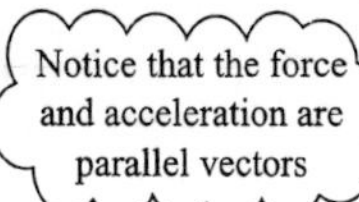

Adding vectors in component form

In component form, addition and subtraction of vectors is simply carried out by adding or subtracting the components of the vectors.

EXAMPLE 5.3

Two vectors **a** and **b** are given by $\mathbf{a} = \begin{pmatrix} 2 \\ 3 \end{pmatrix}$ and $\mathbf{b} = \begin{pmatrix} -1 \\ 4 \end{pmatrix}$.

(i) Find the vectors **a** + **b** and **a** − **b**.

(ii) Verify that your results are the same if you use a scale drawing.

SOLUTION

(i) $$\mathbf{a} + \mathbf{b} = \begin{pmatrix} 2 \\ 3 \end{pmatrix} + \begin{pmatrix} -1 \\ 4 \end{pmatrix} = \begin{pmatrix} 1 \\ 7 \end{pmatrix}$$

$$\mathbf{a} - \mathbf{b} = \begin{pmatrix} 2 \\ 3 \end{pmatrix} - \begin{pmatrix} -1 \\ 4 \end{pmatrix} = \begin{pmatrix} 3 \\ -1 \end{pmatrix}$$

(ii)

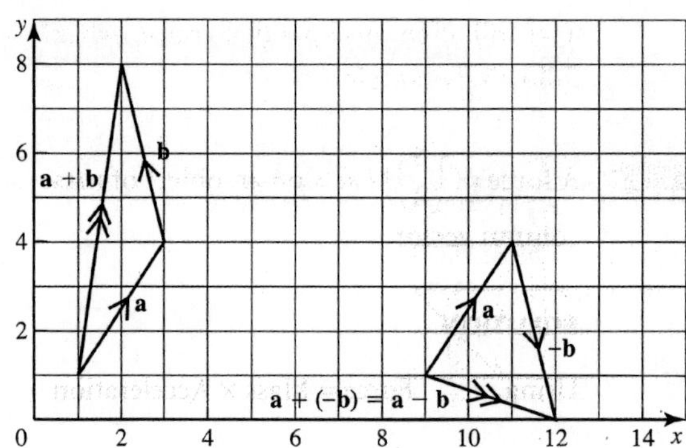

Figure 5.10

From the diagram you can see that $\mathbf{a} + \mathbf{b} = \begin{pmatrix} 1 \\ 7 \end{pmatrix}$

and $\mathbf{a} - \mathbf{b} = \begin{pmatrix} 3 \\ -1 \end{pmatrix}$.

These vectors are the same as those obtained in part **(i)**.

? **a** and **b** are the position vectors of points A and B as shown in the diagram.

How can you write the displacement vector $\overrightarrow{AB}$ in terms of **a** and **b**?

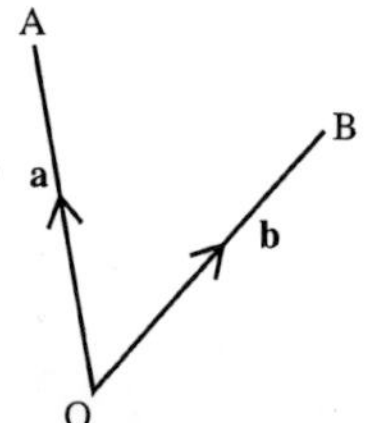

Figure 5.11

EXERCISE 5A

1 The diagram shows a grid of 1 m squares. A person walks first east and then north. How far should the person walk in each of these directions to travel

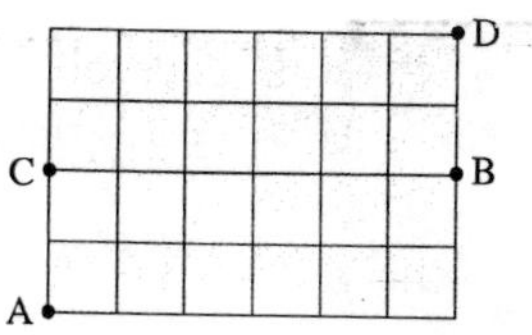

(i) from A to B?

(ii) from B to C?

(iii) from A to D?

2 The diagram shows nine different forces. The units are newtons. Write each of the forces as a column vector.

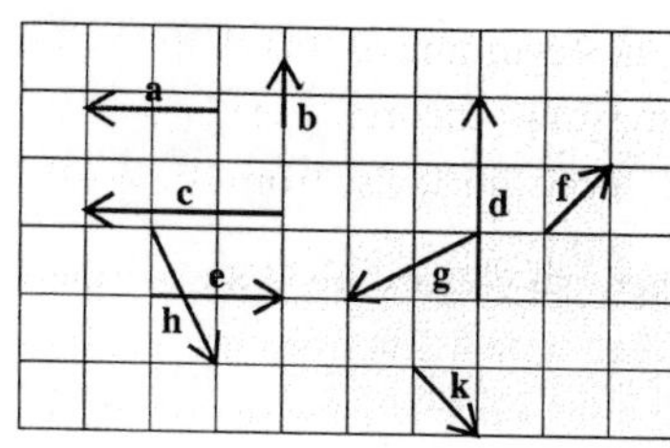

3 Two forces **a** and **b** are given in newtons by $\mathbf{a} = \begin{pmatrix} 2 \\ -1 \end{pmatrix}$ and $\mathbf{b} = \begin{pmatrix} 1 \\ 4 \end{pmatrix}$. Write the force $3\mathbf{a} - 2\mathbf{b}$ as a column vector.

4 Four forces **a**, **b**, **c** and **d** are given in newtons by $\mathbf{a} = \begin{pmatrix} 4 \\ 1 \end{pmatrix}$, $\mathbf{b} = \begin{pmatrix} -1 \\ 0 \end{pmatrix}$, $\mathbf{c} = \begin{pmatrix} -2 \\ -3 \end{pmatrix}$ and $\mathbf{d} = \begin{pmatrix} 2 \\ 6 \end{pmatrix}$. Write each of the following forces as a column vector.

(i) $\mathbf{a} + 2\mathbf{b}$

(ii) $2\mathbf{c} - 3\mathbf{d}$

(iii) $\mathbf{a} + \mathbf{c} - 2\mathbf{b}$

(iv) $-2\mathbf{a} + 3\mathbf{b} + 4\mathbf{d}$.

5 A, B and C are the points (1, 2), (5, 1) and (7, 8).

(i) Write down the position vectors of these three points.

(ii) Find the displacement vectors $\overrightarrow{AB}$, $\overrightarrow{BC}$ and $\overrightarrow{CA}$.

(iii) Draw a diagram to show the position vectors of A, B and C and your answers to part (ii).

6 A, B and C are the points (0, –3), (2, 5) and (3, 9).

(i) Write down the position vectors of these three points.

(ii) Find the displacement vectors $\overrightarrow{AB}$ and $\overrightarrow{BC}$.

(iii) Show that the three points all lie on a straight line.

7 A, B, C and D are the points (4, 2), (1, 3), (0, 10) and (3, *d*).

(i) Find the value of *d* so that DC is parallel to AB.

(ii) Find a relationship between $\overrightarrow{BC}$ and $\overrightarrow{AD}$. What is ABCD?

8 Four forces **a**, **b**, **c** and **d** are given in newtons by $\mathbf{a} = \begin{pmatrix} 1 \\ 1 \end{pmatrix}$, $\mathbf{b} = \begin{pmatrix} 1 \\ 2 \end{pmatrix}$, $\mathbf{c} = \begin{pmatrix} 3 \\ -4 \end{pmatrix}$ and $\mathbf{d} = \begin{pmatrix} 1 \\ 2 \end{pmatrix}$.

A force given by $2\mathbf{a} + 3\mathbf{b} + \mathbf{c} - 8\mathbf{d}$ acts on a particle of mass 3 kg. Find the acceleration of the particle as a column vector and write down its magnitude.

The magnitude and direction of vectors written in component form

At the beginning of this chapter the magnitude of a vector was found by using Pythagoras' theorem (see page 86). The direction was given using bearings, measured clockwise from the north.

When the vectors are in an x–y plane, a mathematical convention is used for direction. Starting from the x axis, angles measured anticlockwise are positive and angles in a clockwise direction are negative as in figure 5.12.

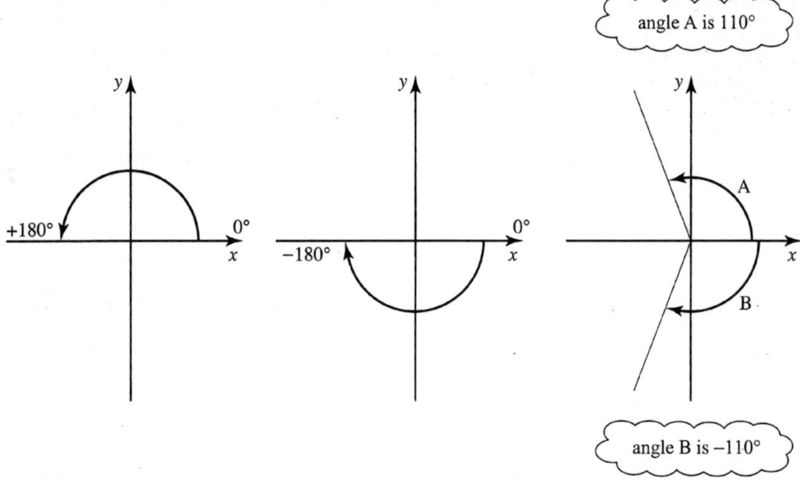

Figure 5.12

Using the notation in figure 5.13, the magnitude and direction can be written in general form.

Magnitude of the vector $\left| \begin{pmatrix} a_1 \\ a_2 \end{pmatrix} \right| = \sqrt{a_1^2 + a_2^2}$

Direction $\tan\theta = \dfrac{a_2}{a_1}$

a
a_2
θ
a_1

Figure 5.13

EXAMPLE 5.4

Find the magnitude and direction of the vectors $\begin{pmatrix} 4 \\ 3 \end{pmatrix}$, $\begin{pmatrix} 4 \\ -3 \end{pmatrix}$, $\begin{pmatrix} -4 \\ 3 \end{pmatrix}$ and $\begin{pmatrix} -4 \\ -3 \end{pmatrix}$.

SOLUTION

First draw diagrams so that you can see which lengths and acute angles to find.

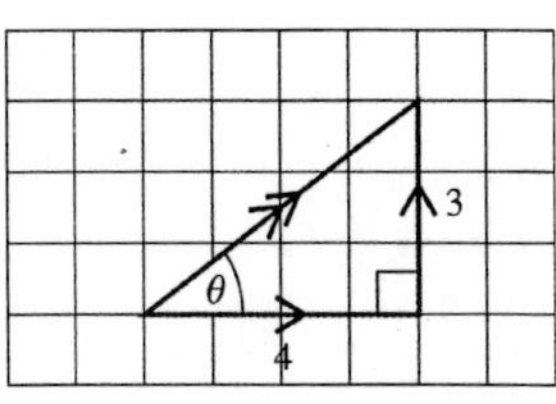

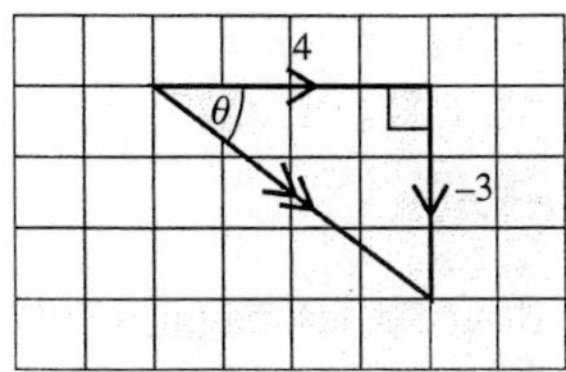

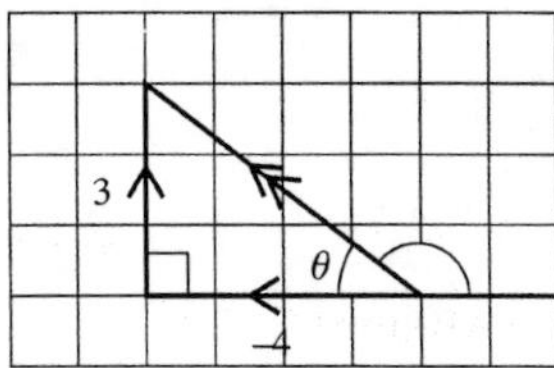

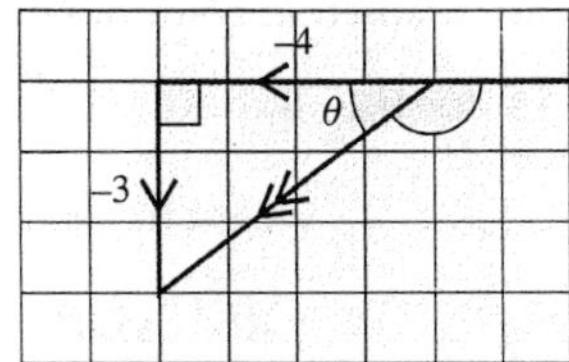

Figure 5.14

The vectors in each of the diagrams have the same magnitude and using Pythagoras' theorem, the resultants all have magnitude $\sqrt{4^2 + 3^2} = 5$.

The angles θ are also the same size in each diagram and can be found using

$$\tan\theta = \tfrac{3}{4} \qquad \theta = 37°$$

The angles the vectors make starting from the x axis specify their directions:

$\begin{pmatrix} 4 \\ 3 \end{pmatrix}$	$37°$
$\begin{pmatrix} 4 \\ -3 \end{pmatrix}$	$-37°$
$\begin{pmatrix} -4 \\ 3 \end{pmatrix}$	$180° - 37° = 143°$
$\begin{pmatrix} -4 \\ -3 \end{pmatrix}$	$-143°$

EXERCISE 5B

Make use of sketches to help you in this exercise.

1 Find the magnitude and direction of

(i) $\begin{pmatrix} 6 \\ -8 \end{pmatrix}$ **(ii)** $\begin{pmatrix} -4 \\ 0 \end{pmatrix}$ **(iii)** $\begin{pmatrix} -1 \\ -2 \end{pmatrix}$.

2 Find the resultant, $\mathbf{F}_1 + \mathbf{F}_2$, of the two forces $\mathbf{F}_1 = \begin{pmatrix} 10 \\ 40 \end{pmatrix}$ and $\mathbf{F}_2 = \begin{pmatrix} 20 \\ -10 \end{pmatrix}$ and then find its magnitude and direction.

3 Find the resultant of the three forces $\mathbf{F}_1 = \begin{pmatrix} -1 \\ 5 \end{pmatrix}$, $\mathbf{F}_2 = \begin{pmatrix} 2 \\ -10 \end{pmatrix}$ and $\mathbf{F}_3 = \begin{pmatrix} -2 \\ 7 \end{pmatrix}$ and then find its magnitude and direction.

❓ **(i)** Show that $\begin{pmatrix} 0.6 \\ 0.8 \end{pmatrix}$ is a unit vector.

(ii) Find unit vectors in the directions of

(a) $\begin{pmatrix} 8 \\ 6 \end{pmatrix}$ **(b)** $\begin{pmatrix} 1 \\ -1 \end{pmatrix}$.

Resolving vectors

A vector has magnitude 10 units and it makes an angle of 60° with the x axis. How can it be represented in component form?

In the diagram:

$$\frac{AC}{AB} = \cos 60° \quad \text{and} \quad \frac{BC}{AB} = \sin 60°$$

$$AC = AB\cos 60° \qquad BC = AB\sin 60°$$

$$= 10\cos 60° \qquad = 10\sin 60°$$

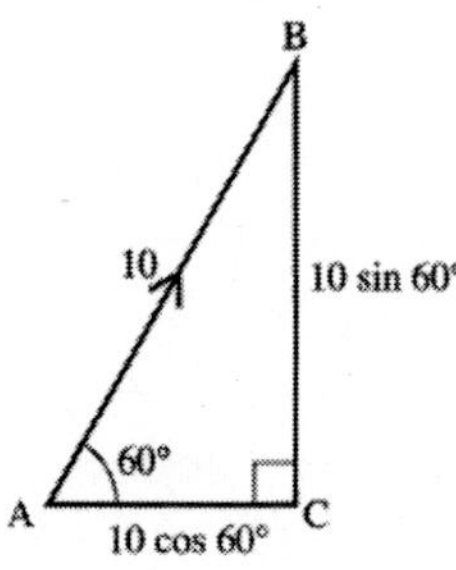

Figure 5.15

The vector can then be written as $\begin{pmatrix} 10\cos 60° \\ 10\sin 60° \end{pmatrix} = \begin{pmatrix} 5 \\ 8.66 \end{pmatrix}$ (to 3 s.f.).

In a similar way, any vector **a** with magnitude a which makes an angle α with the x axis can be written in component form as

$$\mathbf{a} = \begin{pmatrix} a\cos\alpha \\ a\sin\alpha \end{pmatrix}.$$

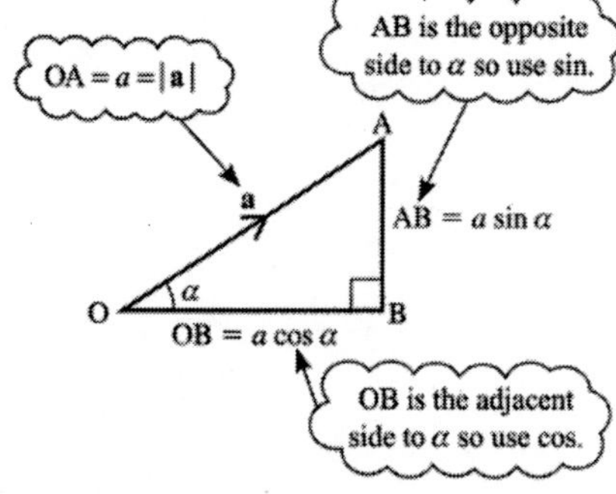

Figure 5.16

When α is an obtuse angle, this expression is still true. For example, when $\alpha = 120°$ and $a = 10$,

$$\mathbf{a} = \begin{pmatrix} a\cos\alpha \\ a\sin\alpha \end{pmatrix}$$

$$= \begin{pmatrix} 10\cos 120° \\ 10\sin 120° \end{pmatrix}$$

$$= \begin{pmatrix} -5 \\ 8.66 \end{pmatrix}$$

However, it is usually easier to write

$$\mathbf{a} = \begin{pmatrix} -10\cos 60° \\ 10\sin 60° \end{pmatrix}$$

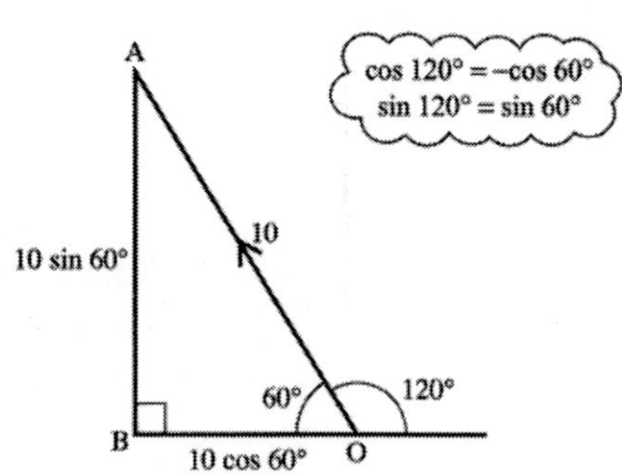

Figure 5.17

EXAMPLE 5.5

Two forces **P** and **Q** have magnitudes 4 and 5 in the directions shown in the diagram.

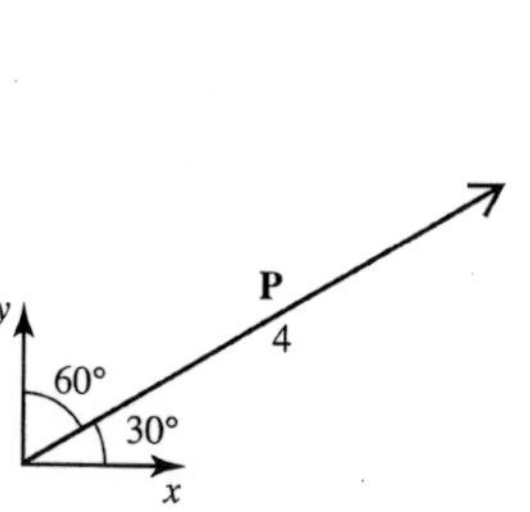

Figure 5.18

Find the magnitude and direction of the resultant force **P** + **Q**.

SOLUTION

$$\mathbf{P} = \begin{pmatrix} 4\cos 30° \\ 4\sin 30° \end{pmatrix} = \begin{pmatrix} 3.46 \\ 2 \end{pmatrix}$$

$$\mathbf{Q} = \begin{pmatrix} -5\cos 60° \\ 5\sin 60° \end{pmatrix} = \begin{pmatrix} -2.5 \\ 4.33 \end{pmatrix}$$

$$\mathbf{P} + \mathbf{Q} = \begin{pmatrix} 3.46 \\ 2 \end{pmatrix} + \begin{pmatrix} -2.5 \\ 4.33 \end{pmatrix} = \begin{pmatrix} 0.96 \\ 6.33 \end{pmatrix}$$

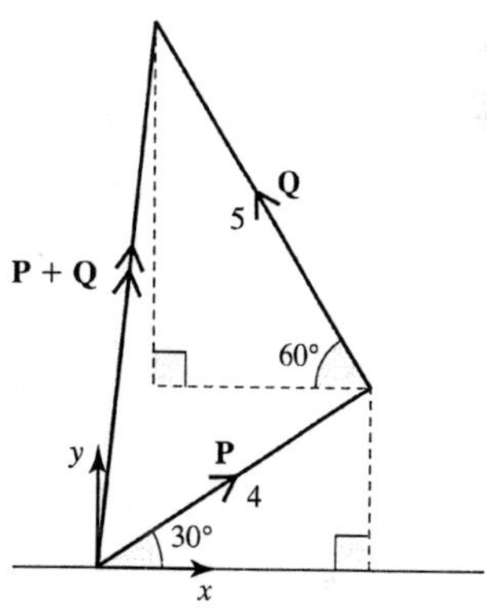

Figure 5.19

This resultant is shown in Figure 5.19.

Magnitude $|\mathbf{P} + \mathbf{Q}| = \sqrt{0.96^2 + 6.33^2}$

$= \sqrt{40.99}$

$= 6.4$

Direction $\tan\theta = \dfrac{6.33}{0.96}$

$= 6.59$

$\theta = 81.4°$

The force **P** + **Q** has magnitude 6.4 and direction 81.4° relative to the positive x direction.

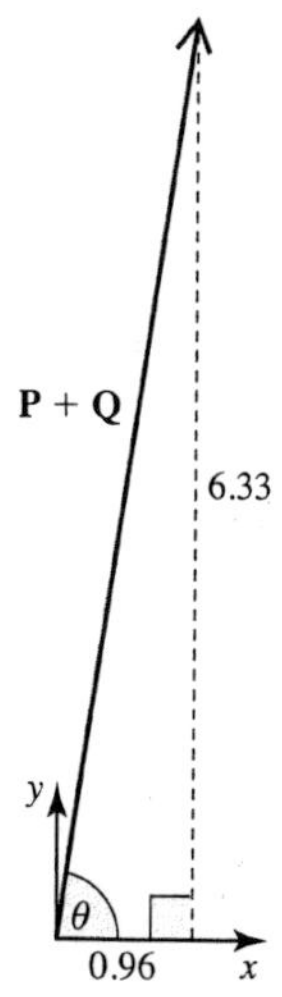

Figure 5.20

EXERCISE 5C

1 Write the following forces as column vectors.

(i)

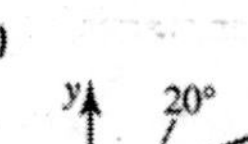
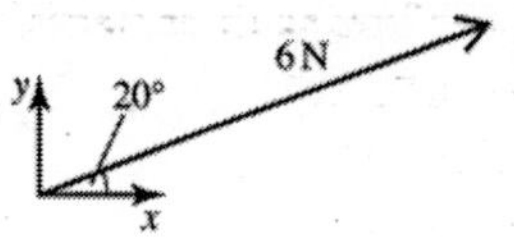

(ii)

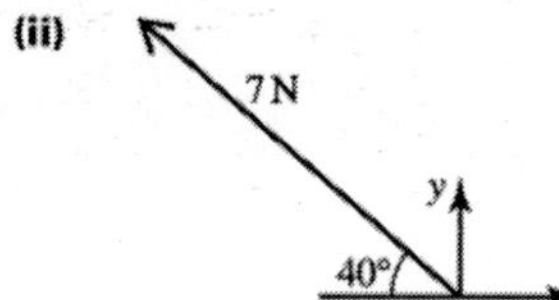

(iii)

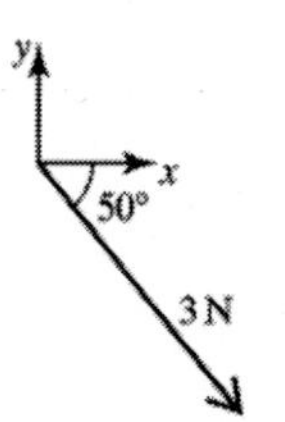

(iv)

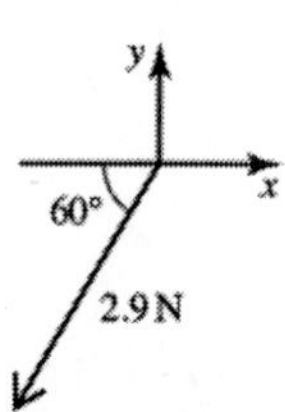

2 Draw a diagram showing each of the following displacements. Write each as a column vector in directions east and north respectively.

(i) 130 km, bearing 060°
(ii) 250 km, bearing 130°
(iii) 400 km, bearing 210°
(iv) 50 km, bearing 300°

3 A boat has a speed of $4\,\text{km h}^{-1}$ in still water and sets its course north-east in an easterly current of $3\,\text{km h}^{-1}$. Write each velocity as a column vector in directions east and north and hence find the magnitude and direction of the resultant velocity.

4 A boy walks 30 m north and then 50 m south-west.

(i) Draw a diagram to show the boy's path.
(ii) Write each displacement using column vectors in directions east and north.
(iii) In which direction should he walk to get directly back to his starting point?

5 **(a)** Write each of the following forces as a column vector.
(b) Find the resultant of each set of vectors.

(i)

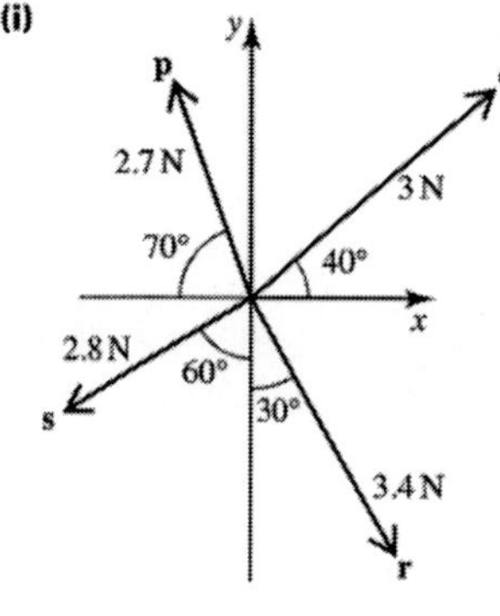

(ii)

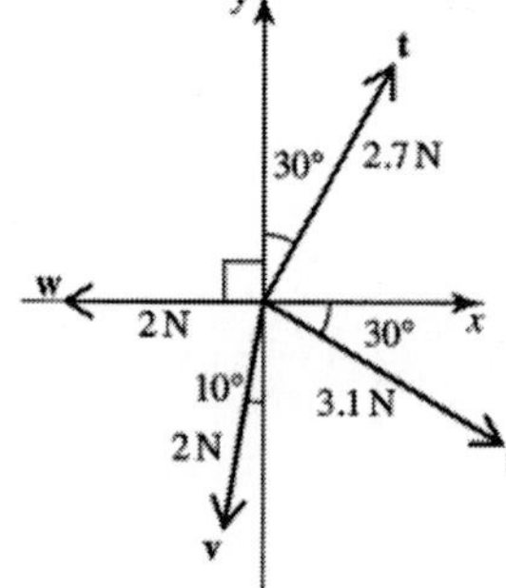

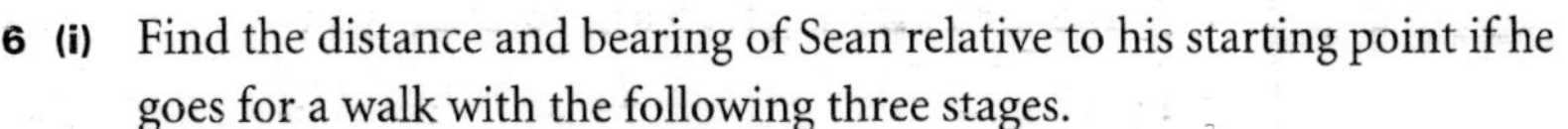

6 (i) Find the distance and bearing of Sean relative to his starting point if he goes for a walk with the following three stages.
Stage 1: 600 m on a bearing 030°
Stage 2: 1 km on a bearing 100°
Stage 3: 700 m on a bearing 340°

(ii) Shona sets off from the same place at the same time as Sean. She walks at the same speed but takes the stages in the order 3–1–2.
How far apart are Sean and Shona at the end of their walks?

7 The diagram shows the journey of a yacht.

Express $\overrightarrow{OA}$, $\overrightarrow{AB}$ and $\overrightarrow{OB}$ as column vectors based on directions east and north respectively.

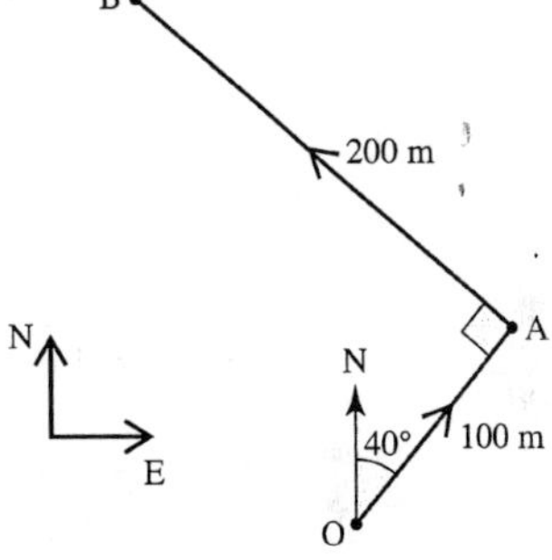

8 The diagram shows the big wheel ride at a fairground. The radius of the wheel is 5 m and the length of the arms that support each carriage is 1 m.

Express the position vector of the carriages A, B, C and D as column vectors.

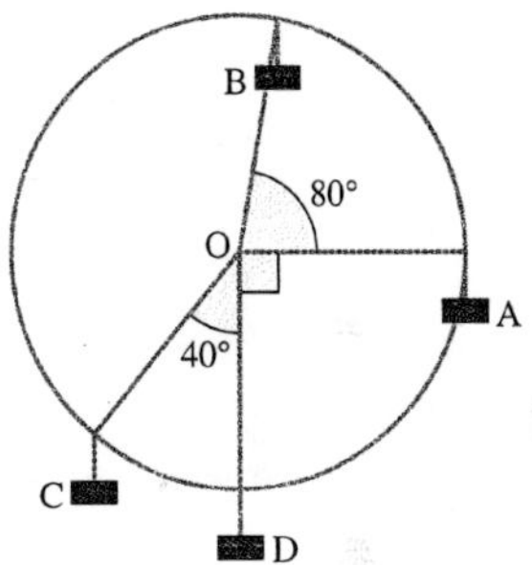

9 Two walkers set off from the same place in different directions. After a period they stop. Their displacements are $\begin{pmatrix} 2 \\ 5 \end{pmatrix}$ and $\begin{pmatrix} -3 \\ 4 \end{pmatrix}$ where the distances are in kilometres and the directions are east and north. On what bearing and for what distance does the second walker have to walk to be reunited with the first (who does not move)?

KEY POINTS

1 A scalar quantity has only magnitude (size).
A vector quantity has both magnitude and direction.
Displacement, velocity, acceleration and force are all vector quantities.

2 Vectors may be represented in either magnitude–direction form or in component form.

Magnitude–direction form | **Component form**

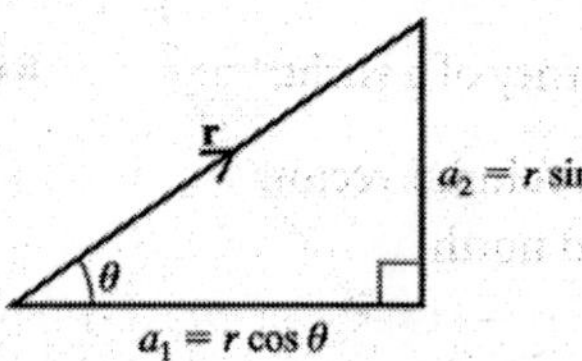

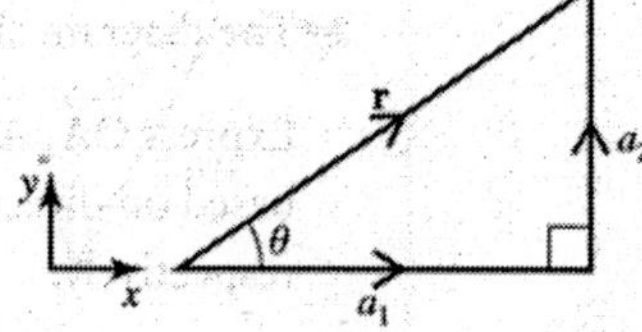

Magnitude, r; direction, θ | $\begin{pmatrix} a_1 \\ a_2 \end{pmatrix}$

where $r = \sqrt{a_1^2 + a_2^2}$ | $a_1 = r\cos\theta$

and $\tan\theta = \frac{a_2}{a_1}$ | $a_2 = r\sin\theta = r\cos(90° - \theta)$

3 When two or more vectors are added, the *resultant* is obtained. Vector addition may be done graphically or algebraically.

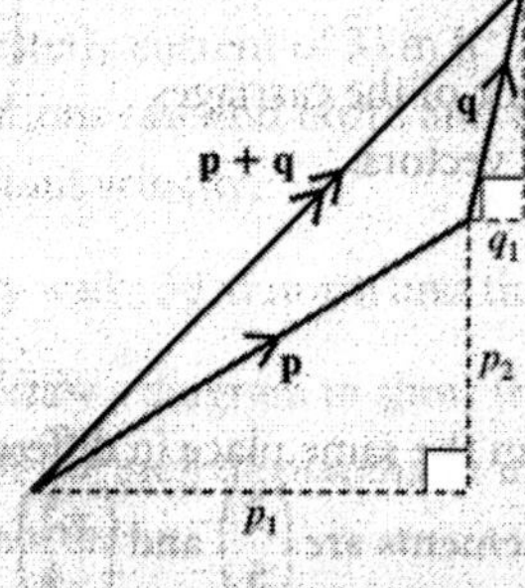

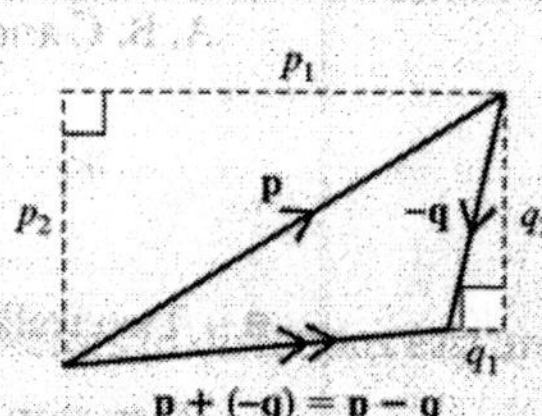

$$\mathbf{p} + \mathbf{q} = \begin{pmatrix} p_1 \\ p_2 \end{pmatrix} + \begin{pmatrix} q_1 \\ q_2 \end{pmatrix} = \begin{pmatrix} p_1 + q_1 \\ p_2 + q_2 \end{pmatrix}$$

4 Multiplication by a scalar: $n\begin{pmatrix} a_1 \\ a_2 \end{pmatrix} = \begin{pmatrix} na_1 \\ na_2 \end{pmatrix}$

5 The *position vector* of a point P is $\overrightarrow{OP}$, its displacement from a fixed origin.

6 When A and B have position vectors **a** and **b**, $\overrightarrow{AB} = \mathbf{b} - \mathbf{a}$.

7 Equal vectors have equal magnitude and are in the same direction.

$$\begin{pmatrix} p_1 \\ p_2 \end{pmatrix} = \begin{pmatrix} q_1 \\ q_2 \end{pmatrix} \Rightarrow p_1 = q_1 \text{ and } p_2 = q_2.$$

8 When $\begin{pmatrix} p_1 \\ p_2 \end{pmatrix}$ and $\begin{pmatrix} q_1 \\ q_2 \end{pmatrix}$ are parallel, $\frac{p_1}{q_1} = \frac{p_2}{q_2}$.

6 Forces in equilibrium and resultant forces

Give me matter and motion and I will construct the Universe.

René Descartes

Finding resultant forces

This cable car is stationary. Are the tensions in the cable greater than the weight of the car?

A child on a sledge is being pulled up a smooth slope of 20° by a rope which makes an angle of 40° with the slope. The mass of the child and sledge together is 20 kg and the tension in the rope is 170 N. Draw a diagram to show the forces acting on the child and sledge together. In what direction is the resultant of these forces?

When the child and sledge are modelled as a particle, all the forces can be assumed to be acting at a point. There is no friction force because the slope is smooth. Here is the force diagram.

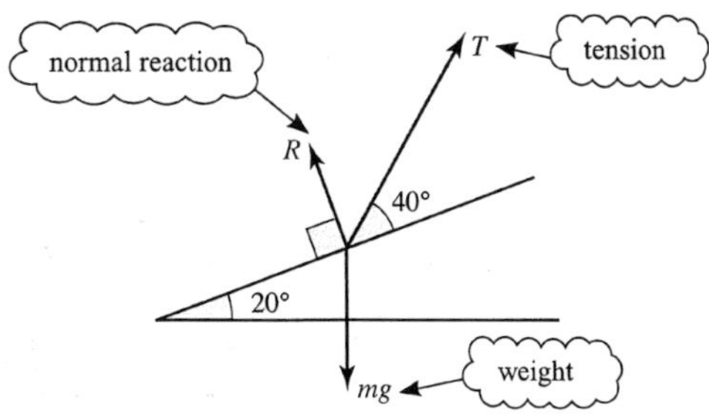

Figure 6.1

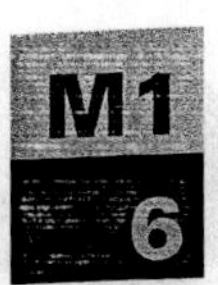

❓ The sledge is sliding along the slope. What direction is the resultant force acting on it?

You can find the normal reaction and the resultant force on the sledge using two methods.

Method 1: Using components

This method involves resolving forces into components in two perpendicular directions as in Chapter 5. It is easiest to use the components of the forces parallel and perpendicular to the slope in the directions shown.

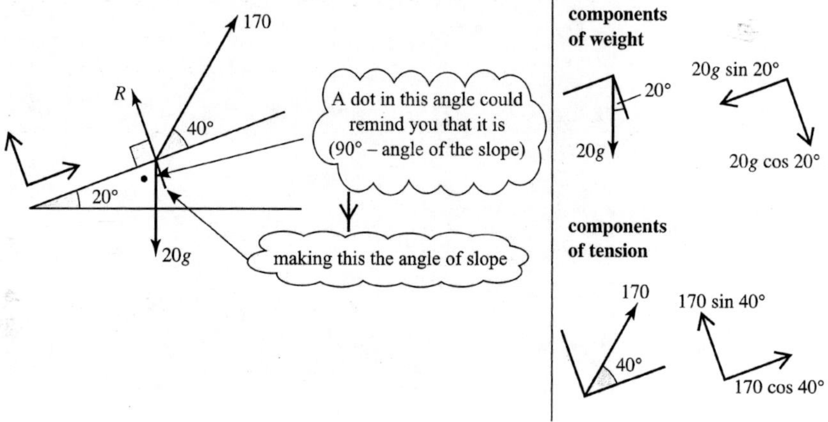

Figure 6.2

Resolve parallel to the slope (↗):

The force R is perpendicular to the slope so it has no component in this direction.

The resultant $\quad F = 170 \cos 40° - 20g \sin 20°$

$= 61.8$ (to 3 s.f.)

$\cos(90° - 20°) = \sin 20°$

Resolve perpendicular to the slope (↖):

$$R + 170 \sin 40° - 20g \cos 20° = 0$$

$$R = 20g \cos 20° - 170 \sin 40°$$

$$= 78.7 \text{ (to 3 s.f.)}$$

There is no resultant in this direction because the motion is parallel to the slope.

The normal reaction is 78.7 N and the resultant is 61.8 N up the slope.

Alternatively, you could have worked in column vectors as follows.

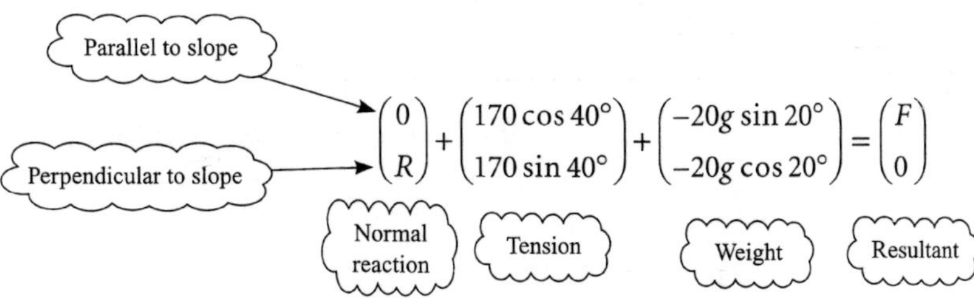

$$\begin{pmatrix} 0 \\ R \end{pmatrix} + \begin{pmatrix} 170\cos 40° \\ 170\sin 40° \end{pmatrix} + \begin{pmatrix} -20g\sin 20° \\ -20g\cos 20° \end{pmatrix} = \begin{pmatrix} F \\ 0 \end{pmatrix}$$

Note

Try resolving horizontally and vertically. You will obtain two equations in the two unknowns R and F. It is perfectly possible to solve these equations, but quite a lot of work. It is much easier to choose to resolve in directions which ensure that one component of at least one of the unknown forces is zero.

Once you know the resultant force, you can work out the acceleration of the sledge using Newton's second law.

$$F = ma$$
$$61.8 = 20a$$

The acceleration is $3.1\,\text{m}\,\text{s}^{-2}$ (correct to 1 d.p.).

Method 2: Scale drawing

An alternative is to draw a scale diagram with the three forces represented by three of the sides of a quadrilateral taken in order (with the arrows following each other) as shown in figure 6.3. The resultant is represented by the fourth side AD. This must be parallel to the slope.

❓ In what order would you draw the lines in the diagram?

From the diagram you can estimate the normal reaction to be about 80 N and the resultant 60 N. This is a reasonable estimate, but components are more precise.

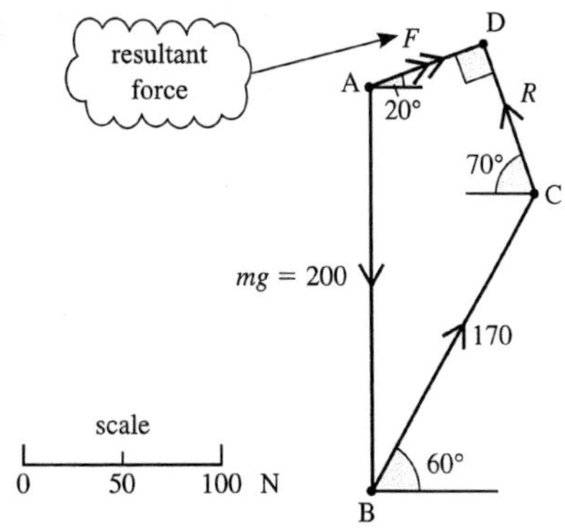

Figure 6.3

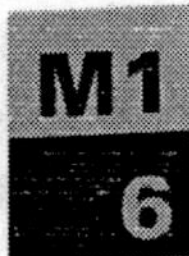

❓ What can you say about the sledge in the cases when

(i) the length AD is not zero?

(ii) the length AD is zero so that the starting point on the quadrilateral is the same as the finishing point?

(iii) BC is so short that the point D is to the left of A as shown in figure 6.4?

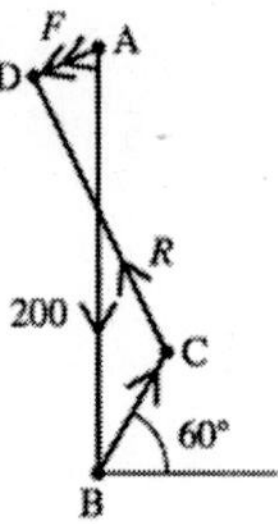

Figure 6.4

EXERCISE 6A

For questions **1** to **6**, carry out the following steps. All forces are in newtons.

(i) Draw a scale diagram to show the polygon of the forces and the resultant.

(ii) State whether you think the forces are in equilibrium and, if not, estimate the magnitude and direction of the resultant.

(iii) Write the forces in component form, using the directions indicated and so obtain the components of the resultant.
Hence find the magnitude and direction of the resultant as on page 95.

(iv) Compare your answers to parts **(ii)** and **(iii)**.

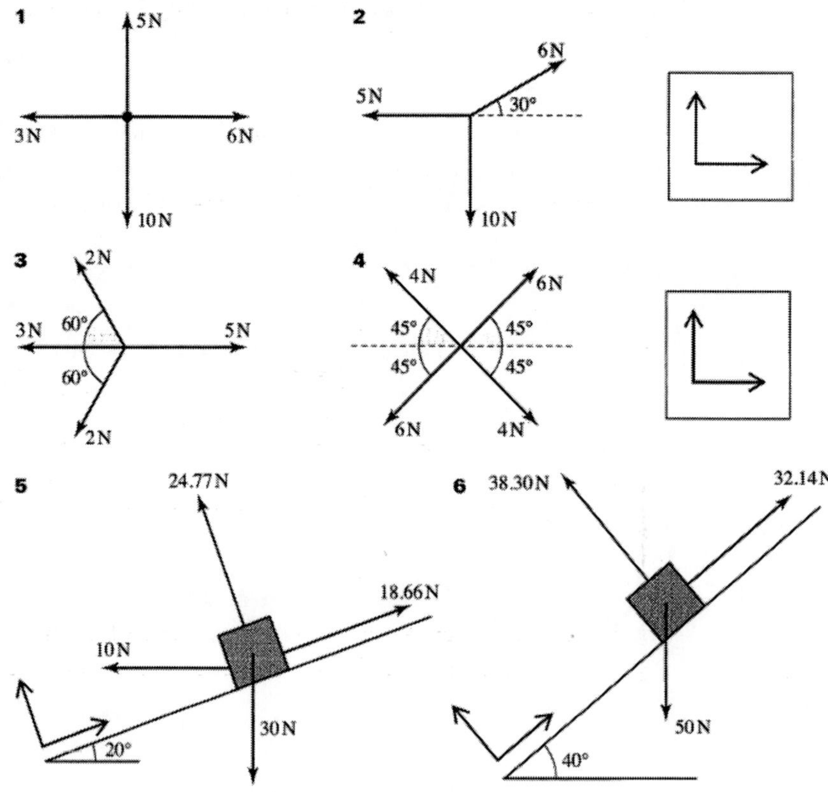

7 Forces of magnitudes 7 N, 10 N and 15 N act on a particle in the directions shown in the diagram.

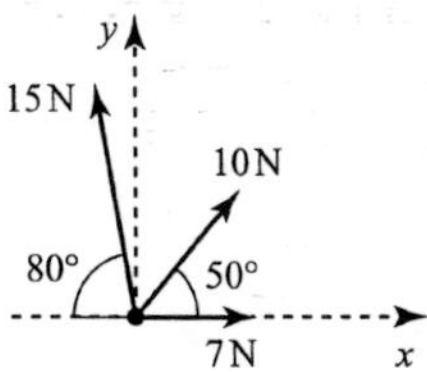

(i) Find the component of the resultant of the three forces

(a) in the x direction,

(b) in the y direction.

(ii) Hence find the direction of the resultant.

[Cambridge AS and A Level Mathematics 9709, Paper 4 Q3 June 2009]

Forces in equilibrium

When forces are in equilibrium their vector sum is zero and the sum of their resolved parts in *any* direction is zero.

EXAMPLE 6.1

A brick of mass 3 kg is at rest on a rough plane inclined at an angle of 30° to the horizontal. Find the friction force F N, and the normal reaction R N of the plane on the brick.

SOLUTION

The diagram shows the forces acting on the brick.

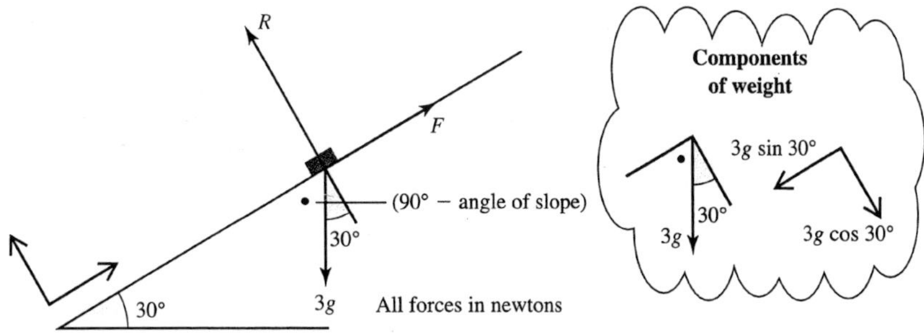

Figure 6.5

Use directions parallel and perpendicular to the plane, as shown. Since the brick is in equilibrium the resultant of the three forces acting on it is zero.

Resolving parallel to the slope: $\quad F - 30\sin 30° = 0 \quad$ ①

$F = 15$ (3g = 30)

Resolving perpendicular to the slope: $\quad R - 30\cos 30° = 0 \quad$ ②

$R = 26.0$ (to 3 s.f.)

Written in vector form the equivalent is

$$\begin{pmatrix} F \\ 0 \end{pmatrix} + \begin{pmatrix} 0 \\ R \end{pmatrix} + \begin{pmatrix} -30\sin 30° \\ -30\cos 30° \end{pmatrix} = \begin{pmatrix} 0 \\ 0 \end{pmatrix}$$

This leads to the equations ① and ②.

The triangle of forces

When there are only three (non-parallel) forces acting and they are in equilibrium, the polygon of forces becomes a closed triangle as shown for the brick on the plane.

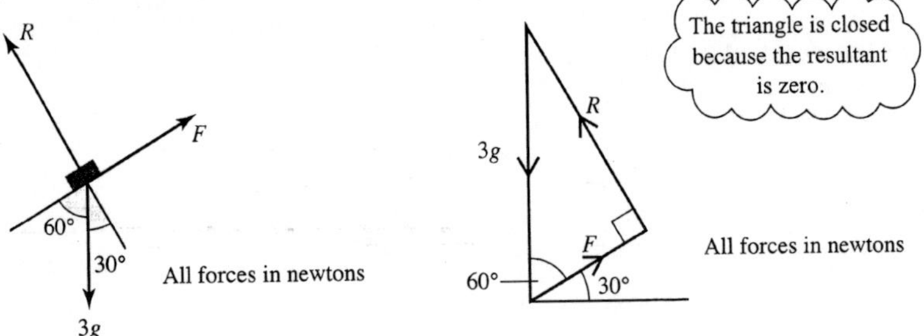

Figure 6.6 **Figure 6.7**

Then $\quad \dfrac{F}{3g} = \cos 60°$

$F = 30\cos 60° = 15\,\text{N}$

and similarly $\quad R = 30\sin 60° = 26.0\,\text{N}$ (to 3 s.f.)

This is an example of the theorem known as the *triangle of forces*.

- When a body is in equilibrium under the action of three non-parallel forces, then

  (i) the forces can be represented in magnitude and direction by the sides of a triangle

  (ii) the lines of action of the forces pass through the same point.

When more than three forces are in equilibrium the first statement still holds but the triangle is then a polygon. The second is not necessarily true.

Often mechanics questions involve the angles 30°, 45° and 60° so that you can use the exact values of cos θ, sin θ and tan θ in your working. Here is a table to remind you of the exact values.

$\theta°$	$\cos\theta°$	$\sin\theta°$	$\tan\theta°$
30°	$\frac{\sqrt{3}}{2}$	$\frac{1}{2}$	$\frac{1}{\sqrt{3}}$
45°	$\frac{1}{\sqrt{2}}$	$\frac{1}{\sqrt{2}}$	1
60°	$\frac{1}{2}$	$\frac{\sqrt{3}}{2}$	$\sqrt{3}$

You can see where these values come from in *Pure Mathematics 1* Chapter 7.

EXAMPLE 6.2

This example illustrates two methods for solving problems involving forces in equilibrium. With experience, you will find it easier to judge which method is best for a particular problem.

A sign of mass 10 kg is to be suspended by two strings arranged as shown in the diagram below. Find the tension in each string.

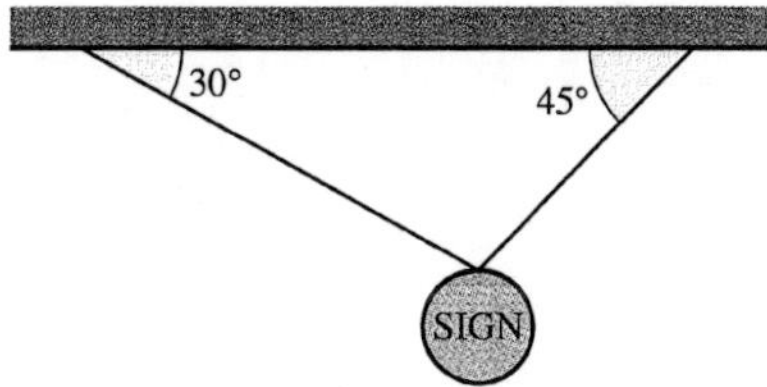

Figure 6.9

SOLUTION

The force diagram for this situation is given below.

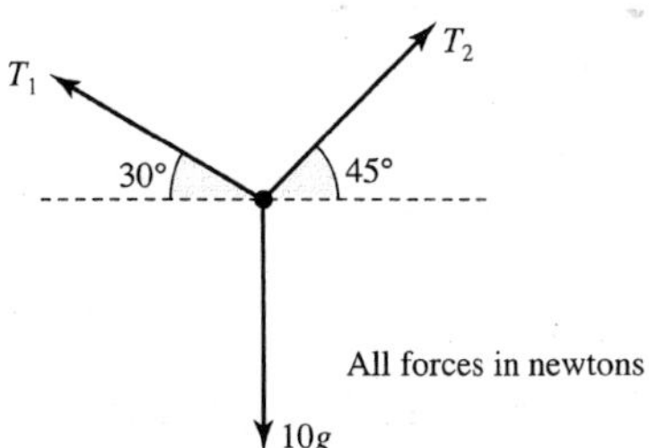

Figure 6.10

Method 1: Resolving forces

Vertically (↑): $T_1 \sin 30° + T_2 \sin 45° - 10g = 0$

$$\frac{1}{2}T_1 + \frac{1}{\sqrt{2}}T_2 = 100 \qquad ①$$

Horizontally (→): $-T_1 \cos 30° + T_2 \cos 45° = 0$

$$-\frac{\sqrt{3}}{2}T_1 + \frac{1}{\sqrt{2}}T_2 = 0 \qquad ②$$

Subtracting ② from ① $\left(\frac{1}{2} + \frac{\sqrt{3}}{2}\right)T_1 = 100$

$$1.366T_1 = 100$$

$$T_1 = 73.2$$

Back substitution gives $T_2 = 89.7$

The tensions are 73.2 N and 89.7 N (to 1 d.p.).

Method 2: Triangle of forces

Since the three forces are in equilibrium they can be represented by the sides of a triangle taken in order.

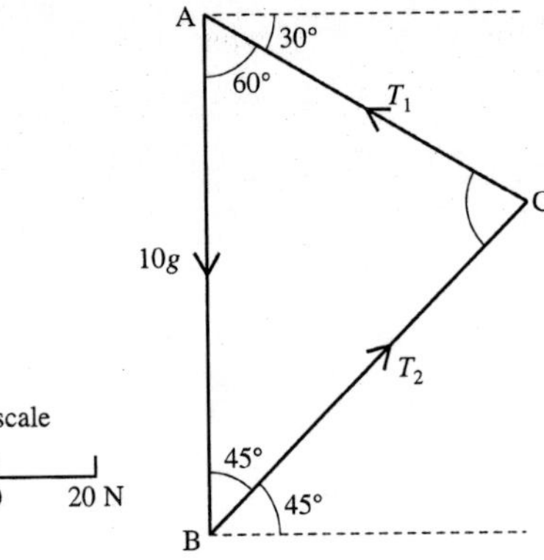

Figure 6.11

In what order would you draw the three lines in this diagram?

You can estimate the tensions by measurement. This will tell you that $T_1 \approx 73$ and $T_2 \approx 90$ in newtons.

Alternatively, you can use the sine rule to calculate T_1 and T_2 accurately.

In the triangle ABC, $\angle CAB = 60°$ and $\angle ABC = 45°$, so $\angle BCA = 75°$.

So $$\frac{T_1}{\sin 45°} = \frac{T_2}{\sin 60°} = \frac{100}{\sin 75°}$$

giving $$T_1 = \frac{100 \sin 45°}{\sin 75°} \quad \text{and} \quad T_2 = \frac{100 \sin 60°}{\sin 75°}$$

As before the tensions are found to be 73.2 N and 89.7 N.

? Lami's theorem states that when three forces acting at a point as shown in the diagram are in equilibrium then

$$\frac{F_1}{\sin\alpha} = \frac{F_2}{\sin\beta} = \frac{F_3}{\sin\gamma}.$$

Sketch a triangle of forces and say how the angles in the triangle are related to α, β and γ. Hence explain why Lami's theorem is true.

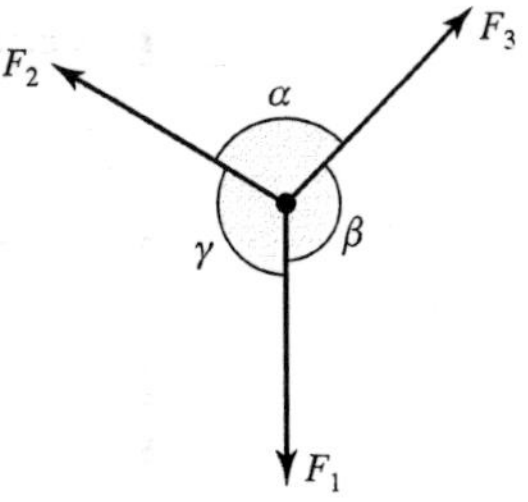

Figure 6.12

EXAMPLE 6.3

The picture shows three men involved in moving a packing case up to the top floor of a warehouse. Brian is pulling on a rope which passes round smooth pulleys at X and Y and is then secured to the point Z at the end of the loading beam.

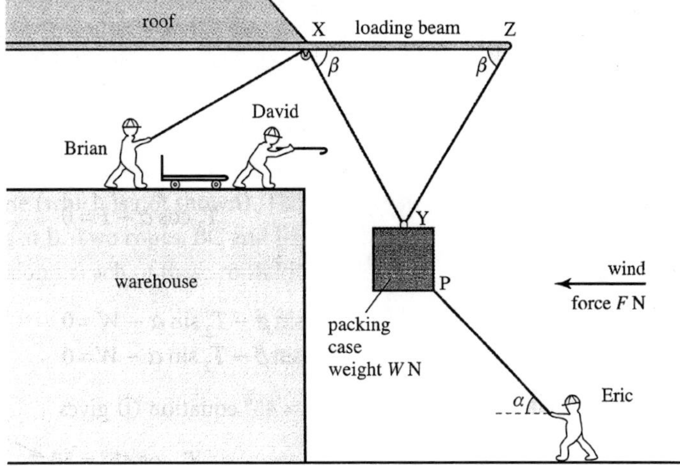

Figure 6.13

The wind is blowing directly towards the building. To counteract this, Eric is pulling on another rope, attached to the packing case at P, with just enough force and in the right direction to keep the packing case central between X and Z.

At the time of the picture the men are holding the packing case motionless.

(i) Draw a diagram showing all the forces acting on the packing case using T_1 and T_2 for the tensions in Brian's and Eric's ropes, respectively.

(ii) Write down equations for the horizontal and vertical equilibrium of the packing case.

In one particular situation, $W = 100$, $F = 50$, $\alpha = 45°$ and $\beta = 75°$.

(iii) Find the tension T_1.

(iv) Explain why Brian has to pull harder if the wind blows more strongly.

[MEI adapted]

SOLUTION

(i) The diagram shows all the forces acting on the packing case and the relevant angles.

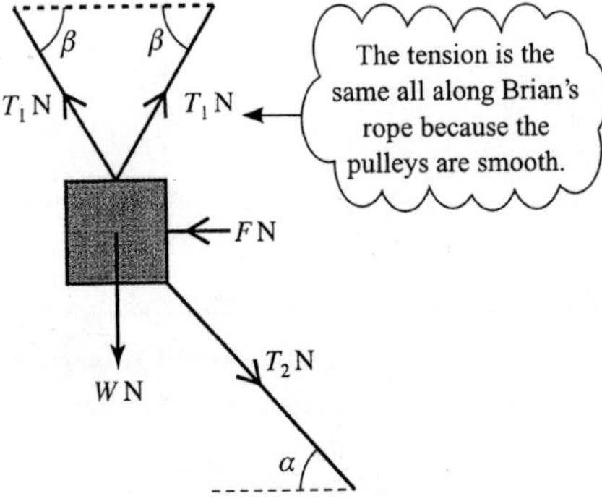

Figure 6.14 *Force diagram*

(ii) Equilibrium equations

Resolving horizontally (→):

$$T_1 \cos\beta + T_2 \cos\alpha - F - T_1 \cos\beta = 0$$

$$T_2 \cos\alpha - F = 0 \quad ①$$

Resolving vertically (↑):

$$T_1 \sin\beta + T_1 \sin\beta - T_2 \sin\alpha - W = 0$$

$$2T_1 \sin\beta - T_2 \sin\alpha - W = 0 \quad ②$$

(iii) When $F = 50$ and $\alpha = 45°$ equation ① gives

$$T_2 \cos 45° = 50$$

$$\Rightarrow T_2 \sin 45° = 50$$

This tells you that T_2 is $\dfrac{50}{\cos 45°}$ but you don't need to work it out because $\cos 45° = \sin 45°$.

Substituting in ② gives $\quad 2T_1 \sin\beta - 50 - W = 0$

So when $W = 100$ and $\beta = 75°$ $\quad 2T_1 \sin 75° = 150$

$$T_1 = \frac{150}{2\sin 75°}$$

The tension in Brian's rope is 77.65 N = 78 N (to the nearest newton).

(iv) When the wind blows more strongly, F increases. Given that all the angles remain unchanged, Eric will have to pull harder so the vertical component of T_2 will increase. This means that T_1 must increase and Brian must pull harder.

Or $F = T_2 \cos\alpha$, so as F increases, T_2 increases $\Rightarrow T_2 \sin\alpha + W$ increases $\Rightarrow 2\,T_1 \sin\beta$ increases. Hence T_1 increases.

EXERCISE 6B

1 The picture shows a boy, Halley, holding onto a post while his two older sisters, Sheuli and Veronica, try to pull him away. Using perpendicular horizontal directions the forces, in newtons, exerted by the two girls are:

Sheuli $\begin{pmatrix} 24 \\ 18 \end{pmatrix}$

Veronica $\begin{pmatrix} 25 \\ 60 \end{pmatrix}$

(i) Calculate the magnitude and direction of the force of each of the girls.

(ii) Use a scale drawing to estimate the magnitude and direction of the resultant of the forces exerted by the two girls.

(iii) Write the resultant as a vector and so calculate (to 3 significant figures) its magnitude and direction.

Check that your answers agree with those obtained by scale drawing in part **(ii)**.

2 The diagram shows a girder CD of mass 20 tonnes being held stationary by a crane (which is not shown). The rope from the crane (AB) is attached to a ring at B. Two ropes, BC and BD, of equal length attach the girder to B; the tension in each of these ropes is T N.

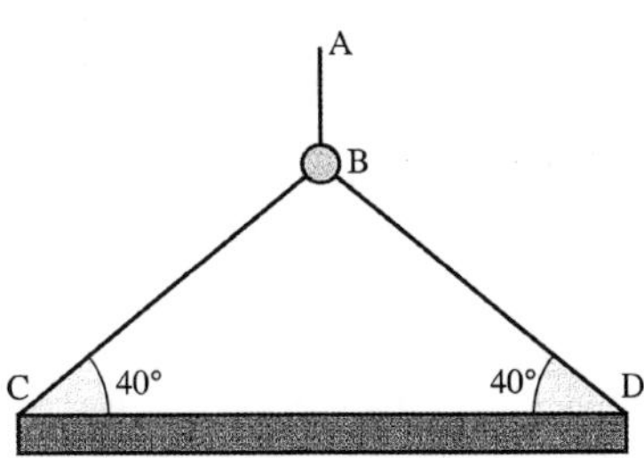

(i) Draw a diagram showing the forces acting on the girder.

(ii) Write down, in terms of T, the horizontal and vertical components of the tensions in the ropes acting at C and D.

(iii) Hence show that the tension in the rope BC is 155.6 kN (to 1 d.p.).

(iv) Draw a diagram to show the three forces acting on the ring at B.

(v) Hence calculate the tension in the rope AB.

(vi) How could you have known the answer to part **(v)** without any calculations?

3 The diagram shows a simple model of a crane. The structure is at rest in a vertical plane. The rod and cables are of negligible mass and the load suspended from the joint at A is 30 N.

(i) Draw a diagram showing the forces acting on

(a) the load

(b) the joint at A.

(ii) Calculate the forces in the rod and cable 1 and state whether they are in compression or in tension.

4 An angler catches a very large fish. When he tries to weigh it he finds that it is more than the 10 kg limit of his spring balance. He borrows another spring balance of exactly the same design and uses the two to weigh the fish, as shown in figure **(A)**. Both balances read 8 kg.

(i) What is the mass of the fish?

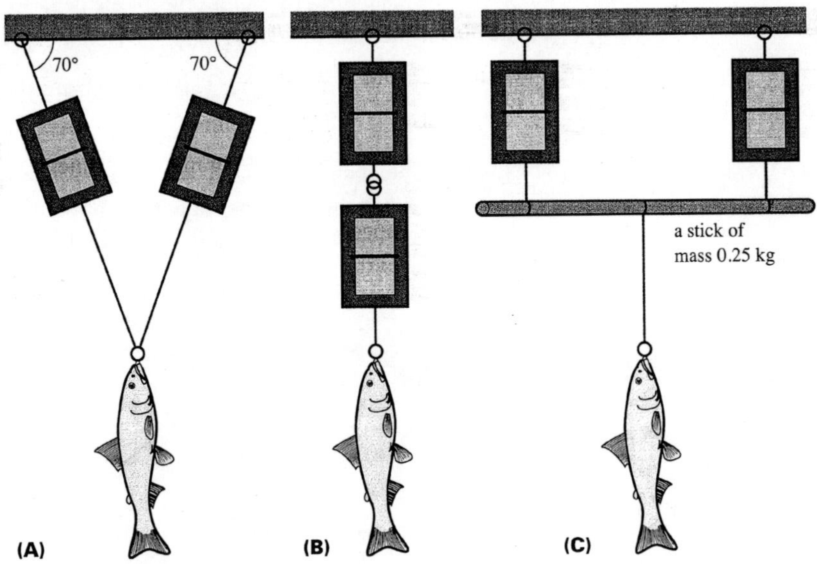

The angler believes the mass of the fish is a record and asks a witness to confirm it. The witness agrees with the measurements but cannot follow the calculations. He asks the angler to weigh the fish in two different positions, still using both balances. These are shown in figures **(B)** and **(C)**.

Assuming the spring balances themselves to have negligible mass, state the readings of the balances as set up in

(ii) figure **(B)**

(iii) figure **(C)**.

(iv) Which of the three methods do you think is the best?

5 The diagram shows a device for crushing scrap cars. The light rod AB is hinged at A and raised by a cable which runs from B round a pulley at D and down to a winch at E. The vertical strut EAD is rigid and strong and AD = AB. A weight of mass 1 tonne is suspended from B by the cable BC. When the weight is correctly situated above the car it is released and falls on to the car.

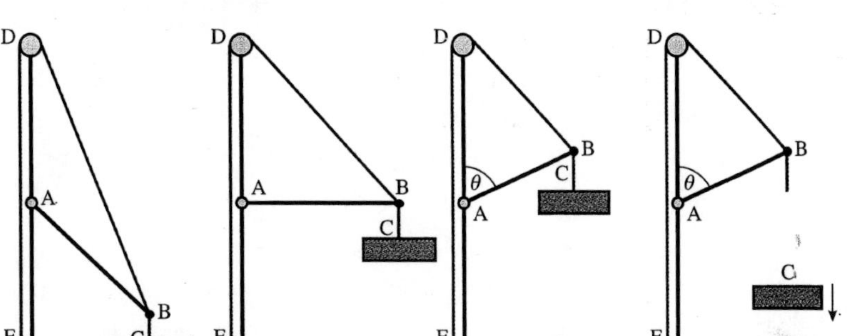

Just before the weight is released the rod AB makes angle θ with the upward vertical AD and the weight is at rest.

(i) Draw a diagram showing the forces acting at point B in this position.

(ii) Explain why the rod AB must be in thrust and not in tension.

(iii) Draw a diagram showing the vector sum of the forces at B (i.e. the polygon of forces).

(iv) Calculate each of the three forces acting at B when

(a) $\theta = 90°$ **(b)** $\theta = 60°$.

6 Four wires, all of them horizontal, are attached to the top of a telegraph pole as shown in the plan view on the right. The pole is in equilibrium and tensions in the wires are as shown.

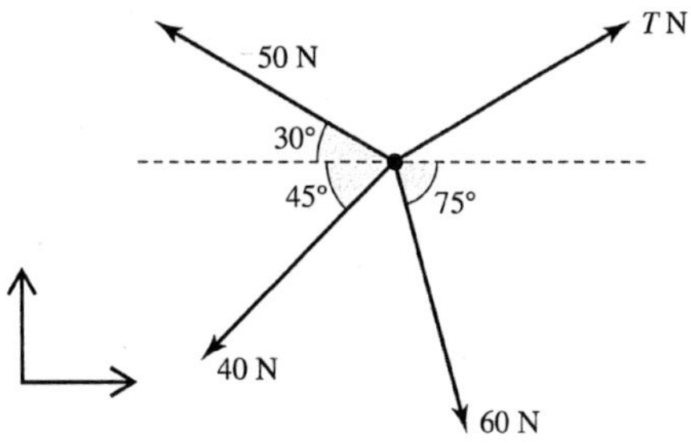

(i) Using perpendicular directions as shown in the diagram, show that the force of 60 N may be written as $\begin{pmatrix} 15.5 \\ -58.0 \end{pmatrix}$ N (to 3 significant figures).

(ii) Find T in both component form and magnitude and direction form.

(iii) The force T is changed to $\begin{pmatrix} 40 \\ 35 \end{pmatrix}$ N. Show that there is now a resultant force on the pole and find its magnitude and direction.

7 A ship is being towed by two tugs. Each tug exerts forces on the ship as indicated. There is also a drag force on the ship.

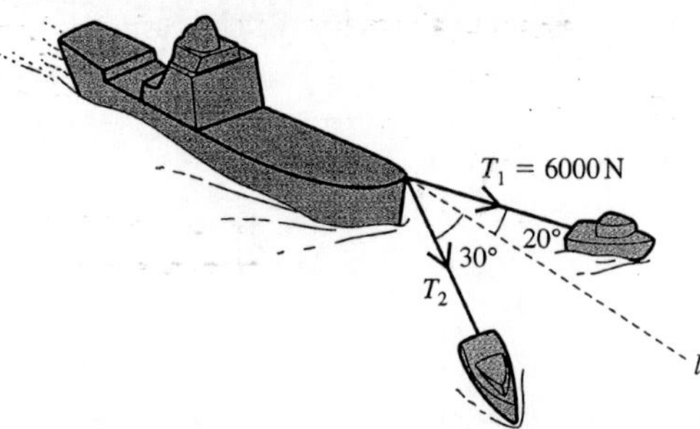

(i) Write down the components of the tensions in the towing cables along and perpendicular to the line of motion, l, of the ship.

(ii) There is no resultant force perpendicular to the line l. Find T_2.

(iii) The ship is travelling with constant velocity along the line l. Find the magnitude of the drag force acting on it.

8 A skier of mass 50 kg is skiing down a 15° slope.

(i) Draw a diagram showing the forces acting on the skier.

(ii) Resolve these forces into components parallel and perpendicular to the slope.

(iii) The skier is travelling at constant speed. Find the normal reaction of the slope on the skier and the resistance force on her.

(iv) The skier later returns to the top of the slope by being pulled up it at constant speed by a rope parallel to the slope. Assuming the resistance on the skier is the same as before, calculate the tension in the rope.

9 The diagram shows a block of mass 5 kg on a rough inclined plane. The block is attached to a 3 kg weight by a light string which passes over a smooth pulley and is on the point of sliding up the slope.

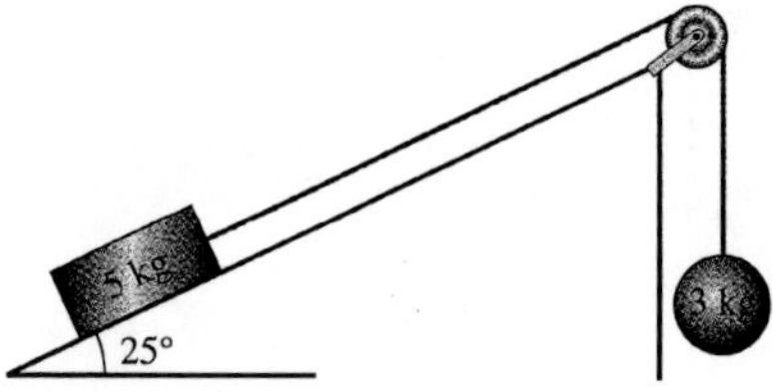

(i) Draw a diagram showing the forces acting on the block.

(ii) Resolve these forces into components parallel and perpendicular to the slope.

(iii) Find the force of resistance to the block's motion.

The 3 kg weight is replaced by one of mass m kg.

(iv) Find the value of m for which the block is on the point of sliding down the slope, assuming the resistance to motion is the same as before.

10 Two husky dogs are pulling a sledge. They both exert forces of 60 N but at different angles to the line of the sledge, as shown in the diagram. The sledge is moving straight forwards.

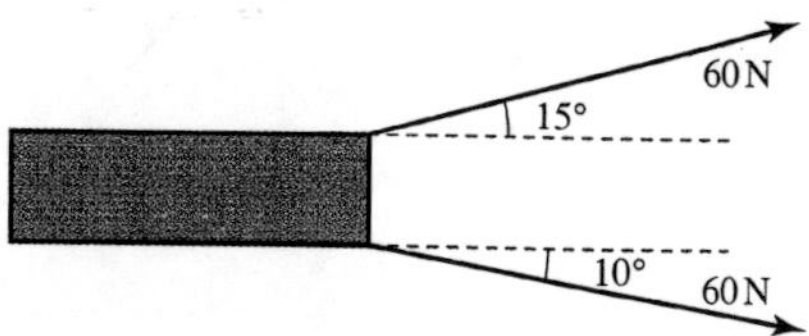

(i) Resolve the two forces into components parallel and perpendicular to the line of the sledge.

(ii) Hence find

(a) the overall forward force from the dogs

(b) the sideways force.

The resistance to motion is 20 N along the line of the sledge but up to 400 N perpendicular to it.

(iii) Find the magnitude and direction of the overall horizontal force on the sledge.

(iv) How much force is lost due to the dogs not pulling straight forwards?

11 One end of a string of length 1 m is fixed to a mass of 1 kg and the other end is fixed to a point A. Another string is fixed to the mass and passes over a frictionless pulley at B which is 1 m horizontally from A but 2 m above it. The tension in the second string is such that the mass is held at the same horizontal level as the point A.

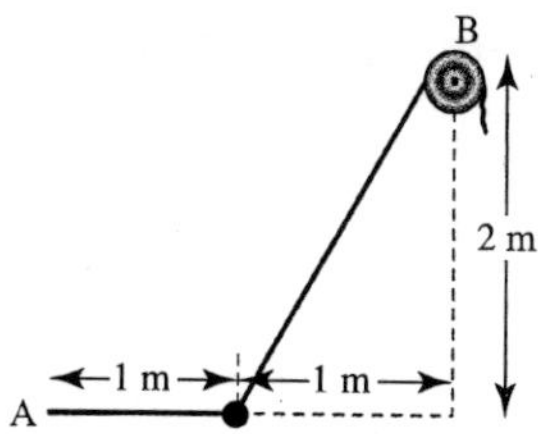

(i) Show that the tension in the horizontal string fixed to the mass and to A is 5 N and find the tension in the string which passes over the pulley at B. Find also the angle that this second string makes with the horizontal.

(ii) If the tension in this second string is slowly increased by drawing more of it over the pulley at B describe the path followed by the mass. Will the point A, the mass, and the point B, ever lie in a straight line? Give reasons for your answer.

12 A particle P is in equilibrium on a smooth horizontal table under the action of horizontal forces of magnitudes F N, F N, G N and 12 N acting in the directions shown. Find the values of F and G.

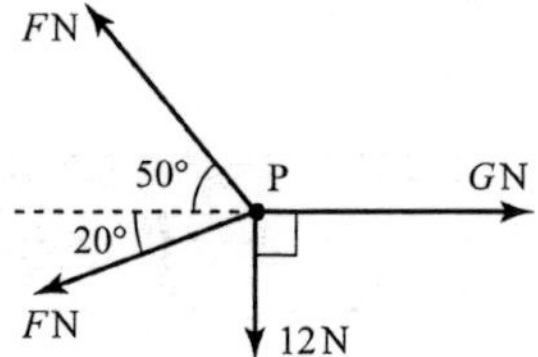

[Cambridge AS and A Level Mathematics 9709, Paper 4 Q3 June 2006]

13 Each of three light strings has a particle attached to one of its ends. The other ends of the strings are tied together at a point A. The strings are in equilibrium with two of them passing over fixed smooth horizontal pegs, and with the particles hanging freely. The weights of the particles, and the angles between the sloping parts of the strings and the vertical, are as shown in the diagram. Find the values of W_1 and W_2.

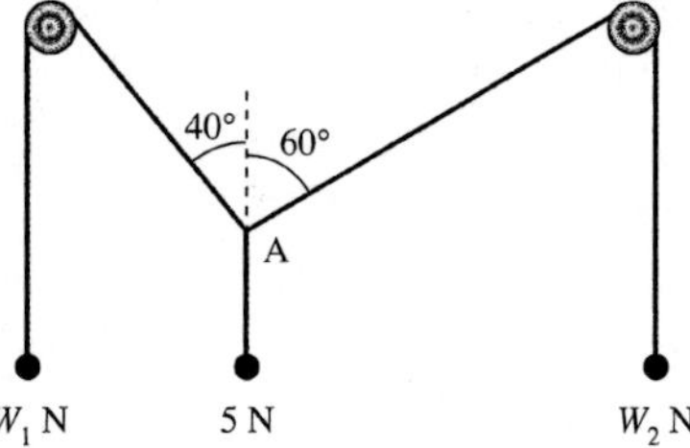

[Cambridge AS and A Level Mathematics 9709, Paper 4 Q3 November 2005]

Newton's second law in two dimensions

When the forces acting on an object are not in equilibrium it will have an acceleration and you can use Newton's second law to solve problems about its motion.

The equation $\mathbf{F} = m\mathbf{a}$ is a vector equation. The resultant force acting on a particle is equal in both magnitude and direction to the mass × acceleration. It can be written in components as

$$\begin{pmatrix} F_1 \\ F_2 \end{pmatrix} = m \begin{pmatrix} a_1 \\ a_2 \end{pmatrix}$$

so that $F_1 = ma_1$ and $F_2 = ma_2$.

What direction is the resultant force acting on a child sliding on a sledge down a smooth straight slope inclined at 15° to the horizontal?

EXAMPLE 6.4

Anna is sledging. Her sister gives her a push at the top of a smooth straight 15° slope and lets go when she is moving at 2 m s^{-1}. She continues to slide for 5 seconds before using her feet to produce a braking force of 95 N parallel to the slope. This brings her to rest. Anna and her sledge have a mass of 30 kg.

How far does she travel altogether?

SOLUTION

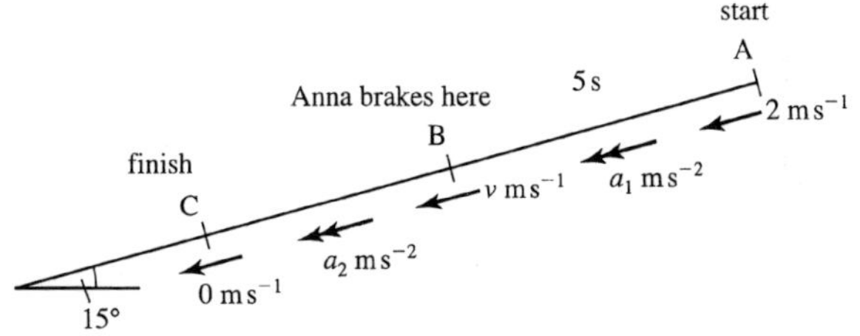

Figure 6.15

To answer this question, you need to know Anna's acceleration for the two parts of her journey. These are constant so you can then use the constant acceleration formulae.

Sliding freely

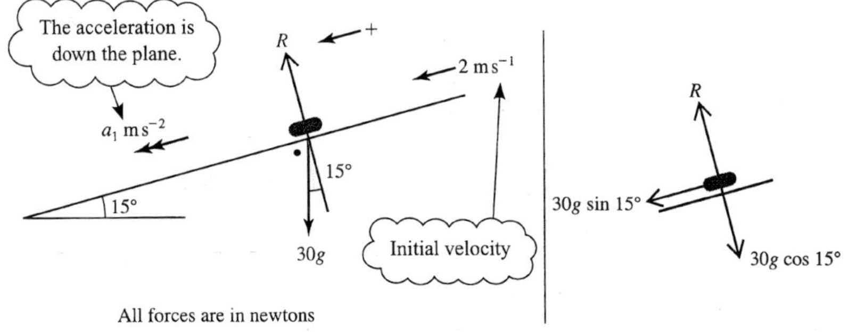

Figure 6.16

Using Newton's second law in the direction of the acceleration gives:

$$30g \sin 15° = 30a_1$$
$$a_1 = 10 \sin 15°$$
$$a_1 = 2.58...$$

Resultant force down the plane = mass × acceleration

Store all the working values in the memory of your calculator so that you avoid rounding errors.

Now you know a_1 you can find how far Anna slides (s_1) and her speed ($v\,\mathrm{m\,s^{-1}}$) before braking.

Given $u = 2$, $t = 5$, $a = 10 \sin 15°$:

$$s = ut + \frac{1}{2}at^2$$
$$s_1 = 2 \times 5 + \frac{1}{2} \times 10 \sin 15° \times 25$$
$$s_1 = 42.352...$$

So Anna slides 42.35 m (to the nearest centimetre).

$$v = u + at$$
$$v = 2 + 10 \sin 15° \times 5$$
$$v = 14.940...$$

So Anna's speed is $14.9\,\mathrm{m\,s^{-1}}$ (to 3 s.f.).

Braking

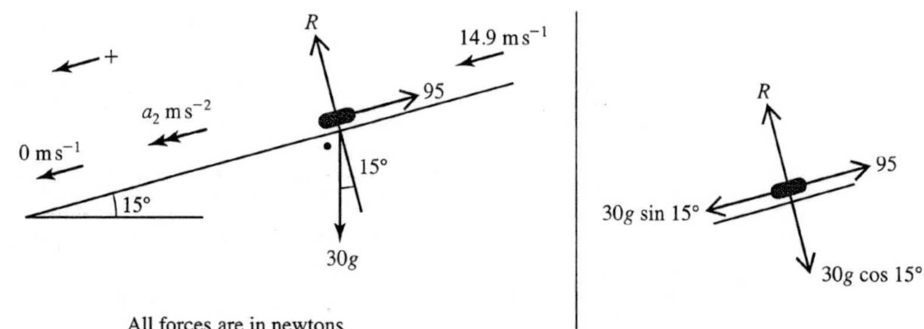

Figure 6.17

By Newton's second law down the plane:

Resultant force = mass × acceleration

$$30g \sin 15° - 95 = 30a_2$$
$$a_2 = -0.578...$$

$$v^2 = u^2 + 2as$$
$$0 = 14.9...^2 - 2 \times 0.578... \times s_2$$

Given $u = 14.9$, $v = 0$, $a = -0.578...$

$$s_2 = \frac{14.9...^2}{2 \times 0.578...} = 192.94...$$

Anna travels a total distance of (42.35... + 192.94...)m = 235 m to the nearest metre.

Make a list of the modelling assumptions used in Example 6.4. What would be the effect of changing these?

EXAMPLE 6.5

A skier is being pulled up a smooth 25° dry ski slope by a rope which makes an angle of 35° with the horizontal. The mass of the skier is 75 kg and the tension in the rope is 350 N. Initially the skier is at rest at the bottom of the slope. The slope is smooth. Find the skier's speed after 5 s and find the distance he has travelled in that time.

SOLUTION

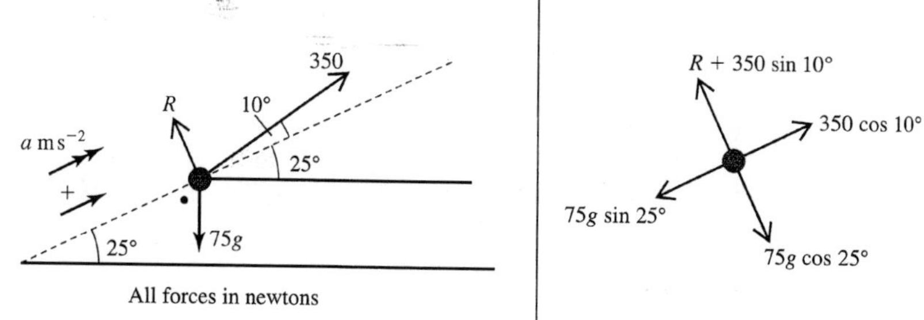

Figure 6.18

In the diagram the skier is modelled as a particle. Since the skier moves parallel to the slope consider motion in that direction.

$$\text{Resultant force} = \text{mass} \times \text{acceleration}$$

$$350 \cos 10° - 75g \sin 25° = 75 \times a$$

Taking g as 10 → $a = \dfrac{27.71...}{75} = 0.369...$

This is a constant acceleration so use the constant acceleration formulae.

$$v = u + at$$

$$v = 0 + 0.369... \times 5$$ ← $u = 0$, $a = 0.369...$, $t = 5$

Speed = 1.85 m s^{-1} (to 2 d.p.).

$$s = ut + \frac{1}{2}at^2$$

$$s = 0 + \frac{1}{2} \times 0.369... \times 25$$

Distance travelled = 4.62 m (to 2 d.p.).

EXAMPLE 6.6

A car of mass 1000 kg including its driver, is being pushed along a horizontal road by three people as indicated in the diagram. The car is moving in the direction PQ.

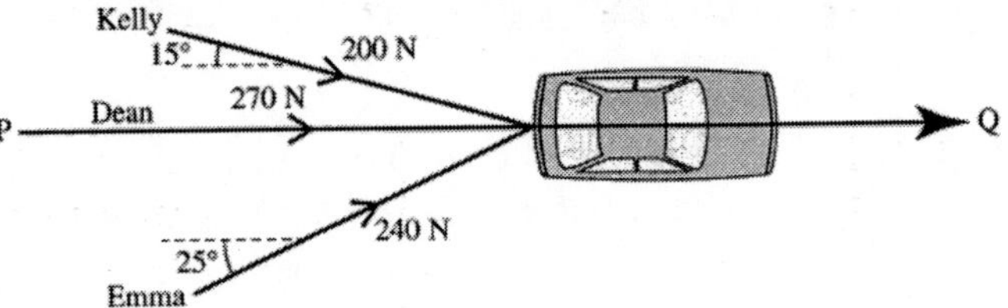

Figure 6.19

(i) Calculate the total force exerted by the three people in the direction PQ.

(ii) Calculate the force exerted overall by the three people in the direction perpendicular to PQ.

(iii) Explain briefly why the car does not move in the direction perpendicular to PQ.

Initially the car is stationary and 5 s later it has a speed of $2\,\text{m s}^{-1}$ in the direction PQ.

(iv) Calculate the force of resistance to the car's movement in the direction PQ assuming the three people continue to push as described above.

[MEI, part]

SOLUTION

(i) Resolving in the direction PQ, the components in newtons are:

Kelly	$200 \cos 15° = 193$
Dean	270
Emma	$240 \cos 25° = 218$

Total force in the direction PQ = 681 N.

(ii) Resolving perpendicular to PQ (↑) the components are:

Kelly	$-200 \sin 15° = -51.8$
Dean	0
Emma	$240 \sin 25° = 101.4$

Total force in the direction perpendicular to PQ = 49.6 N.

(iii) The car does not move perpendicular to PQ because the force in this direction is balanced by a sideways (lateral) friction force between the tyres and the road.

(iv) To find the acceleration, $a\,\text{m s}^{-2}$, of the car:

$$v = u + at$$
$$2 = 0 + 5a \quad \leftarrow \quad u = 0, v = 2, t = 5$$
$$a = 0.4$$

When the resistance to motion in the direction QP is R N, figure 6.20 shows all the horizontal forces acting on the car and its acceleration.

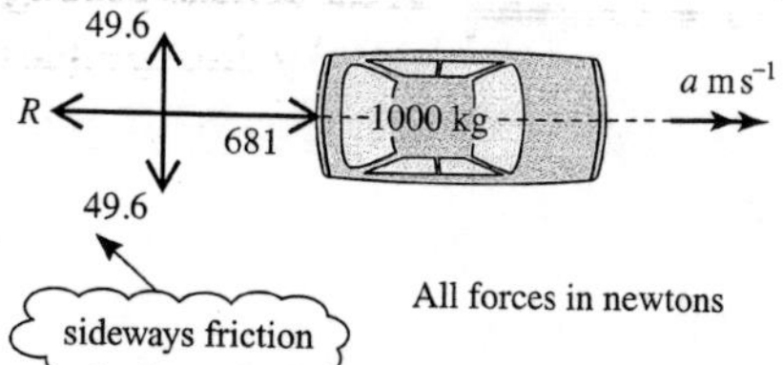

Figure 6.20

The resultant force in the direction PQ is $(681 - R)$ N. So by Newton II

$$681 - R = 1000a$$
$$R = 681 - 400$$

The resistance to motion in the direction Q is 281 N.

EXERCISE 6C

1 The forces $\mathbf{F}_1 = \begin{pmatrix} 4 \\ -5 \end{pmatrix}$ and $\mathbf{F}_2 = \begin{pmatrix} 2 \\ 1 \end{pmatrix}$, in newtons, act on a particle of mass 4 kg.

(i) Find the acceleration of the particle in component form.

(ii) Find the magnitude of the particle's acceleration.

2 Two forces $\mathbf{P}_1$ and $\mathbf{P}_2$ act on a particle of mass 2 kg giving it an acceleration of $\begin{pmatrix} 5 \\ 5 \end{pmatrix}$ (in m s^{-2}).

(i) If $\mathbf{P}_1 = \begin{pmatrix} 6 \\ -1 \end{pmatrix}$ (in newtons), find $\mathbf{P}_2$.

(ii) If instead $\mathbf{P}_1$ and $\mathbf{P}_2$ both act in the same direction but $\mathbf{P}_1$ is four times as big as $\mathbf{P}_2$ find both forces.

3 The diagram shows a girl pulling a sledge at steady speed across level snow-covered ground using a rope which makes an angle of 30° to the horizontal. The mass of the sledge is 8 kg and there is a resistance force of 10 N.

(i) Draw a diagram showing the forces acting on the sledge.

(ii) Find the magnitude of the tension in the rope.

The girl comes to an area of ice where the resistance force on the sledge is only 2 N. She continues to pull the sledge with the same force as before and with the rope still taut at 30°.

(iii) What acceleration must the girl have in order to do this?

(iv) How long will it take to double her initial speed of 0.4 m s^{-1}?

4 The picture shows a situation which has arisen between two anglers, Davies and Jones, standing at the ends of adjacent jetties. Their lines have become entangled under the water with the result that they have both hooked the same fish, which has mass 1.9 kg. Both are reeling in their lines as hard as they can in order to claim the fish.

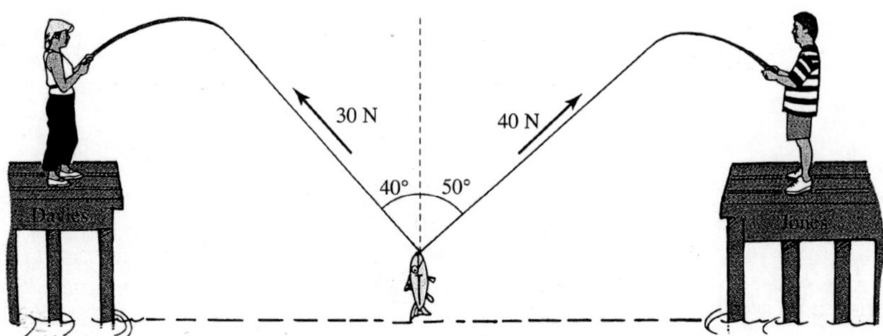

(i) Draw a diagram showing the forces acting on the fish.

(ii) Resolve the tensions in both anglers' lines into horizontal and vertical components and so find the total force acting on the fish.

(iii) Find the magnitude and direction of the acceleration of the fish.

(iv) At this point Davies' line breaks. What happens to the fish?

5 A crate of mass 30 kg is being pulled up a smooth slope inclined at 30° to the horizontal by a rope which is parallel to the slope. The crate has acceleration $0.75\,\mathrm{m\,s^{-2}}$.

(i) Draw a diagram showing the forces acting on the crate and the direction of its acceleration.

(ii) Resolve the forces in directions parallel and perpendicular to the slope.

(iii) Find the tension in the rope.

(iv) The rope suddenly snaps. What happens to the crate?

6 A cyclist of mass 60 kg rides a cycle of mass 7 kg. The greatest forward force that she can produce is 200 N but she is subject to air resistance and friction totalling 50 N.

(i) Draw a diagram showing the forces acting on the cyclist when she is going uphill.

(ii) What is the angle of the steepest slope that she can ascend?

The cyclist reaches a slope of 8° with a speed of $5\,\mathrm{m\,s^{-1}}$ and rides as hard as she can up it.

(iii) Find her acceleration and the distance she travels in 5 s.

(iv) What is her speed now?

7 A builder is demolishing the chimney of a house and slides the old bricks down to the ground on a straight chute 10 m long inclined at 42° to the horizontal. Each brick has mass 3 kg.

(i) Draw a diagram showing the forces acting on a brick as it slides down the chute, assuming the chute to have a flat cross section and a smooth surface.

(ii) Find the acceleration of the brick.

(iii) Find the time the brick takes to reach the ground.

In fact the chute is not smooth and the brick takes 3 s to reach the ground.

(iv) Find the frictional force acting on the brick, assuming it to be constant.

8 A box of mass 80 kg is to be pulled along a horizontal floor by means of a light rope. The rope is pulled with a force of 100 N and the rope is inclined at 20° to the horizontal, as shown in the diagram.

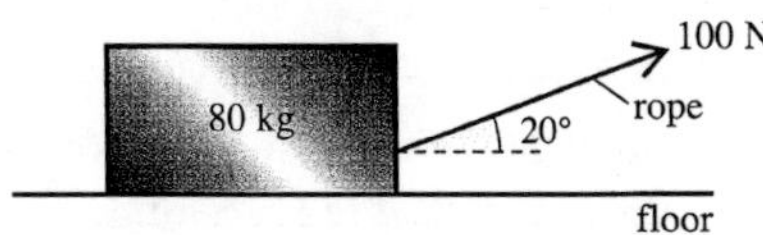

(i) Explain briefly why the box cannot be in equilibrium if the floor is smooth.

In fact the floor is not smooth and the box is in equilibrium.

(ii) Draw a diagram showing all the external forces acting on the box.

(iii) Calculate the frictional force between the box and the floor and also the normal reaction of the floor on the box, giving your answers correct to three significant figures.

The maximum value of the frictional force between the box and the floor is 120 N and the box is now pulled along the floor with the rope always inclined at 20° to the horizontal.

(iv) Calculate the force with which the rope must be pulled for the box to move at a constant speed. Give your answer correct to three significant figures.

(v) Calculate the acceleration of the box if the rope is pulled with a force of 140 N.

[MEI]

9 A block of mass 5 kg is at rest on a plane which is inclined at 30° to the horizontal. A light, inelastic string is attached to the block, passes over a smooth pulley and supports a mass m which is hanging freely. The part of the string between the block and the pulley is parallel to a line of greatest slope of the plane. A friction force of 15 N opposes the motion of the block. The diagram shows the block when it is slipping up the plane at a constant speed.

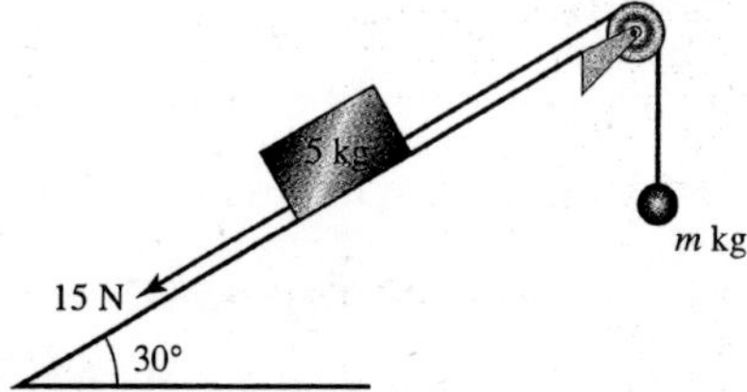

Give your answers correct to two significant figures.

(i) Copy the diagram and mark in all the forces acting on the block and the hanging mass, including the tension in the string.

(ii) Calculate the value of m when the block slides up the plane at a constant speed and find the tension in the string.

(iii) Calculate the acceleration of the system when $m = 6$ kg and find the tension in the string in this case.

[MEI]

KEY POINTS

1 The forces acting on a particle can be combined to form a resultant force using scale drawing or calculation by resolving the forces into their components.

Scale drawing

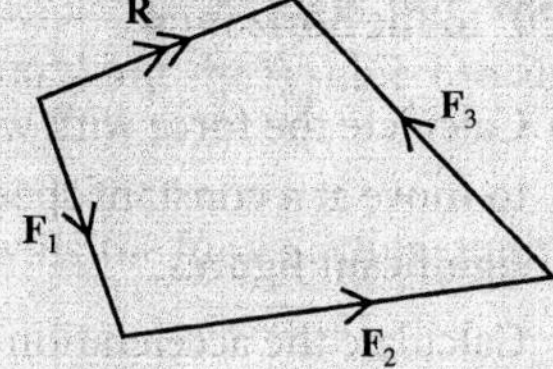

- Draw an accurate diagram, then measure the resultant. This is less accurate than calculation.
- To calculate the resultant, find the components of the various forces and add them. Then find the magnitude and directions of the resultant.

Components

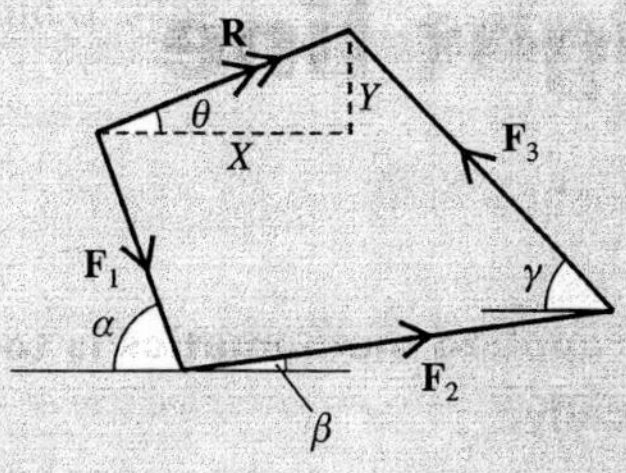

When $\mathbf{R} = \begin{pmatrix} X \\ Y \end{pmatrix}$

$$X = F_1 \cos\alpha + F_2 \cos\beta - F_3 \cos\gamma$$

$$Y = -F_1 \sin\alpha + F_2 \sin\beta + F_3 \sin\gamma$$

$$|\mathbf{R}| = \sqrt{X^2 + Y^2}$$

$$\tan\theta = \frac{Y}{X}$$

2 Equilibrium

When the resultant **R** is zero, the forces are in equilibrium.

3 Triangle of forces

If a body is in equilibrium under three non-parallel forces, their lines of action are concurrent and they can be represented by a triangle.

4 Newton's second law

When the resultant **R** is not zero there is an acceleration **a** and $\mathbf{R} = m\mathbf{a}$.

5 When a particle is on a slope, it is usually helpful to resolve in directions parallel and perpendicular to the slope.

7 General motion in a straight line

The goal of applied mathematics is to understand reality mathematically.

G. G. Hall

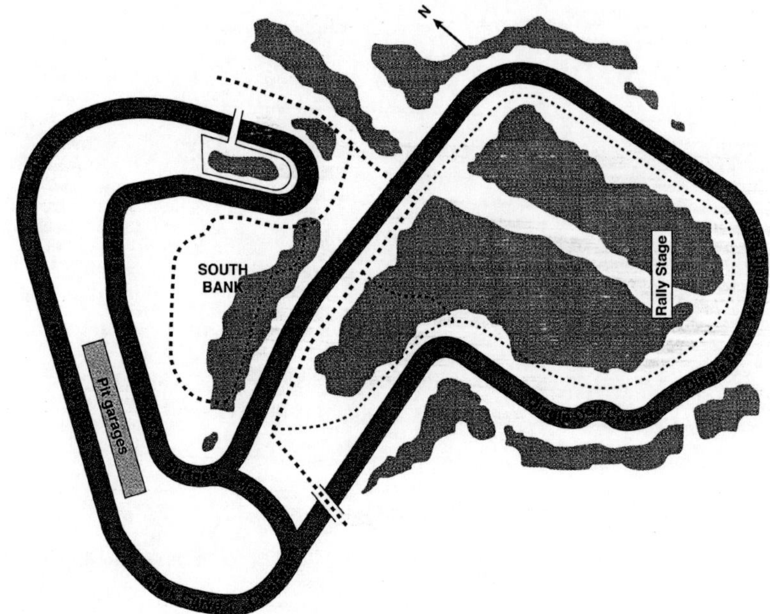

Figure 7.1

So far you have studied motion with constant acceleration in a straight line, but the motion of a car round the Brand's Hatch racing circuit shown in figure 7.1 is much more complex. In this chapter you will see how to deal with variable acceleration.

The equations you have used for constant acceleration do not apply when the acceleration varies. You need to go back to first principles.

Consider how displacement, velocity and acceleration are related to each other. The velocity of an object is the rate at which its position changes with time. When the velocity is not constant the position–time graph is a curve.

The rate of change of the position is the gradient of the tangent to the curve. You can find this by differentiating.

$$v = \frac{ds}{dt} \quad ①$$

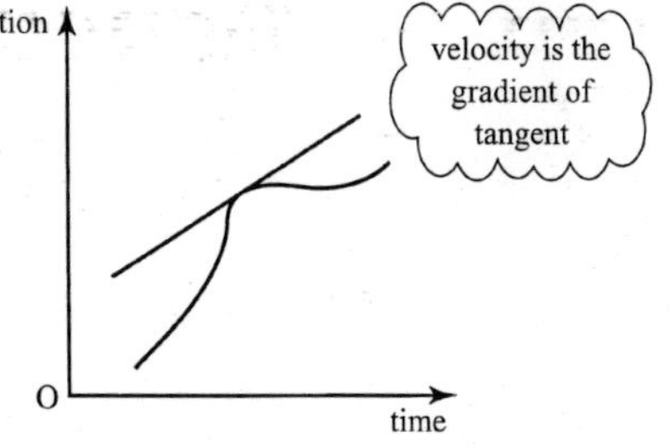

Figure 7.2

Similarly, the acceleration is the rate at which the velocity changes, so

$$a = \frac{dv}{dt} = \frac{d^2s}{dt^2} \quad ②$$

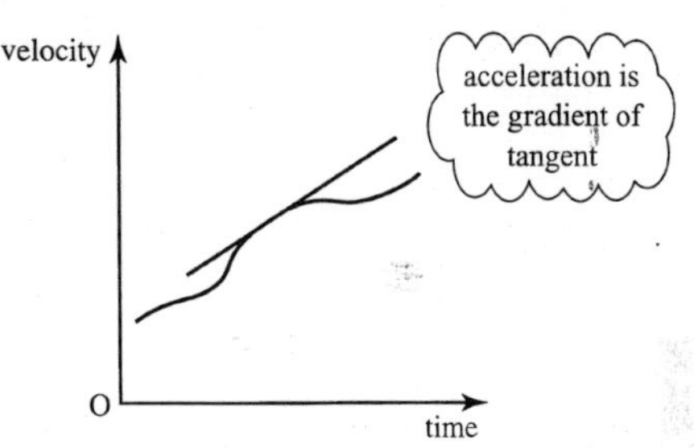

Figure 7.3

Using differentiation

When you are given the position of a moving object in terms of time, you can use equations ① and ② to solve problems even when the acceleration is not constant.

EXAMPLE 7.1

An object moves along a straight line so that its position at time t in seconds is given by

$$x = 2t^3 - 6t \text{ (in metres) } (t \geqslant 0).$$

(i) Find expressions for the velocity and acceleration of the object at time t.

(ii) Find the values of x, v and a when $t = 0$, 1, 2 and 3.

(iii) Sketch the graphs of x, v and a against time.

(iv) Describe the motion of the object.

SOLUTION

(i) Position $\quad x = 2t^3 - 6t \quad ①$

Velocity $\quad v = \frac{dx}{dt} = 6t^2 - 6 \quad ②$

Acceleration $a = \frac{dv}{dt} = 12t \quad ③$

You can now use these three equations to solve problems about the motion of the object.

(ii) When

	$t =$	0	1	2	3
From ①	$x =$	0	−4	4	36
From ②	$v =$	−6	0	18	48
From ③	$a =$	0	12	24	36

(iii) The graphs are drawn under each other so that you can see how they relate.

(iv) The object starts at the origin and moves towards the negative direction, gradually slowing down.

At $t = 1$ it stops instantaneously and changes direction, returning to its initial position at about $t = 1.7$.

It then continues moving in the positive direction with increasing speed.

The acceleration is increasing at a constant rate. This cannot go on for much longer or the speed will become excessive.

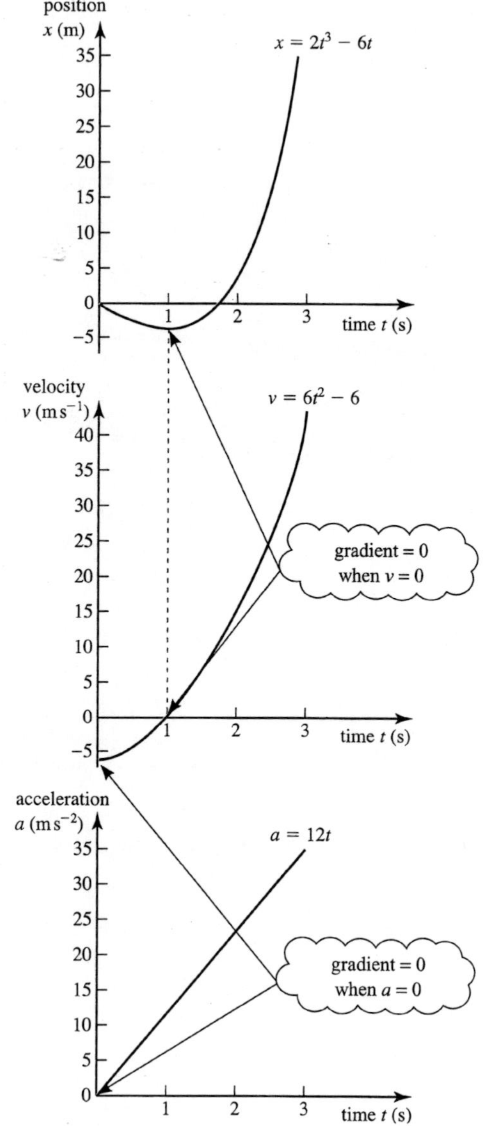

Figure 7.4

EXERCISE 7A

1 In each of the following cases

(a) find an expression for the velocity

(b) use your equations to write down the initial position and velocity

(c) find the time and position when the velocity is zero.

(i) $s = 10 + 2t - t^2$

(ii) $s = -4t + t^2$

(iii) $x = t^3 - 5t^2 + 4$

2 In each of the following cases

(a) find an expression for the acceleration

(b) use your equations to write down the initial velocity and acceleration.

(i) $v = 4t + 3$

(ii) $v = 6t^2 - 2t + 1$

(iii) $v = 7t - 5$

3 The distance travelled by a cyclist is modelled by

$$s = 4t + 0.5t^2 \text{ in S.I. units}$$

Find expressions for the velocity and the acceleration of the cyclist at time t.

4 In each of the following cases

(a) find expressions for the velocity and the acceleration

(b) draw the acceleration–time graph and, below it, the velocity–time graph with the same scale for time and the origins in line

(c) describe how the two graphs for each object relate to each other

(d) describe how the velocity and acceleration change during the motion of each object.

(i) $x = 15t - 5t^2$

(ii) $x = 6t^3 - 18t^2 - 6t + 3$

Finding displacement from velocity

How can you find an expression for the position of an object when you know its velocity in terms of time?

One way of thinking about this is to remember that $v = \dfrac{ds}{dt}$, so you need to do the opposite of differentiation, that is integrate, to find s.

$$s = \int v\,dt$$

The dt indicates that you must write v in terms of t before integrating.

EXAMPLE 7.2

The velocity (in $\mathrm{m\,s^{-1}}$) of a model train which is moving along straight rails is

$$v = 0.3t^2 - 0.5$$

Find its displacement from its initial position

(i) after time t

(ii) after 3 seconds.

SOLUTION

(i) The displacement at any time is

$$\begin{aligned} s &= \int v \,\mathrm{d}t \\ &= \int (0.3t^2 - 0.5)\,\mathrm{d}t \\ &= 0.1t^3 - 0.5t + c \end{aligned}$$

To find the train's displacement from its initial position, put $s = 0$ when $t = 0$.

This gives $c = 0$ and so $s = 0.1t^3 - 0.5t$.

You can use this equation to find the displacement at any time before the motion changes.

(ii) After 3 seconds, $t = 3$ and $s = 2.7 - 1.5$.
The train is 1.2 m from its initial position.

When using integration don't forget the constant. This is very important in mechanics problems and you are usually given some extra information to help you find the value of the constant.

The area under a velocity–time graph

In Chapter 1 you saw that the area under a velocity–time graph represents a displacement. Both the area under the graph and the displacement are found by integrating. To find a particular displacement you calculate the area under the velocity–time graph by integration using suitable limits.

The distance travelled between the times T_1 and T_2 is shown by the shaded area on the graph.

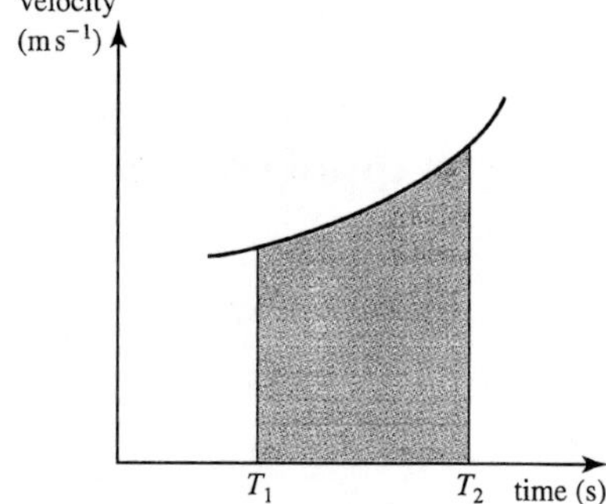

Figure 7.5

$$s = \text{area} = \int_{T_1}^{T_2} v \,\mathrm{d}t$$

EXAMPLE 7.3

A car moves between two sets of traffic lights, stopping at both. Its speed $v\,\mathrm{m\,s^{-1}}$ at time t s is modelled by

$$v = \frac{1}{20}t(40 - t), \qquad 0 \leqslant t \leqslant 40.$$

Find the times at which the car is stationary and the distance between the two sets of traffic lights.

SOLUTION

The car is stationary when $v = 0$. Substituting this into the expression for the speed gives

$$0 = \frac{1}{20}t(40 - t)$$

$$\Rightarrow \qquad t = 0 \text{ or } t = 40.$$

These are the times when the car starts to move away from the first set of traffic lights and stops at the second set.

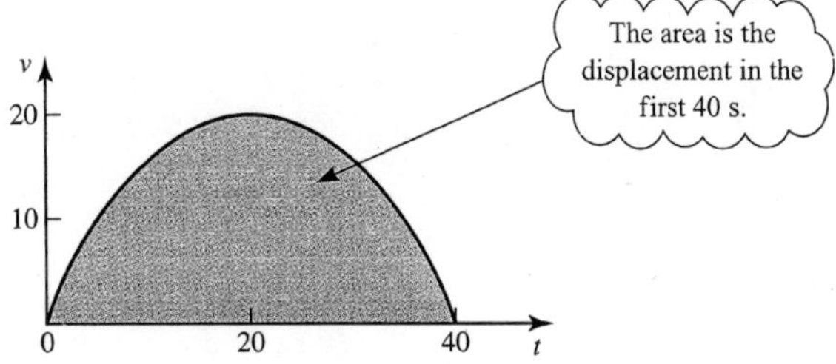

Figure 7.6

The distance between the two sets of lights is given by

$$\text{Distance} = \int_0^{40} \frac{1}{20}t(40 - t)\,\mathrm{d}t$$

$$= \frac{1}{20}\int_0^{40} (40t - t^2)\,\mathrm{d}t$$

$$= \frac{1}{20}\left[20t^2 - \frac{t^3}{3}\right]_0^{40}$$

$$= 533.\dot{3}\,\mathrm{m}$$

Finding velocity from acceleration

You can also find the velocity from the acceleration by using integration.

$$a = \frac{\mathrm{d}v}{\mathrm{d}t}$$

$$\Rightarrow \qquad v = \int a\,\mathrm{d}t$$

The next example shows how you can obtain equations for motion using integration.

EXAMPLE 7.4

The acceleration of a particle (in m s^{-2}) at time t seconds is given by

$$a = 6 - t.$$

The particle is initially at the origin with velocity $-2\,\text{m s}^{-1}$. Find an expression for

(i) the velocity of the particle after t s

(ii) the position of the particle after t s.

Hence find the velocity and position 6 s later.

SOLUTION

The information given may be summarised as follows:

at $t = 0$, $s = 0$ and $v = -2$;

at time t, $a = 6 - t$. ①

(i) $\dfrac{dv}{dt} = a = 6 - t$

Integrating gives

$$v = 6t - \tfrac{1}{2}t^2 + c$$

When $t = 0$, $v = -2$

so $\quad -2 = 0 - 0 + c$

$\quad c = -2$

At time t

$$v = 6t - \tfrac{1}{2}t^2 - 2 \quad ②$$

(ii) 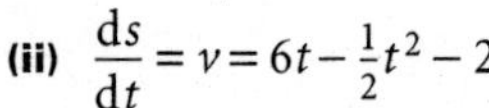

$\dfrac{ds}{dt} = v = 6t - \tfrac{1}{2}t^2 - 2$

Integrating gives

$$s = 3t^2 - \tfrac{1}{6}t^3 - 2t + k$$

When $t = 0$, $s = 0$

so $0 = 0 - 0 - 0 + k$

$\quad k = 0$

At time t

$$s = 3t^2 - \tfrac{1}{6}t^3 - 2t \quad ③$$

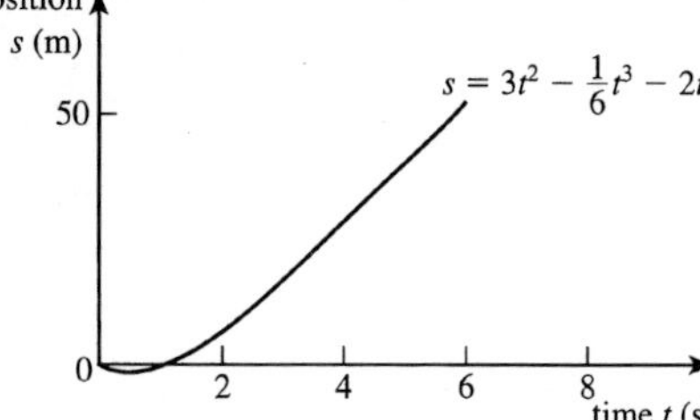

Figure 7.7

Notice that two different arbitrary constants (c and k) are necessary when you integrate twice. You could call them c_1 and c_2 if you wish.

Equations ①, ② and ③ can now be used to give more information about the motion in a similar way to the *suvat* formulae. (The *suvat* formulae only apply when the acceleration is constant.)

When $t = 6$ $\quad v = 36 - 18 - 2 = 16$ $\quad$ from ②

When $t = 6$ $\quad s = 108 - 36 - 12 = 60$ $\quad$ from ③

The particle has a velocity of $+16\,\text{m s}^{-1}$ and is at $+60$ m after 6 s.

EXERCISE 7B

1 Find expressions for the position in each of these cases.

(i) $v = 4t + 3$; initial position 0.
(ii) $v = 6t^3 - 2t^2 + 1$; when $t = 0$, $s = 1$.
(iii) $v = 7t^2 - 5$; when $t = 0$, $s = 2$.

2 The speed of a ball rolling down a hill is modelled by $v = 1.7t$ (in m s^{-1}).

(i) Draw the speed–time graph of the ball.
(ii) How far does the ball travel in 10 s?

3 Until it stops moving, the speed of a bullet t s after entering water is modelled by $v = 216 - t^3$ (in m s^{-1}).

(i) When does the bullet stop moving?
(ii) How far has it travelled by this time?

4 During braking the speed of a car is modelled by $v = 40 - 2t^2$ (in m s^{-1}) until it stops moving.

(i) How long does the car take to stop?
(ii) How far does it move before it stops?

5 In each case below, the object moves along a straight line with acceleration a in m s^{-2}. Find an expression for the velocity v (m s^{-1}) and position x (m) of each object at time t s.

(i) $a = 10 + 3t - t^2$; the object is initially at the origin and at rest.
(ii) $a = 4t - 2t^2$; at $t = 0$, $x = 1$ and $v = 2$.
(iii) $a = 10 - 6t$; at $t = 1$, $x = 0$ and $v = -5$.

The constant acceleration formulae revisited

? In which of the cases in question 1 above is the acceleration constant? Which constant acceleration formulae give the same results for s, v and a in this case? Why would the constant acceleration formulae not apply in the other two cases?

You can use integration to prove the equations for constant acceleration.
When a is constant (and only then)

$$v = \int a\,dt = at + c_1$$

When $t = 0$, $v = u$ $\qquad u = 0 + c_1$

$$\Rightarrow \quad v = u + at \qquad ①$$

You can integrate this again to find $s = ut + \frac{1}{2}at^2 + c_2$

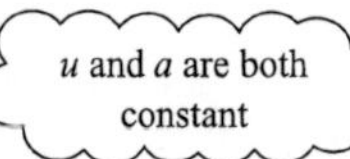

If $s = s_0$ when $t = 0$, $c_2 = s_0$ and $\qquad s = ut + \frac{1}{2}at^2 + s_0 \qquad ②$

❓ How can you use these to derive the other equations for constant acceleration?

$$s = \frac{1}{2}(u + v)\,t + s_0 \qquad ③$$

$$v^2 - u^2 = 2a(s - s_0) \qquad ④$$

$$s = vt - \frac{1}{2}at^2 + s_0 \qquad ⑤$$

EXERCISE 7C

1 A boy throws a ball up in the air from a height of 1.5 m and catches it at the same height. Its height in metres at time t seconds is

$$y = 1.5 + 15t - 5t^2.$$

(i) What is the vertical velocity $v\,\mathrm{m\,s^{-1}}$ of the ball at time t?

(ii) Find the position, velocity and speed of the ball at $t = 1$ and $t = 2$.

(iii) Sketch the position–time, velocity–time and speed–time graphs for $0 \leqslant t \leqslant 3$.

(iv) When does the boy catch the ball?

(v) Explain why the distance travelled by the ball is not equal to $\int_0^3 v\,dt$ and state what information this expression does give.

2 An object moves along a straight line so that its position in metres at time t seconds is given by

$$x = t^3 - 3t^2 - t + 3 \quad (t \geqslant 0).$$

(i) Find the position, velocity and speed of the object at $t = 2$.

(ii) Find the smallest time when

(a) the position is zero

(b) the velocity is zero.

(iii) Sketch position–time, velocity–time and speed–time graphs for $0 \leqslant t \leqslant 3$.

(iv) Describe the motion of the object.

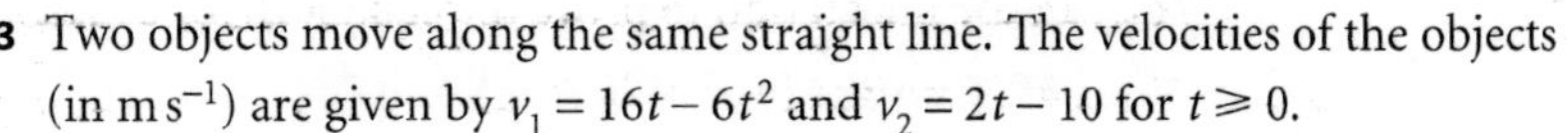

3 Two objects move along the same straight line. The velocities of the objects (in $\mathrm{m\,s^{-1}}$) are given by $v_1 = 16t - 6t^2$ and $v_2 = 2t - 10$ for $t \geqslant 0$.

Initially the objects are 32 m apart. At what time do they collide?

4 An object moves along a straight line so that its acceleration (in $\mathrm{m\,s^{-2}}$) is given by $a = 4 - 2t$. It starts its motion at the origin with speed $4\,\mathrm{m\,s^{-1}}$ in the direction of increasing x.

(i) Find as functions of t the velocity and position of the object.

(ii) Sketch the position–time, velocity–time and acceleration–time graphs for $0 \leqslant t \leqslant 2$.

(iii) Describe the motion of the object.

5 Nick watches a golfer putting her ball 24 m from the edge of the green and into the hole and he decides to model the motion of the ball. Assuming that the ball is a particle travelling along a straight line he models its distance, s metres, from the golfer at time t seconds by

$$s = -\frac{3}{2}t^2 + 12t \qquad 0 \leqslant t \leqslant 4.$$

(i) Find the value of s when $t = 0$, 1, 2, 3 and 4.

(ii) Explain the restriction $0 \leqslant t \leqslant 4$.

(iii) Find the velocity of the ball at time t seconds.

(iv) With what speed does the ball enter the hole?

(v) Find the acceleration of the ball at time t seconds.

6 Andrew and Elizabeth are having a race over 100 m. Their accelerations (in $\mathrm{m\,s^{-2}}$) are as follows:

Andrew		Elizabeth	
$a = 4 - 0.8t$	$0 \leqslant t \leqslant 5$	$a = 4$	$0 \leqslant t \leqslant 2.4$
$a = 0$	$t > 5$	$a = 0$	$t > 2.4$

(i) Find the greatest speed of each runner.

(ii) Sketch the speed–time graph for each runner.

(iii) Find the distance Elizabeth runs while reaching her greatest speed.

(iv) How long does Elizabeth take to complete the race?

(v) Who wins the race, by what time margin and by what distance?

On another day they race over 120 m, both running in exactly the same manner.

(vi) What is the result now?

7 Christine is a parachutist. On one of her descents her vertical speed, $v\,\text{m}\,\text{s}^{-1}$, t s after leaving an aircraft is modelled by

$$\begin{array}{ll} v = 8.5t & 0 \leqslant t \leqslant 10 \\ v = 5 + 0.8(t-20)^2 & 10 < t \leqslant 20 \\ v = 5 & 20 < t \leqslant 90 \\ v = 0 & t > 90 \end{array}$$

(i) Sketch the speed–time graph for Christine's descent and explain the shape of each section.

(ii) How high is the aircraft when Christine jumps out?

(iii) Write down expressions for the acceleration during the various phases of Christine's descent. What is the greatest magnitude of her acceleration?

8 A man of mass 70 kg is standing in a lift which, at a particular time, has an acceleration of $1.6\,\text{m}\,\text{s}^{-2}$ upwards. He is holding a parcel of mass 5 kg by a single string.

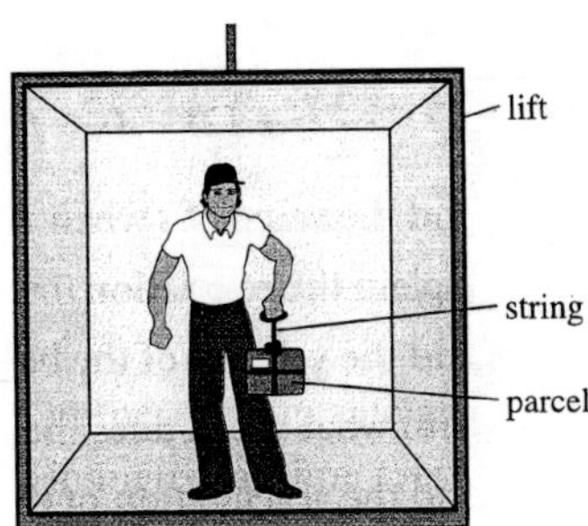

(i) Draw a diagram marking the forces acting on the parcel and the direction of the acceleration.

(ii) Show that the tension in the string is 58 N.

(iii) Calculate the reaction of the lift floor on the man.

During the first two seconds after starting from rest, the lift has acceleration in $\text{m}\,\text{s}^{-2}$ modelled by $3t(2-t)$, where t is in seconds. The maximum tension the string can withstand is 60 N.

(iv) By investigating the maximum acceleration of the system, or otherwise, determine whether the string will break during this time.

[MEI, *adapted*]

9 A bird leaves its nest for a short horizontal flight along a straight line and then returns. Michelle models its distance, s metres, from the nest at time t seconds by

$$s = 25t - \tfrac{5}{2}t^2, \qquad 0 \leqslant t \leqslant 10.$$

(i) Find the value of s when $t = 2$.

(ii) Explain the restriction $0 \leqslant t \leqslant 10$.

(iii) Find the velocity of the bird at time t seconds.

(iv) What is the greatest distance of the bird from the nest?

(v) Michelle's teacher tells her that a better model would be

$$s = 10t^2 - 2t^3 + \frac{1}{10}t^4.$$

Show that the two models agree about the time of the journey and the greatest distance travelled. Compare their predictions about velocity and suggest why the teacher's model is better.

[MEI]

10 A battery-operated toy dog starts at a point O and moves in a straight line. Its motion is modelled by the velocity–time graph below.

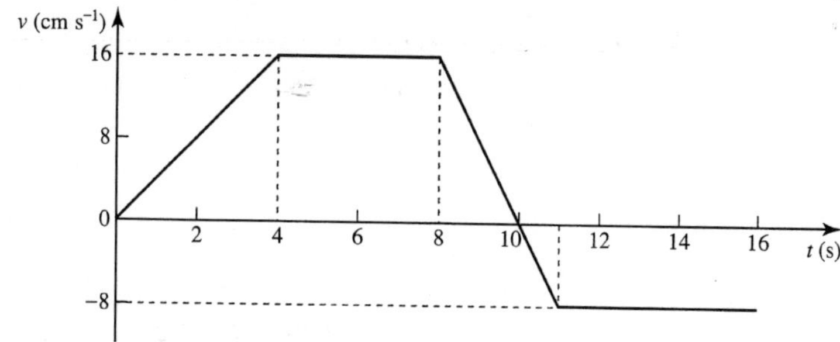

(i) Calculate the displacement from O of the toy

(a) after 10 seconds **(b)** after 16 seconds.

(ii) Write down expressions for the velocity of the toy at time t seconds in the intervals $0 \leqslant t \leqslant 4$ and $4 \leqslant t \leqslant 8$.

(iii) Obtain expressions for the displacement from O of the toy at time t seconds in the intervals $0 \leqslant t \leqslant 4$ and $4 \leqslant t \leqslant 8$.

An alternative model for the motion of the toy in the interval $0 \leqslant t \leqslant 10$ is $v = \frac{2}{3}(10t - t^2)$, where v is the velocity in cm s^{-1}.

(iv) Calculate the difference in the displacement from O after 10 seconds as predicted by the two models.

[MEI]

11 A particle P moves along the x axis in the positive direction. The velocity of P at time t s is $0.03t^2$ m s^{-1}. When $t = 5$ the displacement of P from the origin O is 2.5 m.

(i) Find an expression, in terms of t, for the displacement of P from O.

(ii) Find the velocity of P when its displacement from O is 11.25 m.

[Cambridge AS and A Level Mathematics 9709, Paper 4 Q5 June 2005]

12 A particle P travels in a straight line from A to D, passing through the points B and C. For the section AB the velocity of the particle is $(0.5t - 0.01t^2)$ m s^{-1}, where t s is the time after leaving A.

(i) Given that the acceleration of P at B is $0.1\,\mathrm{m\,s^{-2}}$, find the time taken for P to travel from A to B.

The acceleration of P from B to C is constant and equal to $0.1\,\mathrm{m\,s^{-2}}$.

(ii) Given that P reaches C with speed $14\,\mathrm{m\,s^{-1}}$, find the time taken for P to travel from B to C.

P travels with constant deceleration $0.3\,\mathrm{m\,s^{-2}}$ from C to D. Given that the distance CD is 300 m, find

(iii) the speed with which P reaches D,

(iv) the distance AD.

[Cambridge AS and A Level Mathematics 9709, Paper 4 Q7 June 2009]

13 A particle P starts from rest at the point A and travels in a straight line, coming to rest again after 10 s. The velocity–time graph for P consists of two straight line segments (see diagram). A particle Q starts from rest at A at the same instant as P and travels along the same straight line as P. The velocity of Q is given by $v = 3t - 0.3t^2$ for $0 \leqslant t \leqslant 10$. The displacements from A of P and Q are the same when $t = 10$.

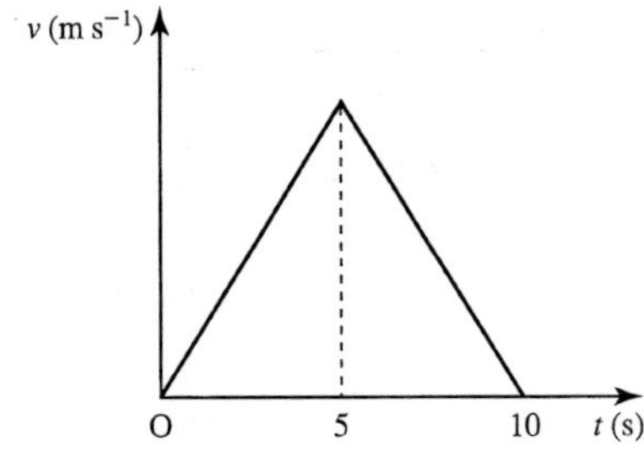

(i) Show that the greatest velocity of P during its motion is $10\,\mathrm{m\,s^{-1}}$.

(ii) Find the value of t, in the interval $0 < t < 5$, for which the acceleration of Q is the same as the acceleration of P.

[Cambridge AS and A Level Mathematics 9709, Paper 4 Q6 June 2007]

14 (i) A man walks in a straight line from A to B with constant acceleration $0.004\,\mathrm{m\,s^{-2}}$. His speed at A is $1.8\,\mathrm{m\,s^{-1}}$ and his speed at B is $2.2\,\mathrm{m\,s^{-1}}$. Find the time taken for the man to walk from A to B, and find the distance AB.

(ii) A woman cyclist leaves A at the same instant as the man. She starts from rest and travels in a straight line to B, reaching B at the same instant as the man. At time t s after leaving A the cyclist's speed is $k(200t - t^2)\,\mathrm{m\,s^{-1}}$, where k is a constant. Find

(a) the value of k,

(b) the cyclist's speed at B.

(iii) Sketch, using the same axes, the velocity–time graphs for the man's motion and the woman's motion from A to B.

[Cambridge AS and A Level Mathematics 9709, Paper 4 Q6 November 2007]

15 A particle P starts from rest at the point A at time $t = 0$, where t is in seconds, and moves in a straight line with constant acceleration $a\,\text{m}\,\text{s}^{-2}$ for 10 s. For $10 \leqslant t \leqslant 20$, P continues to move along the line with velocity $v\,\text{m}\,\text{s}^{-1}$, where $v = \dfrac{800}{t^2} - 2$. Find

(i) the speed of P when $t = 10$, and the value of a,

(ii) the value of t for which the acceleration of P is $-a\,\text{m}\,\text{s}^{-2}$,

(iii) the displacement of P from A when $t = 20$.

[Cambridge AS and A Level Mathematics 9709, Paper 41 Q7 November 2009]

16 A vehicle is moving in a straight line. The velocity $v\,\text{m}\,\text{s}^{-1}$ at time t s after the vehicle starts is given by

$$v = A(t - 0.05t^2) \qquad \text{for } 0 \leqslant t \leqslant 15,$$
$$v = \frac{B}{t^2} \qquad \text{for } t \geqslant 15,$$

where A and B are constants. The distance travelled by the vehicle between $t = 0$ and $t = 15$ is 225 m.

(i) Find the value of A and show that $B = 3375$.

(ii) Find an expression in terms of t for the total distance travelled by the vehicle when $t \geqslant 15$.

(iii) Find the speed of the vehicle when it has travelled a total distance of 315 m.

[Cambridge AS and A Level Mathematics 9709, Paper 41 Q7 June 2010]

KEY POINTS

1 Relationships between the variables describing motion

Position →	Velocity →	Acceleration
	differentiate →	
s	$v = \dfrac{\mathrm{d}s}{\mathrm{d}t}$	$a = \dfrac{\mathrm{d}v}{\mathrm{d}t} = \dfrac{\mathrm{d}^2s}{\mathrm{d}t^2}$

Acceleration →	Velocity →	Position
	← *integrate*	
a	$v = \int a\,\mathrm{d}t$	$s = \int v\,\mathrm{d}t$

2 Acceleration may be due to change in direction or change in speed or both.

A model for friction

Theories do not have to be 'right' to be useful.

Alvin Toffler

This statement about a road accident was offered to a magistrate's court by a solicitor.

'Briefly the circumstances of the accident are that our client was driving his Porsche motor car. He had just left work at the end of the day. He was stationary at the junction with Victoria Road when a motorcyclist travelling north down Victoria Road lost control of his motorcycle due to excessive speed and collided with the front offside of our client's motor car.

'The motorcyclist was braking when he lost control and left a 26-metre skid mark on the road. Our advice from an expert witness is that the motorcyclist was exceeding the speed limit of $50\,\text{km}\,\text{h}^{-1}$.'

? It is the duty of a court to decide whether the motorcyclist was innocent or guilty. Is it possible to deduce his speed from the skid mark? Draw a sketch map and make a list of the important factors that you would need to consider when modelling this situation.

A model for friction

Clearly the key information is provided by the skid marks. To interpret it, you need a model for how friction works; in this case between the motorcycle's tyres and the road.

As a result of experimental work, Coulomb formulated a model for friction between two surfaces. The following laws are usually attributed to him.

1 Friction always opposes relative motion between two surfaces in contact.

2 Friction is independent of the relative speed of the surfaces.

3 The magnitude of the frictional force has a maximum which depends on the normal reaction between the surfaces and on the roughness of the surfaces in contact.

4 If there is no sliding between the surfaces

$$F \leqslant \mu R$$

where F is the force due to friction and R is the normal reaction. μ is called the *coefficient of friction*.

5 When sliding is just about to occur, friction is said to be *limiting* and $F = \mu R$.

6 When sliding occurs $F = \mu R$.

According to Coulomb's model, μ is a constant for any pair of surfaces. Typical values and ranges of values for the coefficient of friction μ are given in this table.

Surfaces in contact	μ
wood sliding on wood	0.2–0.6
metal sliding on metal	0.15–0.3
normal tyres on dry road	0.8
racing tyres on dry road	1.0
sandpaper on sandpaper	2.0
skis on snow	0.02

How fast was the motorcyclist going?

You can now proceed with the problem. As an initial model, you might make the following assumptions:

1 that the road is level;

2 that the motorcycle was at rest just as it hit the car. (Obviously it was not, but this assumption allows you to estimate a minimum initial speed for the motorcycle);

3 that the motorcycle and rider may be treated as a particle, subject to Coulomb's laws of friction with $\mu = 0.8$ (i.e. dry road conditions).

The calculation then proceeds as follows.

Taking the direction of travel as positive, let the motorcycle and rider have acceleration $a\,\mathrm{m\,s^{-2}}$ and mass m kg. You have probably realised that the acceleration will be negative. The forces (in N) and acceleration are shown in figure 8.1.

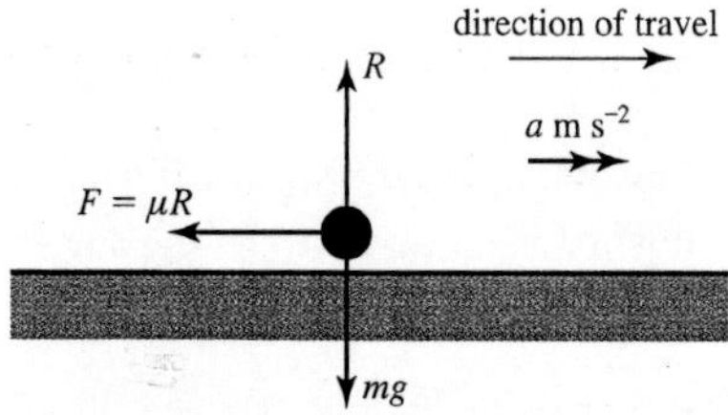

Figure 8.1

Applying Newton's second law:

perpendicular to the road, since there is no vertical acceleration we have

$$R - mg = 0; \qquad ①$$

parallel to the road, there is a constant force $-\mu R$ from friction, so we have

$$-\mu R = ma. \qquad ②$$

Solving for a gives

From ① $R = mg$

$$a = -\frac{\mu R}{m} = -\frac{\mu mg}{m} = -\mu g.$$

Taking $g = 10\ \mathrm{m\,s^{-2}}$ and $\mu = 0.8$ gives $a = -8\,\mathrm{m\,s^{-2}}$.

The constant acceleration equation

$$v^2 = u^2 + 2as$$

can be used to calculate the initial speed of the motorcycle. Substituting $s = 26$, $v = 0$ and $a = 8$ gives

$$u = \sqrt{2 \times 8 \times 26} = 20.4\ \mathrm{m\,s^{-1}}.$$

Convert this figure to kilometres per hour:

$$\text{speed} = \frac{20.4 \times 3600}{1000}$$
$$= 73.4\,\mathrm{km\,h^{-1}}.$$

So this first simple model suggests that the motorcycle was travelling at a speed of at least $73.4\,\mathrm{km\,h^{-1}}$ before skidding began.

 How good is this model and would you be confident in offering the answer as evidence in court? Look carefully at the three assumptions. What effect do they have on the estimate of the initial speed?

Modelling with friction

Whilst there is always some frictional force between two sliding surfaces its magnitude is often very small. In such cases we ignore the frictional force and describe the surfaces as *smooth.*

In situations where frictional forces cannot be ignored we describe the surface(s) as *rough.* Coulomb's law is the standard model for dealing with such cases.

Frictional forces are essential in many ways. For example, a ladder leaning against a wall would always slide if there were no friction between the foot of the ladder and the ground. The absence of friction in icy conditions causes difficulties for road users: pedestrians slip over, cars and motorcycles skid.

Remember that friction always opposes sliding motion.

 In what direction is the frictional force between the back wheel of a cycle and the road?

Historical note

Charles Augustin de Coulomb was born in Angoulême in France in 1736 and is best remembered for his work on electricity rather than for that on friction. The unit for electric charge is named after him.

Coulomb was a military engineer and worked for many years in the West Indies, eventually returning to France in poor health not long before the revolution. He worked in many fields, including the elasticity of metal, silk fibres and the design of windmills. He died in Paris in 1806.

EXAMPLE 8.1

A horizontal rope is attached to a crate of mass 70 kg at rest on a flat surface. The coefficient of friction between the floor and the crate is 0.6. Find the maximum force that the rope can exert on the crate without moving it.

SOLUTION

The forces (in N) acting on the crate are shown in figure 8.2. Since the crate does not move, it is in equilibrium.

Horizontal forces: $T = F$

Vertical forces: $R = mg$

$$= 70 \times 10 = 700$$

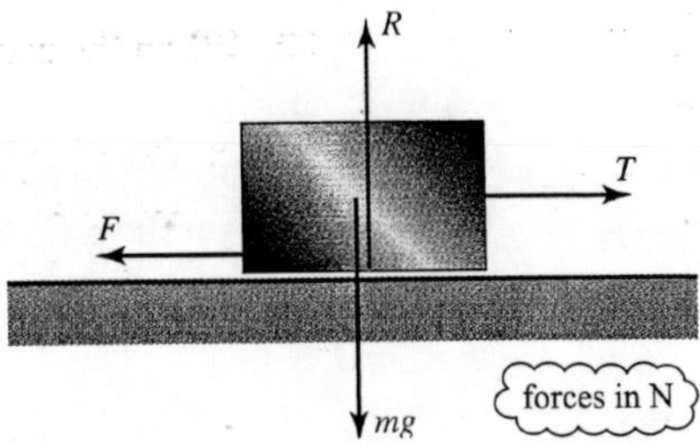

Figure 8.2

The law of friction states that

$$F \leqslant \mu R$$

for objects at rest.

So in this case

$$F \leqslant 0.6 \times 700$$
$$F \leqslant 420$$

The maximum frictional force is 420 N. As the tension in the rope and the force of friction are the only forces which have horizontal components, the crate will remain in equilibrium unless the tension in the rope is greater than 420 N.

EXAMPLE 8.2

Figure 8.3 shows a block of mass 5 kg on a rough table. It is connected by light inextensible strings passing over smooth pulleys to masses of 4 kg and 7 kg which hang vertically. The coefficient of friction between the block and the table is 0.4.

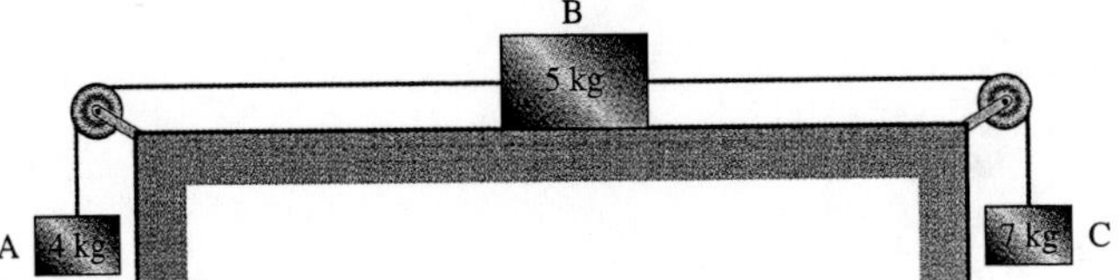

Figure 8.3

(i) Draw a diagram showing the forces acting on the three blocks and the direction of acceleration if the system moves.

(ii) Show that acceleration does take place.

(iii) Find the acceleration of the system and the tensions in the strings.

SOLUTION

(i)

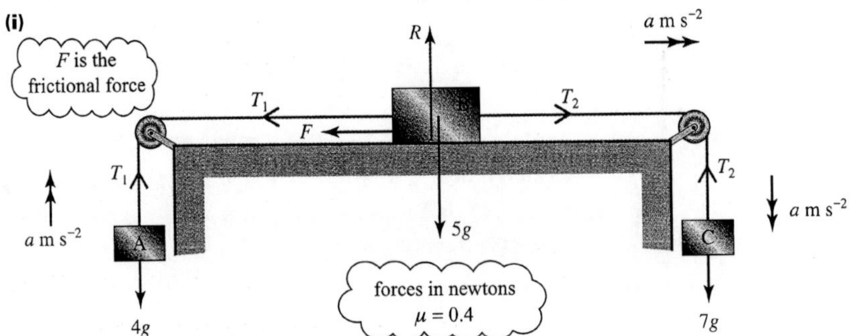

Figure 8.4

If acceleration takes place it is in the direction shown and $a > 0$.

(ii) When the acceleration is $a\ \text{m s}^{-2}$ ($\geqslant 0$), Newton's second law gives

for B, horizontally: $T_2 - T_1 - F = 5a$ ①
for A, vertically upwards: $T_1 - 4g = 4a$ ②
for C, vertically downwards: $7g - T_2 = 7a$ ③
Adding ①, ② and ③, $3g - F = 16a$ ④

B has no vertical acceleration so $R = 5g$
The maximum possible value of F is $\mu R = 0.4 \times 5g = 2g$.

In ④, a can be zero only if $F = 3g$, so $a > 0$ and sliding occurs.

(iii) When sliding occurs, you can replace F by $\mu R = 2g$

Then ④ gives
$$g = 16a$$
$$a = 0.625$$

Back-substituting gives $T_1 = 42.5$ and $T_2 = 65.625$.
The acceleration is $0.625\ \text{m s}^{-2}$ and the tensions are 42.5 N and 65.6 N.

EXAMPLE 8.3

Angus is pulling a sledge of mass 12 kg at steady speed across level snow by means of a rope which makes an angle of 20° with the horizontal. The coefficient of friction between the sledge and the ground is 0.15. What is the tension in the rope?

SOLUTION

Since the sledge is travelling at steady speed, the forces acting on it are in equilibrium. They are shown in figure 8.5.

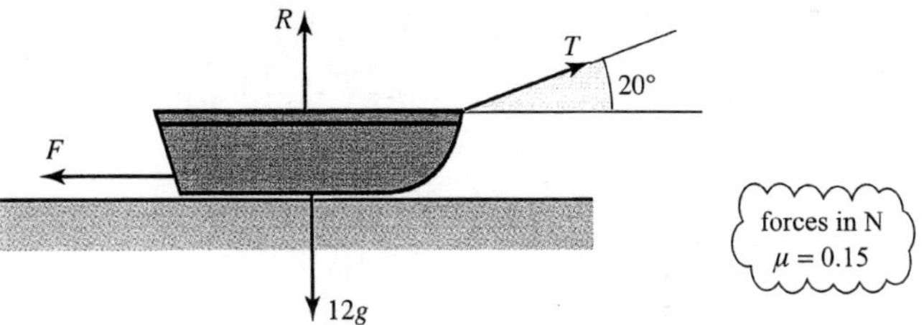

Figure 8.5

Horizontally: $T\cos 20° = F$
$= 0.15R$ ← F = μR when the sledge slides

Vertically: $T\sin 20° + R = 12g$
$R = 12 \times 10 - T\sin 20°$

Combining these gives
$$T\cos 20° = 0.15\,(12 \times 10 - T\sin 20°)$$
$$T\,(\cos 20° + 0.15\sin 20°) = 0.15 \times 12 \times 10$$
$$T = 18.2 \text{ (to 3 s.f.)}$$

The tension is 18.2 N.

Notice that the normal reaction is reduced when the rope is pulled in an upward direction. This has the effect of reducing the friction and making the sledge easier to pull.

EXAMPLE 8.4

A ski slope is designed for beginners. Its angle to the horizontal is such that skiers will either remain at rest on the point of moving or, if they are pushed off, move at constant speed. The coefficient of friction between the skis and the slope is 0.35. Find the angle that the slope makes with the horizontal.

SOLUTION

Figure 8.6 shows the forces on the skier.

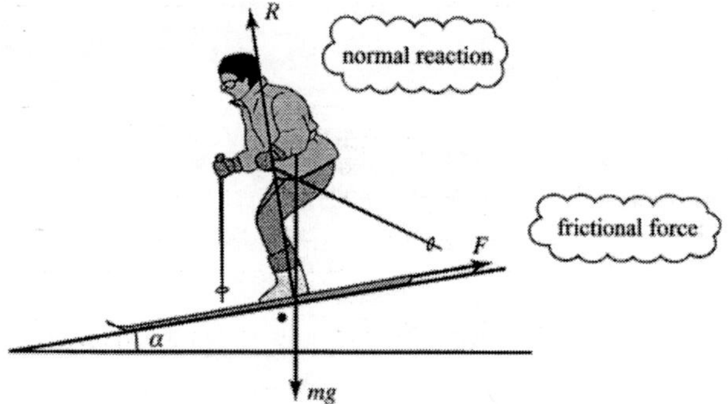

Figure 8.6

The weight mg can be resolved into components $mg\cos\alpha$ perpendicular to the slope and $mg\sin\alpha$ parallel to the slope.

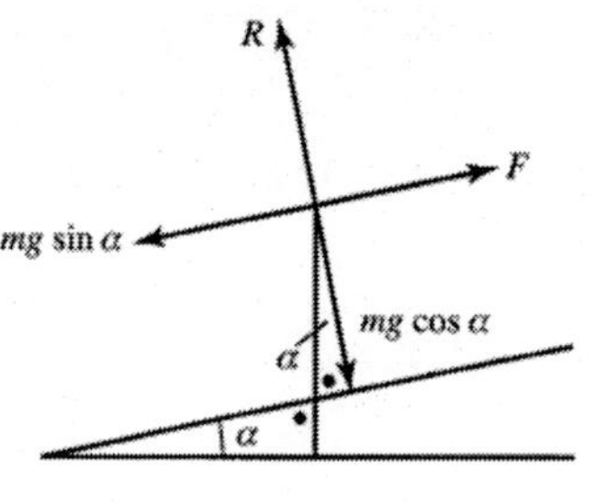

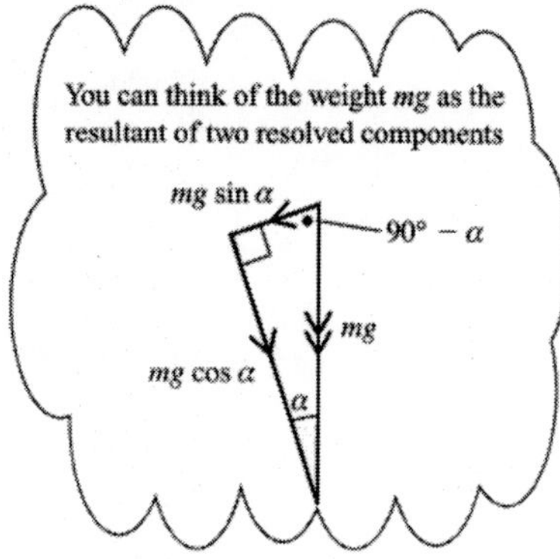

Figure 8.7

Since the skier is in equilibrium (at rest or moving with constant speed) applying Newton's second law:

Parallel to slope: $mg\sin\alpha - F = 0$

$\Rightarrow F = mg\sin\alpha$ ①

Perpendicular to slope: $R - mg\cos\alpha = 0$

$\Rightarrow R = mg\cos\alpha$ ②

In limiting equilibrium or moving at constant speed,

$$F = \mu R$$

$$mg\sin\alpha = \mu\, mg\cos\alpha$$

Substituting for F and R from ① and ②

$$\Rightarrow \quad \mu = \frac{\sin\alpha}{\cos\alpha} = \tan\alpha.$$

In this case $\mu = 0.35$, so $\tan\alpha = 0.35$ and $\alpha = 19.3°$.

Notes

1 The result is independent of the mass of the skier. This is often found in simple mechanics models. For example, two objects of different mass fall to the ground with the same acceleration. However when such models are refined, for example to take account of air resistance, mass is often found to have some effect on the result.

2 The angle for which the skier is about to slide down the slope is called the angle of friction. The angle of friction is often denoted by λ (lambda) and defined by $\tan\lambda = \mu$.

When the angle of the slope (α) is equal to the angle of the friction (λ), it is just possible for the skier to stand on the slope without sliding. If the slope is slightly steeper, the skier will slide immediately, and if it is less steep he or she will find it difficult to slide at all without using the ski poles.

EXERCISE 8A

You will find it helpful to draw diagrams when answering these questions.

1 A block of mass 10 kg is resting on a horizontal surface. It is being pulled by a horizontal force T (in N), and is on the point of sliding. Draw a diagram showing the forces acting and find the coefficient of friction when

(i) $T = 10$

(ii) $T = 5$.

2 In each of the following situations, use the equation of motion for each object to decide whether the block moves. If so, find the magnitude of the acceleration and if not, write down the magnitude of the frictional force.

(i) $\mu = \frac{1}{2}$

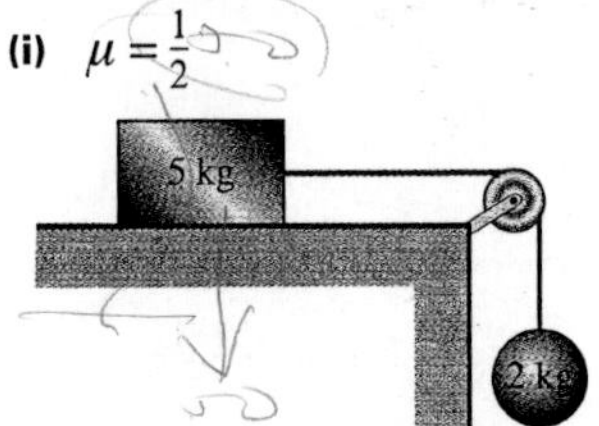

(ii) $\mu = \frac{1}{4}$

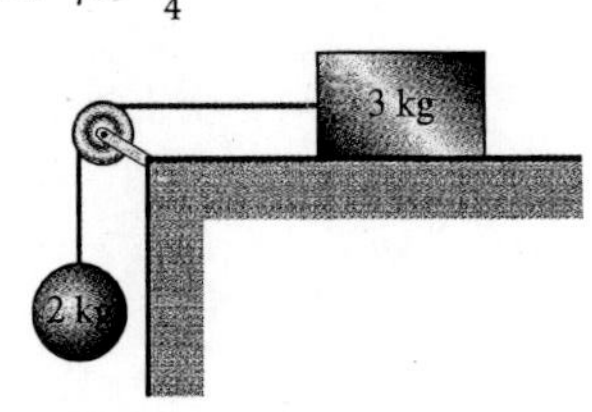

(iii) $\mu = 0.3$

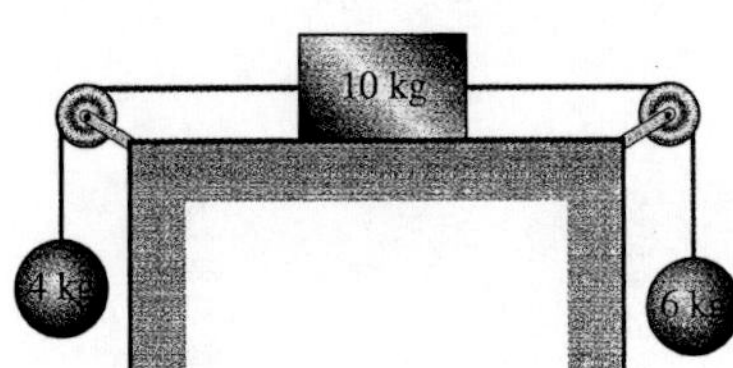

(iv) $\mu = \frac{1}{4}$

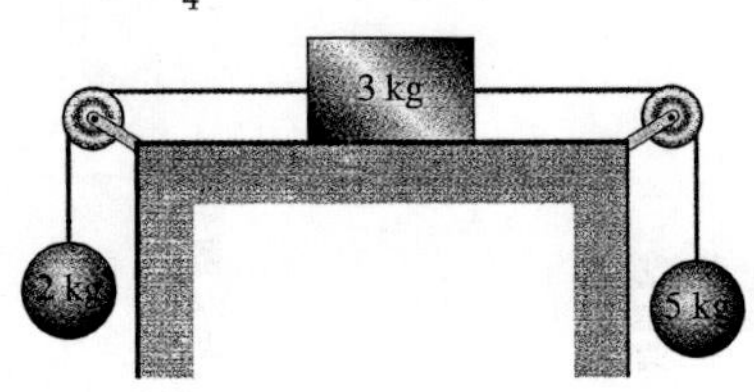

3 The brakes on a caravan of mass 700 kg have seized so that the wheels will not turn. What force must be exerted on the caravan to make it move horizontally? (The coefficient of friction between the tyres and the road is 0.7.)

4 A boy slides a piece of ice of mass 100 g across the surface of a frozen lake. Its initial speed is $10\,\text{m s}^{-1}$ and it takes 49 m to come to rest.

(i) Find the deceleration of the piece of ice.

(ii) Find the frictional force acting on the piece of ice.

(iii) Find the coefficient of friction between the piece of ice and the surface of the lake.

(iv) How far will a 200 g piece of ice travel if it, too, is given an initial speed of $10\,\text{m s}^{-1}$?

5 Jasmine is cycling at $12\,\text{m s}^{-1}$ when her bag falls off the back of her cycle. The bag slides a distance of 9 m before coming to rest. Calculate the coefficient of friction between the bag and the road.

6 A box of mass 50 kg is being moved across a room. To help it to slide a suitable mat is placed underneath the box.

(i) Explain why the mat makes it easier to slide the box.

A force of 100 N is needed to slide the mat at a constant velocity.

(ii) What is the value of the coefficient of friction between the mat and the floor?

A child of mass 20 kg climbs onto the box.

(iii) What force is now needed to slide the mat at constant velocity?

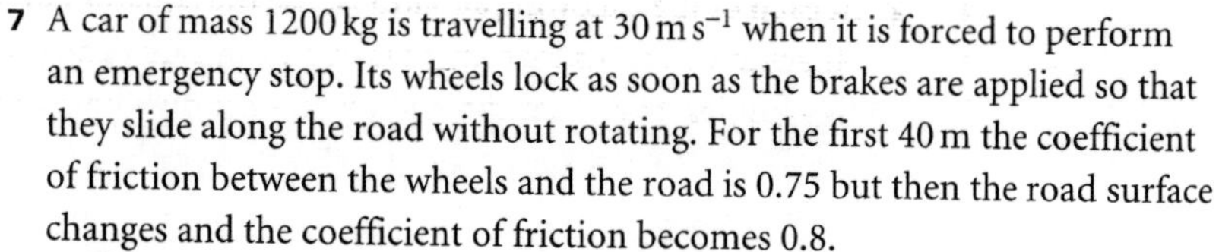

7 A car of mass 1200 kg is travelling at $30\,\text{m}\,\text{s}^{-1}$ when it is forced to perform an emergency stop. Its wheels lock as soon as the brakes are applied so that they slide along the road without rotating. For the first 40 m the coefficient of friction between the wheels and the road is 0.75 but then the road surface changes and the coefficient of friction becomes 0.8.

(i) Find the deceleration of the car immediately after the brakes are applied.

(ii) Find the speed of the car when it comes to the change of road surface.

(iii) Find the total distance the car travels before it comes to rest.

8 Shona, whose mass is 30 kg, is sitting on a sledge of mass 10 kg which is being pulled at constant speed along horizontal ground by her older brother, Aloke. The coefficient of friction between the sledge and the snow-covered ground is 0.15. Find the tension in the rope from Aloke's hand to the sledge when

(i) the rope is horizontal;

(ii) the rope makes an angle of 30° with the horizontal.

9 In each of the following situations a brick is about to slide down a rough inclined plane. Find the unknown quantity.

(i) The plane is inclined at 30° to the horizontal and the brick has mass 2 kg: find μ.

(ii) The brick has mass 4 kg and the coefficient of friction is 0.7: find the angle of the slope.

(iii) The plane is at 65° to the horizontal and the brick has mass 5 kg: find μ.

(iv) The brick has mass 6 kg and μ is 1.2: find the angle of slope.

10 The diagram shows a boy on a simple playground slide. The coefficient of friction between a typically clothed child and the slide is 0.25 and it can be assumed that no speed is lost when changing direction at B. The section AB is 3 m long and makes an angle of 40° with the horizontal. The slide is designed so that a child, starting from rest, stops at just the right moment of arrival at C.

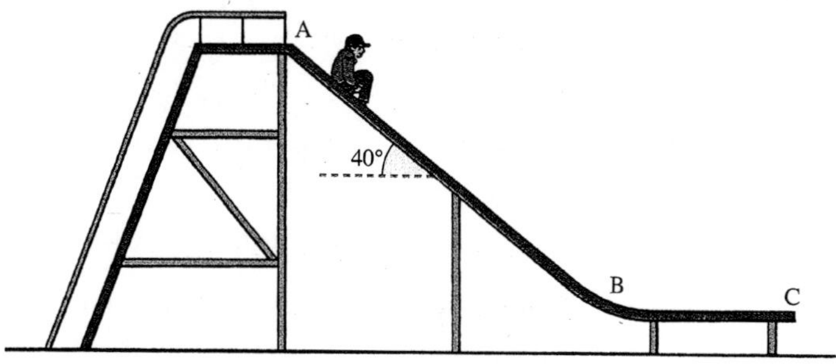

(i) Draw a diagram showing the forces acting on the boy when on the sloping section AB.

(ii) Calculate the acceleration of the boy when on the section AB.

(iii) Calculate the speed on reaching B.

(iv) Find the length of the horizontal section BC.

11 A chute at a water sports centre has been designed so that swimmers first slide down a steep part which is 10 m long and at an angle of 40° to the horizontal. They then come to a 20 m section with a gentler slope, 11° to the horizontal, where they travel at constant speed.

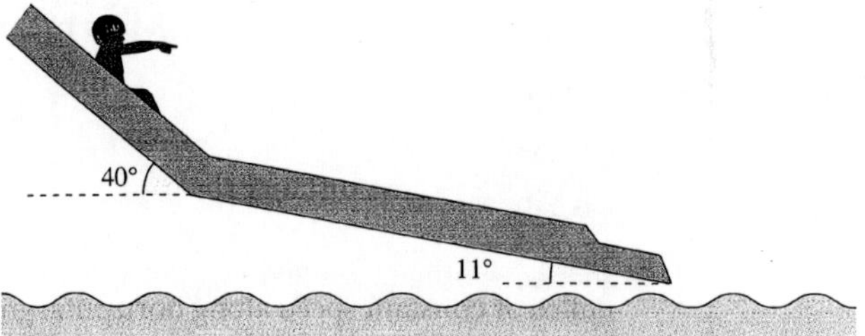

(i) Find the coefficient of friction between a swimmer and the chute.

(ii) Find the acceleration of a swimmer on the steep part.

(iii) Find the speed at the end of the chute of a swimmer who starts at rest. (You may assume that no speed is lost at the point where the slope changes.)

An alternative design of chute has the same starting and finishing points but has a constant gradient.

(iv) With what speed do swimmers arrive at the end of this chute?

12 One winter day, Veronica is pulling a sledge up a hill with slope 30° to the horizontal at a steady speed. The weight of the sledge is 40 N. Veronica pulls the sledge with a rope inclined at 15° to the slope of the hill. The tension in the rope is 24 N.

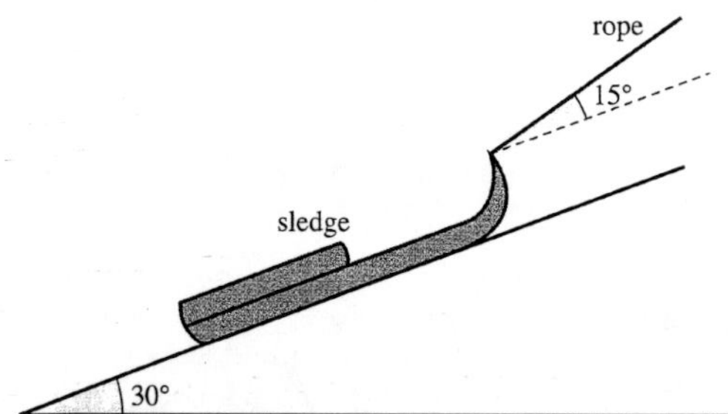

(i) Draw a force diagram showing the forces on the sledge and find the values of the normal reaction of the ground and the frictional force on the sledge.

(ii) Show that the coefficient of friction is slightly more than 0.1.

Veronica stops and when she pulls the rope to start again it breaks and the sledge begins to slide down the hill. The coefficient of friction is now 0.1.

(iii) Find the new value of the frictional force and the acceleration down the slope.

[MEI, *adapted*]

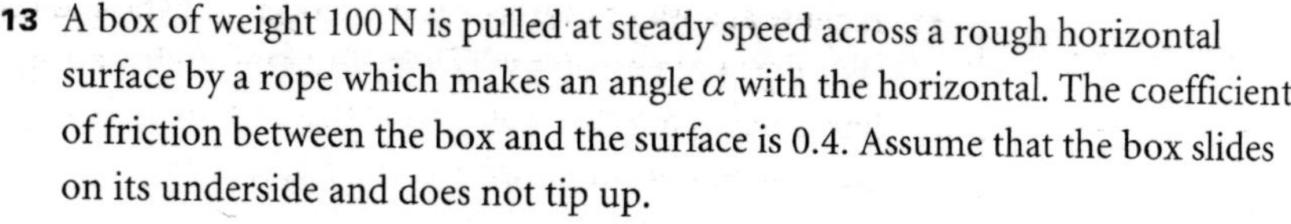

13 A box of weight 100 N is pulled at steady speed across a rough horizontal surface by a rope which makes an angle α with the horizontal. The coefficient of friction between the box and the surface is 0.4. Assume that the box slides on its underside and does not tip up.

(i) Find the tension in the string when the value of α is

(a) $10°$

(b) $20°$

(c) $30°$

(ii) Find an expression for the value of T for any angle α.

(iii) For what value of α is T a minimum?

14 A and B are points on the same line of greatest slope of a rough plane inclined at $30°$ to the horizontal. A is higher up the plane than B and the distance AB is 2.25 m. A particle P, of mass m kg, is released from rest at A and reaches B 1.5 s later. Find the coefficient of friction between P and the plane.

[Cambridge AS and A Level Mathematics 9709, Paper 4 Q3 June 2005]

15 Particles P and Q are attached to opposite ends of a light inextensible string. P is at rest on a rough horizontal table. The string passes over a small smooth pulley which is fixed at the edge of the table. Q hangs vertically below the pulley (see diagram). The force exerted on the string by the pulley has magnitude $4\sqrt{2}$ N. The coefficient of friction between P and the table is 0.8.

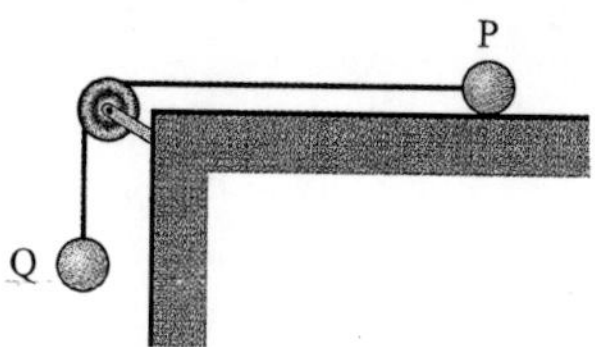

(i) Show that the tension in the string is 4 N and state the mass of Q.

(ii) Given that P is on the point of slipping, find its mass.

A particle of mass 0.1 kg is now attached to Q and the system starts to move.

(iii) Find the tension in the string while the particles are in motion.

[Cambridge AS and A Level Mathematics 9709, Paper 4 Q5 June 2006]

16 Two light strings are attached to a block of mass 20 kg. The block is in equilibrium on a horizontal surface AB with the strings taut. The strings make angles of 60° and 30° with the horizontal, on either side of the block, and the tensions in the strings are T N and 75 N respectively (see diagram).

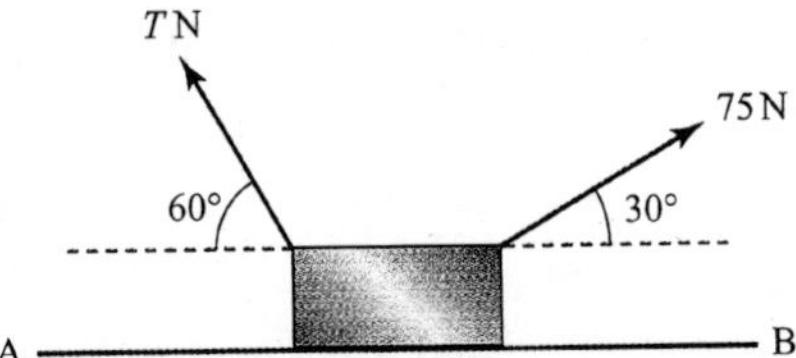

(i) Given that the surface is smooth, find the value of T and the magnitude of the contact force acting on the block.

(ii) It is given instead that the surface is rough and that the block is on the point of slipping. The frictional force on the block has magnitude 25 N and acts towards A. Find the coefficient of friction between the block and the surface.

[Cambridge AS and A Level Mathematics 9709, Paper 4 Q7 June 2007]

17 A block of mass 8 kg is at rest on a plane inclined at 20° to the horizontal. The block is connected to a vertical wall at the top of the plane by a string. The string is taut and parallel to a line of greatest slope of the plane (see diagram).

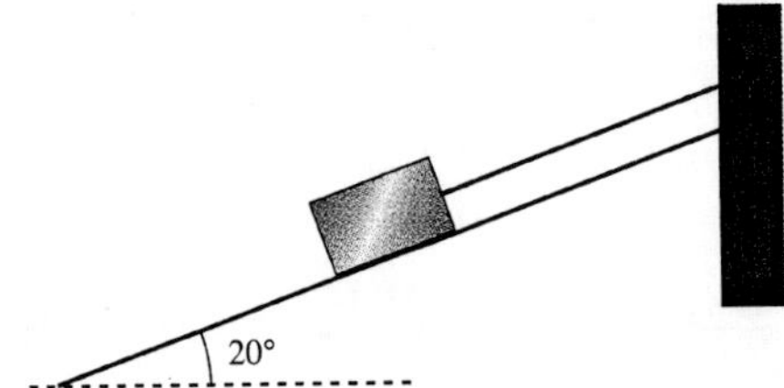

(i) Given that the tension in the string is 13 N, find the frictional and normal components of the force exerted on the block by the plane.

The string is cut; the block remains at rest, but is on the point of slipping down the plane.

(ii) Find the coefficient of friction between the block and the plane.

[Cambridge AS and A Level Mathematics 9709, Paper 4 Q4 June 2009]

18 A stone slab of mass 320 kg rests in equilibrium on rough horizontal ground. A force of magnitude X N acts upwards on the slab at an angle of θ to the vertical, where $\tan\theta = \frac{7}{24}$ (see diagram).

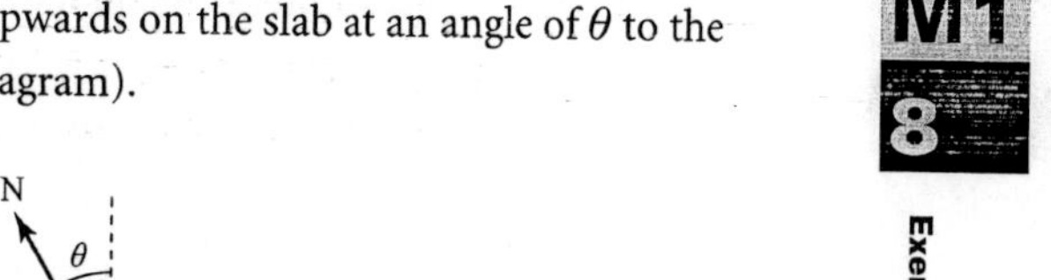
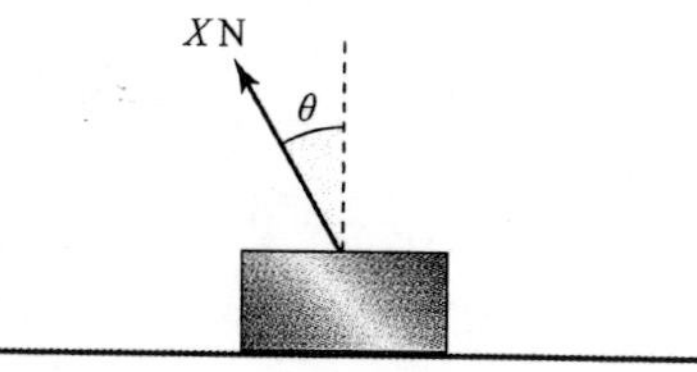

(i) Find, in terms of X, the normal component of the force exerted on the slab by the ground.

(ii) Given that the coefficient of friction between the slab and the ground is $\frac{3}{8}$, find the value of X for which the slab is about to slip.

[Cambridge AS and A Level Mathematics 9709, Paper 4 Q4 November 2005]

19 A rough inclined plane of length 65 cm is fixed with one end at a height of 16 cm above the other end. Particles P and Q, of masses 0.13 kg and 0.11 kg respectively, are attached to the ends of a light inextensible string which passes over a small smooth pulley at the top of the plane. Particle P is held at rest on the plane and particle Q hangs vertically below the pulley (see diagram). The system is released from rest and P starts to move up the plane.

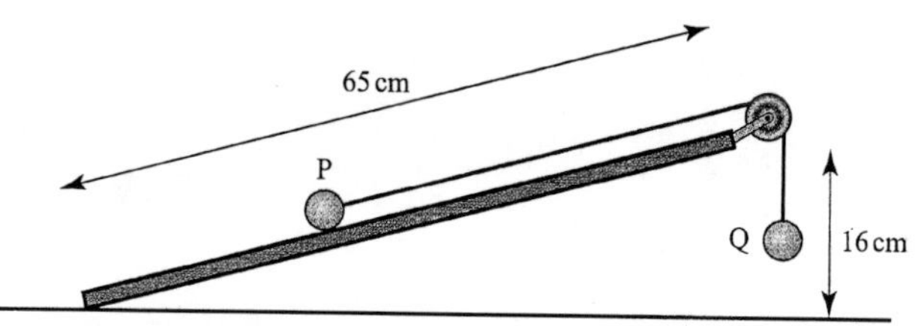

(i) Draw a diagram showing the forces acting on P during its motion up the plane.

(ii) Show that $T - F > 0.32$, where T N is the tension in the string and F N is the magnitude of the frictional force on P.

The coefficient of friction between P and the plane is 0.6.

(iii) Find the acceleration of P.

[Cambridge AS and A Level Mathematics 9709, Paper 4 Q7 November 2007]

20 A block of mass 20 kg is at rest on a plane inclined at 10° to the horizonal. A force acts on the block parallel to a line of greatest slope of the plane. The coefficient of friction between the block and the plane is 0.32. Find the least magnitude of the force necessary to move the block,

(i) given that the force acts up the plane,

(ii) given instead that the force acts down the plane.

[Cambridge AS and A Level Mathematics 9709, Paper 4 Q2 November 2008]

21 A particle P of mass 0.6 kg moves upwards along a line of greatest slope of a plane inclined at 18° to the horizontal. The deceleration of P is $4\,\text{m}\,\text{s}^{-2}$.

(i) Find the frictional and normal components of the force exerted on P by the plane. Hence find the coefficient of friction between P and the plane, correct to 2 significant figures.

After P comes to instantaneous rest it starts to move down the plane with acceleration $a\,\text{m}\,\text{s}^{-2}$.

(ii) Find the value of a.

[Cambridge AS and A Level Mathematics 9709, Paper 41 Q5 November 2009]

22 A small ring of mass 0.8 kg is threaded on a rough rod which is fixed horizontally. The ring is in equilibrium, acted on by a force of magnitude 7 N pulling upwards at 45° to the horizontal (see diagram).

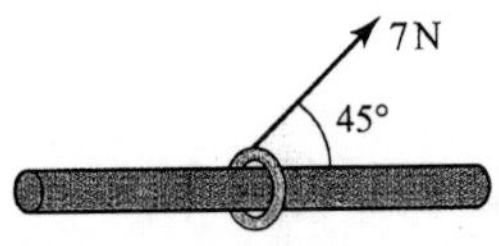

(i) Show that the normal component of the contact force acting on the ring has magnitude 3.05 N, correct to 3 significant figures.

(ii) The ring is in limiting equilibrium. Find the coefficient of friction between the ring and the rod.

[Cambridge AS and A Level Mathematics 9709, Paper 41 Q3 June 2010]

INVESTIGATION

The sliding ruler

Hold a metre ruler horizontally across your two index fingers and slide your fingers smoothly together, fairly slowly. What happens?

Use the laws of friction to investigate what you observe.

Optimum angle

A packing case is pulled across rough ground by means of a rope making an angle θ with the horizontal. Investigate how the tension can be minimised by varying the angle between the rope and the horizontal.

KEY POINTS

Coulomb's laws

1 The frictional force, F, between two surfaces is given by

$F < \mu R$ when there is no sliding except in limiting equilibrium
$F = \mu R$ in limiting equilibrium
$F = \mu R$ when sliding occurs

where R is the normal reaction of one surface on the other and μ is the coefficient of friction between the surfaces.

2 The frictional force always acts in the direction to oppose sliding.

3 Remember that the value of the normal reaction is affected by a force which has a component perpendicular to the direction of sliding.

Energy, work and power

I like work: it fascinates me. I can sit and look at it for hours.

Jerome K. Jerome

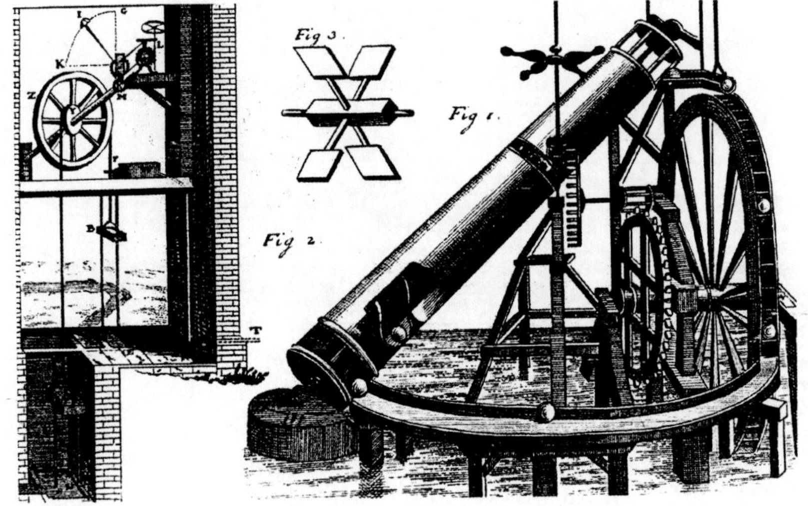

This is a picture of a perpetual motion machine. What does this term mean and will this one work?

Energy and momentum

When describing the motion of objects in everyday language the words *energy* and *momentum* are often used quite loosely and sometimes no distinction is made between them. In mechanics they must be defined precisely.

For an object of mass m moving with velocity $\mathbf{v}$:

- *Kinetic energy* $= \frac{1}{2}mv^2$ (this is the energy it has due to its motion)
- *Momentum* $= m\mathbf{v}$

Notice that kinetic energy is a scalar quantity with magnitude only, but momentum is a vector in the same direction as the velocity.

Both the kinetic energy and the momentum are liable to change when a force acts on a body and you will learn more about how the energy is changed in this chapter.

Work and energy

In everyday life you encounter many forms of energy such as heat, light, electricity and sound. You are familiar with the conversion of one form of energy to another: from chemical energy stored in wood to heat energy when you burn it; from electrical energy to the energy of a train's motion, and so on. The S.I. unit for energy is the joule, J.

Mechanical energy and work

In mechanics two forms of energy are particularly important.

Kinetic energy is the energy which a body possesses because of its motion.

- *The kinetic energy of a moving object* $= \frac{1}{2} \times$ *mass* $\times$ (*speed*)^2.

Potential energy is the energy which a body possesses because of its position. It may be thought of as stored energy which can be converted into kinetic or other forms of energy. You will meet this again on page 163.

The energy of an object is usually changed when it is acted on by a force. When a force is applied to an object which moves in the direction of its line of action, the force is said to do *work*. For a constant force this is defined as follows.

- The work done by a constant force = force × distance moved in the direction of the force.

The following examples illustrate how to use these ideas.

EXAMPLE 9.1

A brick, initially at rest, is raised by a force averaging 40 N to a height 5 m above the ground where it is left stationary. How much work is done by the force?

SOLUTION

The work done by the force raising the brick is

$$40 \times 5 = 200\,\text{J}.$$

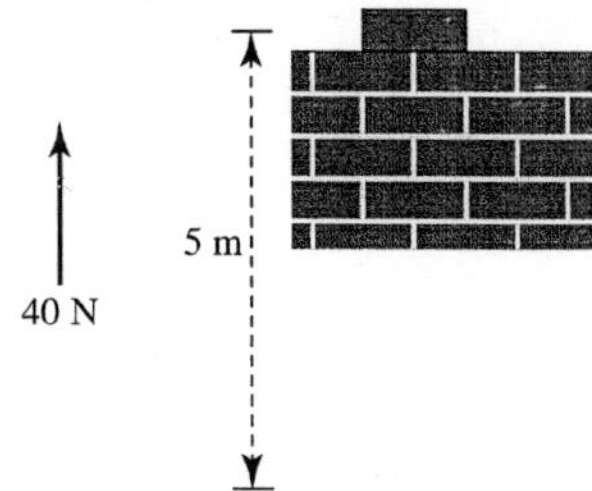

Figure 9.1

Examples 9.2 and 9.3 show how the work done by a force can be related to the change in kinetic energy of an object.

EXAMPLE 9.2

A train travelling on level ground is subject to a resisting force (from the brakes and air resistance) of 250 kN for a distance of 5 km. How much kinetic energy does the train lose?

SOLUTION

The forward force is $-250\,000$ N.

Work and energy have the same units

The work done by it is $-250\,000 \times 5000 = -1250\,000\,000$ J.

Hence $-1250\,000\,000$ J of kinetic energy are gained by the train, in other words $+1250\,000\,000$ J of kinetic energy are lost and the train slows down. This energy is converted to other forms such as heat and perhaps a little sound.

EXAMPLE 9.3

A car of mass m kg is travelling at u m s^{-1} when the driver applies a constant driving force of F N. The ground is level and the road is straight and air resistance can be ignored. The speed of the car increases to v m s^{-1} in a period of t s over a distance of s m. Find the relationship between F, s, m, u and v.

SOLUTION

Treating the car as a particle and applying Newton's second law:

$$F = ma$$
$$a = \frac{F}{m}$$

Since F is assumed constant, the acceleration is constant also, so using $v^2 = u^2 + 2as$

$$v^2 = u^2 + \frac{2Fs}{m}$$
$$\Rightarrow \quad \tfrac{1}{2}mv^2 = \tfrac{1}{2}mu^2 + Fs$$
$$Fs = \tfrac{1}{2}mv^2 - \tfrac{1}{2}mu^2$$

Thus

- work done by force = final kinetic energy − initial kinetic energy of car.

The work–energy principle

Examples 9.4 and 9.5 illustrate the *work–energy principle* which states that:

- The total work done by the forces acting on a body is equal to the increase in the kinetic energy of the body.

EXAMPLE 9.4

A sledge of total mass 30 kg, initially moving at $2\,\text{m s}^{-1}$, is pulled 14 m across smooth horizontal ice by a horizontal rope in which there is a constant tension of 45 N. Find its final velocity.

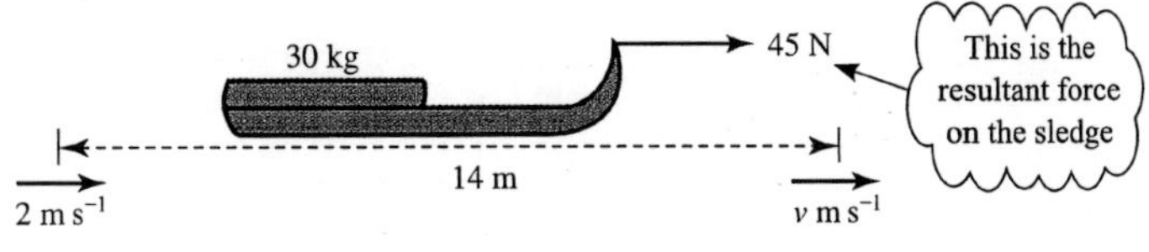

Figure 9.2

SOLUTION

Since the ice is smooth, the work done by the force is all converted into kinetic energy and the final velocity can be found using

work done by the force = final kinetic energy – initial kinetic energy

$$45 \times 14 = \tfrac{1}{2} \times 30 \times v^2 - \tfrac{1}{2} \times 30 \times 2^2$$

So $v^2 = 46$ and the final velocity of the sledge is $6.8\,\text{m s}^{-1}$ (to 2 s.f.).

EXAMPLE 9.5

The combined mass of a cyclist and her bicycle is 65 kg. She accelerated from rest to $8\,\text{m s}^{-1}$ in 80 m along a horizontal road.

(i) Calculate the work done by the net force in accelerating the cyclist and her bicycle.

(ii) Hence calculate the net forward force (assuming the force to be constant).

SOLUTION

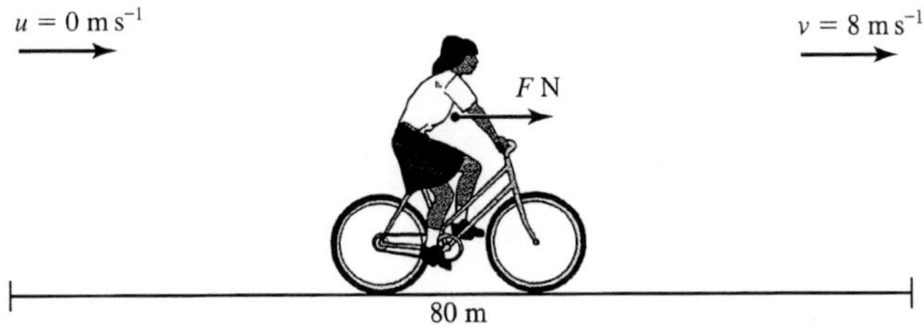

Figure 9.3

(i) The work done by the net force F is given by

$$\begin{aligned} \text{work} &= \text{final K.E.} - \text{initial K.E.} \\ &= \tfrac{1}{2}mv^2 - \tfrac{1}{2}mu^2 \\ &= \tfrac{1}{2} \times 65 \times 8^2 - 0 \\ &= 2080\text{ J} \end{aligned}$$

The work done is 2080 J.

(ii) Work done $= Fs$

$$= F \times 80$$

So $\quad 80F = 2080$

$$F = 26$$

The net forward force is 26 N.

Work

It is important to realise that:

- work is done by a force
- work is only done when there is movement
- a force only does work on an object when it has a component in the direction of motion of the object.

It is quite common to speak of the work done by a person, say in pushing a lawn mower. In fact this is the work done by the force of the person on the lawn mower.

Notice that if you stand holding a brick stationary above your head, painful though it may be, the force you are exerting on it is doing no work. Nor is this vertical force doing any work if you walk round the room keeping the brick at the same height. However, once you start climbing the stairs, a component of the brick's movement is in the direction of the upward force that you are exerting on it, so the force is now doing some work.

When applying the work–energy principle, you have to be careful to include *all* the forces acting on the body. In the example of a brick of weight 40 N being raised 5 m vertically, starting and ending at rest, the change in kinetic energy is clearly 0.

This seems paradoxical when it is clear that the force which raised the brick has done $40 \times 5 = 200$ J of work. However, the brick was subject to another force, namely its weight, which did $-40 \times 5 = 200$ J of work on it, giving a total of $200 + (-200) = 0$ J.

Conservation of mechanical energy

The net forward force on the cyclist in Example 9.5 is the girl's driving force minus resistive forces such as air resistance and friction in the bearings. In the absence of such resistive forces, she would gain more kinetic energy; also the work she does against them is lost, it is dissipated as heat and sound. Contrast this with the work a cyclist does against gravity when going uphill. This work can be recovered as kinetic energy on a downhill run. The work done against the force of gravity is conserved and gives the cyclist potential energy (see page 163).

Forces such as friction which result in the dissipation of mechanical energy are called *dissipative forces.* Forces which conserve mechanical energy are called *conservative forces.* The force of gravity is a conservative force and so is the tension in an elastic string; you can test this using an elastic band.

EXAMPLE 9.6 A bullet of mass 25 g is fired at a wooden barrier 3 cm thick. When it hits the barrier it is travelling at $200\,\text{m s}^{-1}$. The barrier exerts a constant resistive force of 5000 N on the bullet.

(i) Does the bullet pass through the barrier and if so with what speed does it emerge?

(ii) Is energy conserved in this situation?

SOLUTION

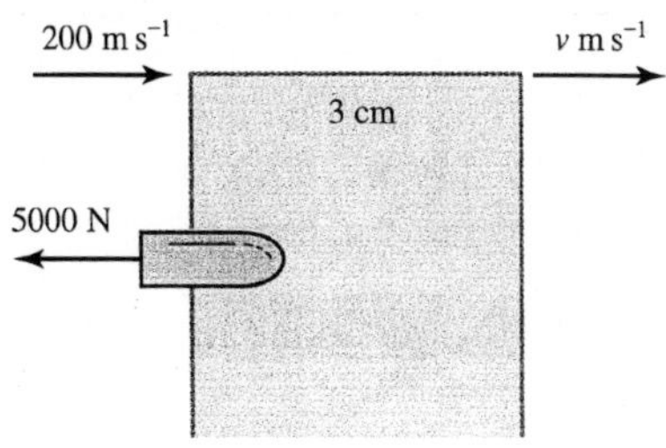

Figure 9.4

(i) The work done *by* the force is defined as the product of the force and the distance moved *in the direction of the force.* Since the bullet is moving in the direction opposite to the net resistive force, the work done by this force is negative.

$$\begin{aligned}\text{Work done} &= -5000 \times 0.03\\ &= 150\,\text{J}\end{aligned}$$

The initial kinetic energy of the bullet is

$$\begin{aligned}\text{Initial K.E.} &= \tfrac{1}{2}mu^2\\ &= \tfrac{1}{2} \times 0.025 \times 200^2\\ &= 500\,\text{J}\end{aligned}$$

A loss in energy of 150 J will not reduce kinetic energy to zero, so the bullet will still be moving on exit.

Since the work done is equal to the change in kinetic energy,

$$-150 = \tfrac{1}{2}mv^2 - 500$$

Solving for v

$$\begin{aligned}\tfrac{1}{2}mv^2 &= 500 - 150\\ v^2 &= \frac{2 \times (500 - 150)}{0.025}\\ v &= 167 \text{ (to nearest whole number)}\end{aligned}$$

So the bullet emerges from the barrier with a speed of $167\,\text{m s}^{-1}$.

(ii) Total energy is conserved but there is a loss of mechanical energy of $\tfrac{1}{2}mu^2 - \tfrac{1}{2}mv^2 = 150\,\text{J}$. This energy is converted into non-mechanical forms such as heat and sound.

EXAMPLE 9.7

An aircraft of mass m kg is flying at a constant velocity $v\,\text{m s}^{-1}$ horizontally. Its engines are providing a horizontal driving force F N.

(i) Draw a diagram showing the driving force, the lift force L N, the air resistance (drag force) R N and the weight of the aircraft.
(ii) State which of these forces are equal in magnitude.
(iii) State which of the forces are doing no work.
(iv) In the case when $m = 100\,000$, $v = 270$ and $F = 350\,000$, find the work done in a 10-second period by those forces which are doing work, and show that the work–energy principle holds in this case.

At a later time the pilot increases the thrust of the aircraft's engines to 400 000 N. When the aircraft has travelled a distance of 30 km, its speed has increased to $300\,\text{m s}^{-1}$.

(v) Find the work done against air resistance during this period, and the average resistance force.

SOLUTION

(i)

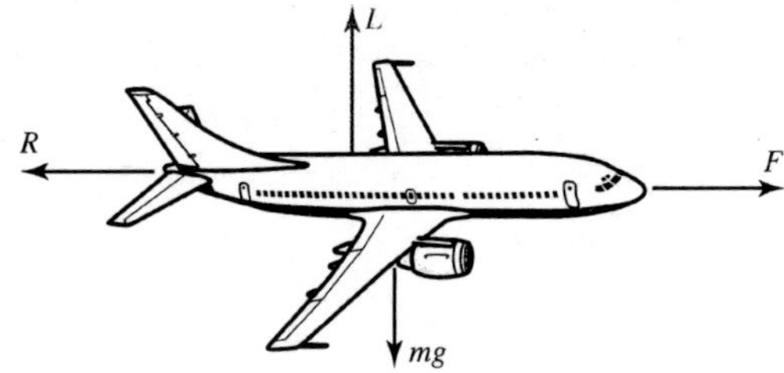

Figure 9.5

(ii) Since the aircraft is travelling at constant velocity it is in equilibrium.

Horizontal forces: $F = R$
Vertical forces: $L = mg$

(iii) Since the aircraft's velocity has no vertical component, the vertical forces, L and mg, are doing no work.

(iv) In 10 s at $270\,\text{m s}^{-1}$ the aircraft travels 2700 m.

Work done by force $F = 350\,000 \times 2700 = 9\,450\,000$ J
Work done by force $R = 350\,000 \times -2700 = -9\,450\,000$ J

The work–energy principle states that in this situation

work done by F + work done by R = change in kinetic energy.

Now work done by F + work done by $R = (9\,450\,000 - 9\,450\,000) = 0$ J, and change in kinetic energy = 0 (since velocity is constant), so the work–energy principle does indeed hold in this case.

(v) Final K.E. – initial K.E.
$$= \tfrac{1}{2}mv^2 - \tfrac{1}{2}mu^2$$
$$= \tfrac{1}{2} \times 100\,000 \times 300^2 - \tfrac{1}{2} \times 100\,000 \times 270^2$$
$$= 855 \times 10^6\ \text{J}$$

$$\text{Work done by driving force} = 400\,000 \times 30\,000$$
$$\text{Total work done} = \text{K.E. gained}$$
$$= 12\,000 \times 10^6\,\text{J}$$

$$\text{Work done by resistance force} + 12\,000 \times 10^6 = 855 \times 10^6$$
$$\text{Work done by resistance force} = 11\,145 \times 10^6\,\text{J}$$
$$\text{Average force} \times \text{distance} = \text{work done by force}$$
$$\text{Average force} \times 30\,000 = 11\,145 \times 10^6$$

$\Rightarrow$ The average resistance force is 371 500 N (in the negative direction).

Note

When an aircraft is in flight, most of the work done by the resistance force results in air currents and the generation of heat. A typical large jet cruising at 35 000 feet has a body temperature about 30°C above the surrounding air temperature. For supersonic flight the temperature difference is much greater. Concorde used to fly with a skin temperature more than 200°C above that of the surrounding air.

EXERCISE 9A

1 Find the kinetic energy of the following objects.

(i) An ice skater of mass 50 kg travelling with speed 10 m s^{-1}.

(ii) An elephant of mass 5 tonnes moving at 4 m s^{-1}.

(iii) A train of mass 7000 tonnes travelling at 40 m s^{-1}.

(iv) The moon, mass 7.4×10^{22} kg, travelling at 1000 m s^{-1} in its orbit round the earth.

(v) A bacterium of mass 2×10^{-16} g which has speed 1 mm s^{-1}.

2 Find the work done by a man in the following situations.

(i) He pushes a packing case of mass 35 kg a distance of 5 m across a rough floor against a resistance of 200 N. The case starts and finishes at rest.

(ii) He pushes a packing case of mass 35 kg a distance of 5 m across a rough floor against a resistance force of 200 N. The case starts at rest and finishes with a speed of 2 m s^{-1}.

(iii) He pushes a packing case of mass 35 kg a distance of 5 m across a rough floor against a resistance force of 200 N. Initially the case has speed 2 m s^{-1} but it ends at rest.

(iv) He is handed a packing case of mass 35 kg. He holds it stationary, at the same height, for 20 s and then someone else takes it from him.

3 A sprinter of mass 60 kg is at rest at the beginning of a race and accelerates to 12 m s^{-1} in a distance of 30 m. Assume air resistance to be negligible.

(i) Calculate the kinetic energy of the sprinter at the end of the 30 m.

(ii) Write down the work done by the sprinter over this distance.

(iii) Calculate the forward force exerted by the sprinter, assuming it to be constant, using work = force $\times$ distance.

(iv) Using force = mass $\times$ acceleration and the constant acceleration formulae, show that this force is consistent with the sprinter having speed 12 m s^{-1} after 30 m.

4 A sports car of mass 1.2 tonnes accelerates from rest to $30\,\mathrm{m\,s^{-1}}$ in a distance of 150 m. Assume air resistance to be negligible.

(i) Calculate the work done in accelerating the car. Does your answer depend on an assumption that the driving force is constant?

(ii) If the driving force is in fact constant, what is its magnitude?

5 A car of mass 1600 kg is travelling at speed $25\,\mathrm{m\,s^{-1}}$ when the brakes are applied so that it stops after moving a further 75 m.

(i) Find the work done by the brakes.

(ii) Find the retarding force from the brakes, assuming that it is constant and that other resistive forces may be neglected.

6 The forces acting on a hot air balloon of mass 500 kg are its weight and the total uplift force.

(i) Find the total work done when the speed of the balloon changes from

(a) $2\,\mathrm{m\,s^{-1}}$ to $5\,\mathrm{m\,s^{-1}}$ **(b)** $8\,\mathrm{m\,s^{-1}}$ to $3\,\mathrm{m\,s^{-1}}$.

(ii) If the balloon rises 100 m vertically while its speed changes calculate in each case the work done by the uplift force.

7 A bullet of mass 20 g, found at the scene of a police investigation, had penetrated 16 cm into a wooden post. The speed for that type of bullet is known to be $80\,\mathrm{m\,s^{-1}}$.

(i) Find the kinetic energy of the bullet before it entered the post.

(ii) What happened to this energy when the bullet entered the wooden post?

(iii) Write down the work done in stopping the bullet.

(iv) Calculate the resistive force on the bullet, assuming it to be constant.

Another bullet of the same mass and shape had clearly been fired from a different and unknown type of gun. This bullet had penetrated 20 cm into the post.

(v) Estimate the speed of this bullet before it hit the post.

8 The UK Highway Code give the braking distance for a car travelling at $22\,\mathrm{m\,s^{-1}}$ (50 mph) to be 38 m (125 ft). A car of mass 1300 kg is brought to rest in just this distance. It may be assumed that the only resistance forces come from the car's brakes.

(i) Find the work done by the brakes.

(ii) Find the average force exerted by the brakes.

(iii) What happened to the kinetic energy of the car?

(iv) What happens when you drive a car with the handbrake on?

9 A car of mass 1200 kg experiences a constant resistance force of 600 N. The driving force from the engine depends upon the gear, as shown in the table.

Gear	1	2	3	4
Force (N)	2800	2100	1400	1000

Starting from rest, the car is driven 20 m in first gear, 40 m in second, 80 m in third and 100 m in fourth. How fast is the car travelling at the end?

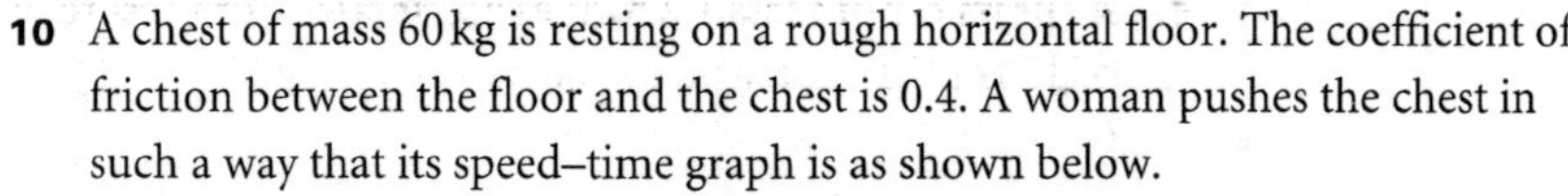

10 A chest of mass 60 kg is resting on a rough horizontal floor. The coefficient of friction between the floor and the chest is 0.4. A woman pushes the chest in such a way that its speed–time graph is as shown below.

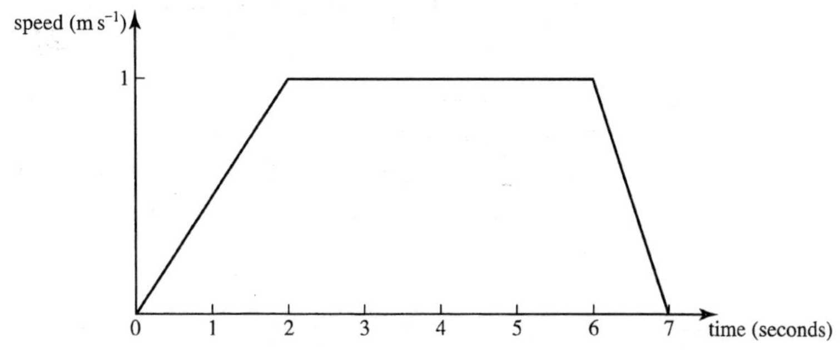

(i) Find the force of frictional resistance acting on the chest when it moves.

(ii) Use the speed–time graph to find the total distance travelled by the chest.

(iii) Find the total work done by the woman.

(iv) Find the acceleration of the chest in the first 2 s of its motion and hence the force exerted by the woman during this time, and the work done.

(v) In the same way find the work done by the woman during the time intervals 2 s to 6 s, and 6 s to 7 s.

(vi) Show that your answers to parts **(iv)** and **(v)** are consistent with your answer to part **(iii)**.

Gravitational potential energy

As you have seen, kinetic energy (K.E.) is the energy that an object has because of its motion. Potential energy (P.E.) is the energy an object has because of its position. The units of potential energy are the same as those of kinetic energy or any other form of energy, namely joules.

One form of potential energy is *gravitational potential energy*. The gravitational potential energy of the object in figure 9.6 of mass m kg at height h m above a fixed reference level, 0, is mgh J. If it falls to the reference level, the force of gravity does mgh J of work and the body loses mgh J of potential energy.

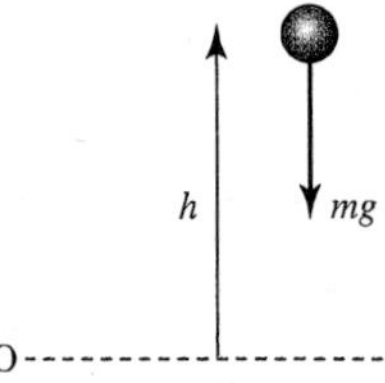

Figure 9.6

A loss in gravitational potential energy is an alternative way of accounting for the work done by the force of gravity.

If a mass m kg is raised through a distance h m, the gravitational potential energy *increases* by mgh J. If a mass m kg is *lowered* through a distance h m the gravitational potential energy *decreases* by mgh J.

EXAMPLE 9.8

Calculate the gravitational potential energy, relative to the ground, of a ball of mass 0.15 kg at a height of 2 m above the ground.

SOLUTION

Mass $m = 0.15$, height $h = 2$.

$$\begin{aligned}\text{Gravitational potential energy} &= mgh \\ &= 0.15 \times 10 \times 2 \\ &= 3\,\text{J}.\end{aligned}$$

Note

If the ball falls:

$$\begin{aligned}\text{loss in P.E.} &= \text{work done by gravity} \\ &= \text{gain in K.E.}\end{aligned}$$

There is no change in the total energy (P.E. + K.E.) of the ball.

Using conservation of mechanical energy

When gravity is the only force which does work on a body, mechanical energy is conserved. When this is the case, many problems are easily solved using energy. This is possible even when the acceleration is not constant.

EXAMPLE 9.9

A skier slides down a smooth ski slope 400 m long which is at an angle of 30° to the horizontal. Find the speed of the skier when he reaches the bottom of the slope.

At the foot of the slope the ground becomes horizontal and is made rough in order to help him to stop. The coefficient of friction between his skis and the ground is $\frac{1}{4}$.

(i) Find how far the skier travels before coming to rest.

(ii) In what way is your model unrealistic?

SOLUTION

The skier is modelled as a particle.

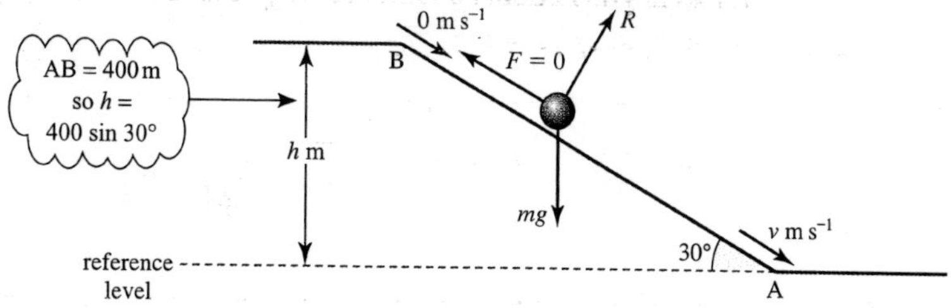

Figure 9.7

(i) Since in this case the slope is smooth, the frictional force is zero. The skier is subject to two external forces: his weight mg and the normal reaction from the slope.

The normal reaction between the skier and the slope does no work because the skier does not move in the direction of this force. The only force which does work is gravity, so mechanical energy is conserved.

$$\begin{aligned}\text{Total mechanical energy at B} &= mgh + \tfrac{1}{2}mu^2\\ &= m \times 10 \times 400 \sin 30° + 0\\ &= 2000m\,\text{J}\end{aligned}$$

Total mechanical energy at A $= (0 + \frac{1}{2}mv^2)$ J

Since mechanical energy is conserved,

$$\tfrac{1}{2}mv^2 = 2000m \qquad ①$$
$$v^2 = 4000$$
$$v = 63.2...$$

The skier's speed at the bottom of the slope is 63.2 m s⁻¹ (to 3 s.f.).

Notice that the mass of the skier cancels out. Using this model, all skiers should arrive at the bottom of the slope with the same speed. Also the slope could be curved so long as the total height lost is the same.

For the horizontal part there is some friction. Suppose that the skier travels a further distance s m before stopping.

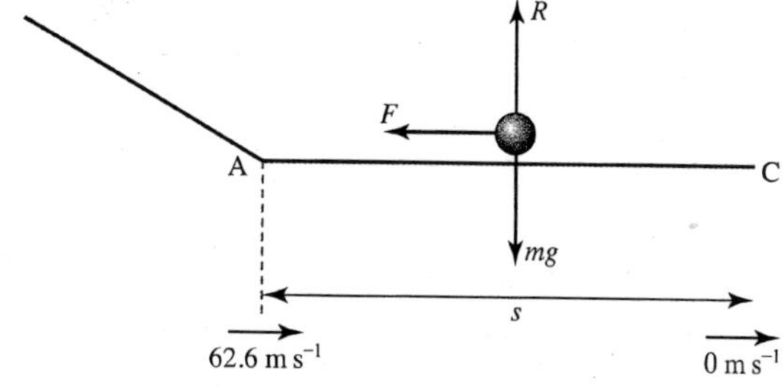

Figure 9.8

Coulomb's law of friction gives $\quad F = \mu R = \frac{1}{4}R.$

Since there is no vertical acceleration we can also say $R = mg$.

So $\quad F = \frac{1}{4}mg.$

Work done by the friction force $= F \times (-s) = -\frac{1}{4}mgs$.

Negative because the motion is in the opposite direction to the force.

The increase in kinetic energy between A and C $= (0 - \frac{1}{2}mv^2)$ J.

Using the work–energy principle

$$-\tfrac{1}{4}mgs = -\tfrac{1}{2}mv^2 = -2000m \text{ from ①}$$

Solving for s gives $s = 800$.

So the distance the skier travels before stopping is 800 m.

(ii) The assumptions made in solving this problem are that friction on the slope and air resistance are negligible, and that the slope ends in a smooth curve at A. Clearly the speed of $63.2\,\text{m s}^{-1}$ is very high, so the assumption that friction and air resistance are negligible must be suspect.

EXAMPLE 9.10

Ama, whose mass is 40 kg, is taking part in an assault course. The obstacle shown in figure 9.9 is a river at the bottom of a ravine 8 m wide which she has to cross by swinging on a rope 5 m long secured to a point on the branch of a tree, immediately above the centre of the ravine.

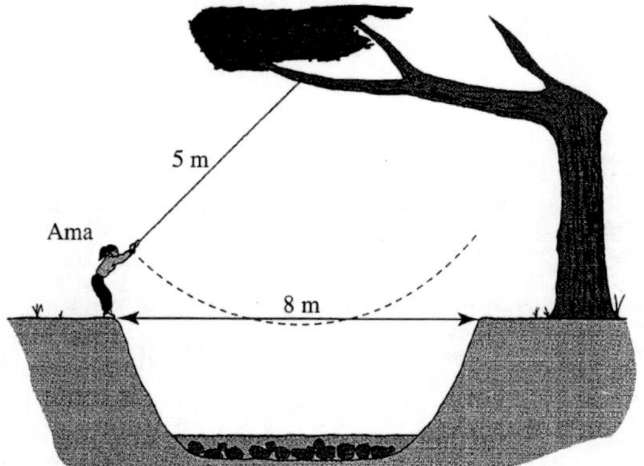

Figure 9.9

(i) Find how fast Ama is travelling at the lowest point of her crossing
 (a) if she starts from rest
 (b) if she launches herself off at a speed of $1\,\text{m s}^{-1}$.

(ii) Will her speed be $1\,\text{m s}^{-1}$ faster throughout her crossing?

SOLUTION

(i) (a) The vertical height Ama loses is HB in the diagram.

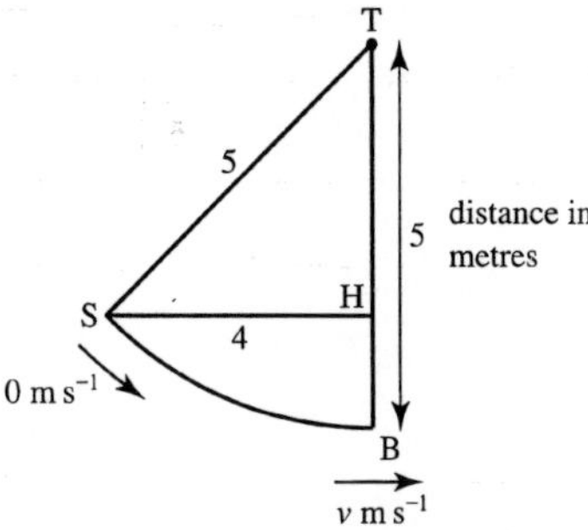

Figure 9.10

$$\begin{aligned} \text{Using Pythagoras}\qquad \text{TH} &= \sqrt{5^2 - 4^2} = 3 \\ \text{HB} &= 5 - 3 = 2 \\ \text{P.E. lost} &= mgh \\ &= 40g \times 2 \\ \text{K.E. gained} &= \tfrac{1}{2}mv^2 - 0 \\ &= \tfrac{1}{2} \times 40 \times v^2 \end{aligned}$$

By conservation of energy, K.E. gained = P.E. lost

$$\begin{aligned} \tfrac{1}{2} \times 40 \times v^2 &= 40 \times 10 \times 2 \\ v &= 6.32 \end{aligned}$$

Ama is travelling at 6.32 $\mathrm{m\,s^{-1}}$.

(b) If she has initial speed 1 $\mathrm{m\,s^{-1}}$ at S and speed v $\mathrm{m\,s^{-1}}$ at B, her initial K.E. is $\frac{1}{2} \times 40 \times 1^2$ and her K.E. at B is $\frac{1}{2} \times 40 \times v^2$.

Using conservation of energy,

$$\begin{aligned} \tfrac{1}{2} \times 40 \times v^2 - \tfrac{1}{2} \times 40 \times 1^2 &= 40 \times 10 \times 2 \\ v &= 6.40 \end{aligned}$$

(ii) Ama's speed at the lowest point is only 0.08 $\mathrm{m\,s^{-1}}$ faster in part **(i)(b)** compared with that in part **(i)(a)**, so she clearly will not travel 1 $\mathrm{m\,s^{-1}}$ faster throughout in part **(i)(b)**.

Historical note

James Joule was born in Salford in Lancashire on Christmas Eve 1818. He studied at Manchester University at the same time as the famous chemist, Dalton.

Joule spent much of his life conducting experiments to measure the equivalence of heat and mechanical forms of energy to ever-increasing degrees of accuracy. Working with William Thomson, he also discovered that a gas cools when it expands without doing work against external forces. It was this discovery that paved the way for the development of refrigerators.

Joule died in 1889 but his contribution to science is remembered with the S.I. unit for energy named after him.

Work and kinetic energy for two-dimensional motion

? Imagine that you are cycling along a level winding road in a strong wind. Suppose that the strength and direction of the wind are constant, but because the road is winding sometimes the wind is directly against you but at other times it is from your side.

How does the work you do in travelling a certain distance – say 1 m – change with your direction?

Work done by a force at an angle to the direction of motion

You have probably deduced that as a cyclist you would do work against the component of the wind force that is directly against you. The sideways component does not resist your forward progress.

Suppose that you are sailing and the angle between the force, F, of the wind on your sail and the direction of your motion is θ. In a certain time you travel a distance d in the direction of F, see figure 9.11, but during that time you actually travel a distance s along the line OP.

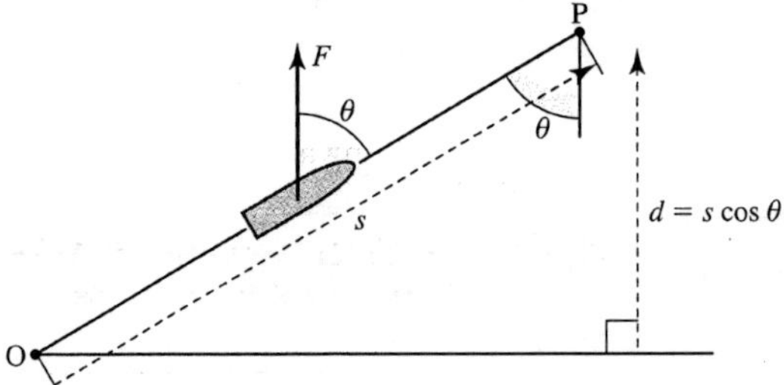

Figure 9.11

Work done by $F = Fd$

Since $d = s\cos\theta$, the work done by the force F is $Fs\cos\theta$. This can also be written as the product of the component of F along OP, $F\cos\theta$, and the distance moved along OP, s.

$$F \times s\cos\theta = F\cos\theta \times s$$

(Notice that the direction of F is not necessarily the same as the direction of the wind, it depends on how you have set your sails.)

EXAMPLE 9.11 As a car of mass m kg drives up a slope at an angle α to the horizontal it experiences a constant resistive force F N and a driving force D N. What can be deduced about the work done by D as the car moves a distance d m uphill if:

(i) the car moves at constant speed?
(ii) the car slows down?
(iii) the car gains speed?

The initial and final speeds of the car are denoted by u m s^{-1} and v m s^{-1} respectively.

(iv) Write v^2 in terms of the other variables.

SOLUTION

The diagram shows the forces acting on the car. The table shows the work done by each force. The normal reaction, R, does no work as the car moves no distance in the direction of R.

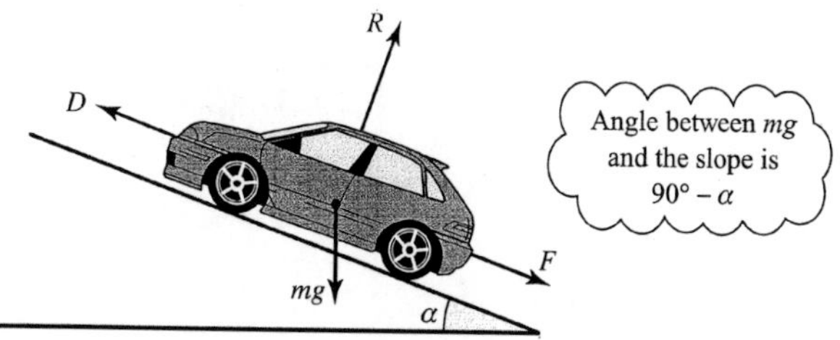

Figure 9.12

Force	Work done
Resistance F	$-Fd$
Normal reaction R	0
Force of gravity mg	$-mgd\cos(90° - \alpha) = -mgd\sin\alpha$
Driving force D	Dd
Total work done	$Dd - Fd - mgd\sin\alpha$

(i) If the car moves at a constant speed there is no change in kinetic energy so the total work done is zero, giving

Work done by D is

$$Dd = Fd + mgd\sin\alpha.$$

(ii) If the car slows down the total work done by the forces is negative, hence

Work done by D is

$$Dd < Fd + mgd\sin\alpha.$$

(iii) If the car gains speed the total work done by the forces is positive so

Work done by D is

$$Dd > Fd + mgd \sin \alpha.$$

(iv) Total work done = final K.E. – initial K.E.

$$\Rightarrow \quad Dd - Fd - mgd \sin \alpha = \tfrac{1}{2}mv^2 - \tfrac{1}{2}mu^2$$

Multiplying by $\dfrac{2}{m}$

$$\Rightarrow \quad v^2 = u^2 + \frac{2d}{m}(D - F) - 2gd \sin \alpha$$

EXERCISE 9B

1 Calculate the gravitational potential energy, relative to the reference level OA, for each of the objects shown.

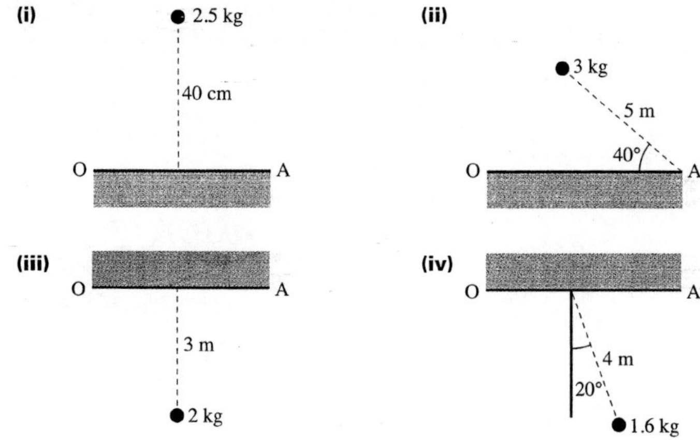

2 Calculate the change in gravitational potential energy when each object moves from A to B in the situations shown below. State whether the change is an increase or a decrease.

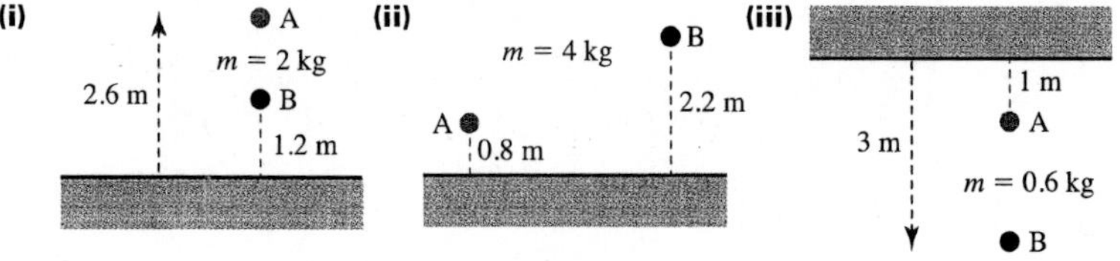

3 A vase of mass 1.2 kg is lifted from ground level and placed on a shelf at a height of 1.5 m. Find the work done against the force of gravity.

4 Find the increase in gravitational potential energy of a woman of mass 60 kg who climbs to the twelfth floor of a block of flats. The distance between floors is 3.3 m.

5 A car of mass 0.9 tonnes is driven 200 m up a slope inclined at 5° to the horizontal. There is a resistance force of 100 N.

(i) Find the work done by the car against gravity.

(ii) Find the work done against the resistance force.

(iii) When asked to work out the total work done by the car, a student replied '$(900g + 100) \times 200$ J'. Explain the error in this answer.

(iv) If the car slows down from $12\,\mathrm{m\,s^{-1}}$ to $8\,\mathrm{m\,s^{-1}}$, what is the total work done by the engine?

6 A sledge of mass 10 kg is being pulled across level ground by a rope which makes an angle of 20° with the horizontal. The tension in the rope is 80 N and there is a resistance force of 14 N.

(i) Find the work done while the sledge moves a distance of 20 m by

(a) the tension in the rope

(b) the resistance force.

(ii) Find the speed of the sledge after it has moved 20 m

(a) if it starts at rest

(b) if it starts at $4\,\mathrm{m\,s^{-1}}$.

7 A bricklayer carries a hod of bricks of mass 25 kg up a ladder of length 10 m inclined at an angle of 60° to the horizontal.

(i) Calculate the increase in the gravitational potential energy of the bricks.

(ii) If instead he had raised the bricks vertically to the same height, using a rope and pulleys, would the increase in potential energy be **(a)** less, **(b)** the same, or **(c)** more than in part **(i)**?

8 A girl of mass 45 kg slides down a smooth water chute of length 6 m inclined at an angle of 40° to the horizontal.

(i) Find

(a) the decrease in her potential energy

(b) her speed at the bottom.

(ii) How are answers to part **(i)** affected if the slide is not smooth?

9 A gymnast of mass 50 kg swings on a rope of length 10 m. Initially the rope makes an angle of 50° with the vertical.

(i) Find the decrease in her potential energy when the rope has reached the vertical.

(ii) Find her kinetic energy and hence her speed when the rope is vertical, assuming that air resistance may be neglected.

(iii) The gymnast continues to swing. What angle will the rope make with the vertical when she is next temporarily at rest?

(iv) Explain why the tension in the rope does no work.

10 A stone of mass 0.2 kg is dropped from the top of a building 80 m high. After t s it has fallen a distance x m and has speed $v\,\text{m s}^{-1}$.

(i) What is the gravitational potential energy of the stone relative to ground level when it is at the top of the building?

(ii) What is the potential energy of the stone t s later?

(iii) Show that, for certain values of t, $v^2 = 20x$ and state the range of values of t for which it is true.

(iv) Find the speed of the stone when it is half-way to the ground.

(v) At what height will the stone have half its final speed?

11 Wesley, whose mass is 70 kg, inadvertently steps off a bridge 50 m above water. When he hits the water, Wesley is travelling at $25\,\text{m s}^{-1}$.

(i) Calculate the potential energy Wesley has lost and the kinetic energy he has gained.

(ii) Find the size of the resistance force acting on Wesley while he is in the air, assuming it to be constant.

Wesley descends to a depth of 5 m below the water surface, then returns to the surface.

(iii) Find the total upthrust (assumed constant) acting on him while he is moving downwards in the water.

12 A hockey ball of mass 0.15 kg is hit from the centre of a pitch. Its position vector (in m), t s later is modelled by

$$\begin{pmatrix} x \\ y \end{pmatrix} = \begin{pmatrix} 10t \\ 10t - 4.9t^2 \end{pmatrix}$$

where the directions are along the line of the pitch and vertically upwards.

(i) What value of g is used in this model?

(ii) Find an expression for the gravitational potential energy of the ball at time t. For what values of t is your answer valid?

(iii) What is the maximum height of the ball? What is its velocity at that instant?

(iv) Find the initial velocity, speed and kinetic energy of the ball.

(v) Show that according to this model mechanical energy is conserved and state what modelling assumption is implied by this. Is it reasonable in this context?

13 A ski-run starts at altitude 2471 m and ends at 1863 m.

(i) If all resistance forces could be ignored, what would the speed of the skier be at the end of the run?

A particular skier of mass 70 kg actually attains a speed of $42\,\text{m s}^{-1}$. The length of the run is 3.1 km.

(ii) Find the average force of resistance acting on a skier.

Two skiers are equally skilful.

(iii) Which would you expect to be travelling faster by the end of the run, the heavier or the lighter?

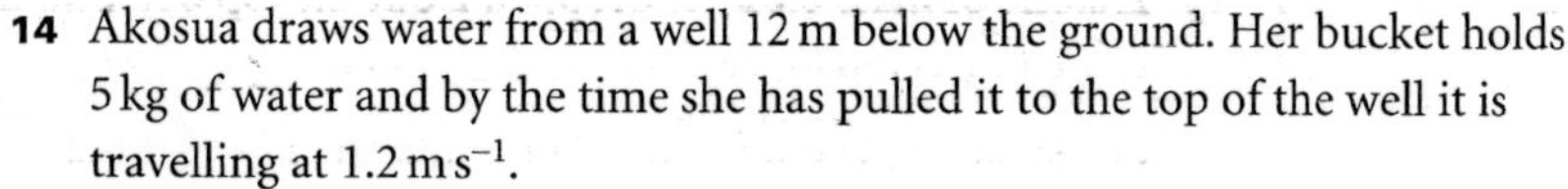

14 Akosua draws water from a well 12 m below the ground. Her bucket holds 5 kg of water and by the time she has pulled it to the top of the well it is travelling at $1.2\,\text{m s}^{-1}$.

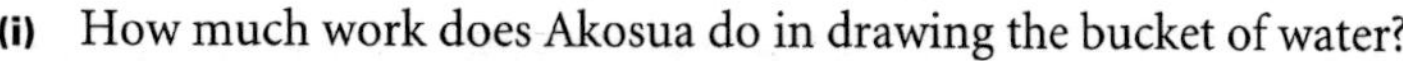

(i) How much work does Akosua do in drawing the bucket of water?

On an average day 150 people in the village each draw six such buckets of water. One day a new electric pump is installed that takes water from the well and fills an overhead tank 5 m above ground level every morning. The flow rate through the pump is such that the water has speed $2\,\text{m s}^{-1}$ on arriving in the tank.

(ii) Assuming that the villagers' demand for water remains unaltered, how much work does the pump do in one day?

It takes the pump one hour to fill the tank each morning.

(iii) At what rate does the pump do work, in joules per second (watts)?

15 A block of mass 50 kg is pulled up a straight hill and passes through points A and B with speeds $7\,\text{m s}^{-1}$ and $3\,\text{m s}^{-1}$ respectively. The distance AB is 200 m and B is 15 m higher than A. For the motion of the block from A to B, find

(i) the loss in kinetic energy of the block,

(ii) the gain in potential energy of the block.

The resistance to motion of the block has magnitude 7.5 N.

(iii) Find the work done by the pulling force acting on the block.

The pulling force acting on the block has constant magnitude 45 N and acts at an angle $\alpha°$ upwards from the hill.

(iv) Find the value of α.

[Cambridge AS and A Level Mathematics 9709, Paper 4 Q6 June 2006]

16 A lorry of mass 12 500 kg travels along a road that has a straight horizontal section AB and a straight inclined section BC. The length of BC is 500 m. The speeds of the lorry at A, B and C are $17\,\text{m s}^{-1}$, $25\,\text{m s}^{-1}$ and $17\,\text{m s}^{-1}$ respectively (see diagram).

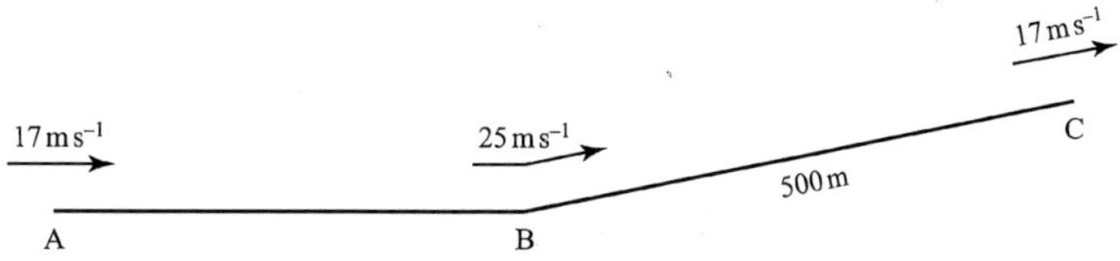

(i) The work done against the resistance to motion of the lorry, as it travels from A to B, is 5000 kJ. Find the work done by the driving force as the lorry travels from A to B.

(ii) As the lorry travels from B to C, the resistance to motion is 4800 N and the work done by the driving force is 3300 kJ. Find the height of C above the level of AB.

[Cambridge AS and A Level Mathematics 9709, Paper 4 Q5 June 2007]

17 A crate of mass 50 kg is dragged along a horizontal floor by a constant force of magnitude 400 N acting at an angle $\alpha°$ upwards from the horizontal. The total resistance to motion of the crate has constant magnitude 250 N. The crate starts from rest at the point O and passes the point P with a speed of 2 m s^{-1}. The distance OP is 20 m. For the crate's motion from O to P, find

(i) the increase in kinetic energy of the crate,

(ii) the work done against the resistance to the motion of the crate,

(iii) the value of α.

[Cambridge AS and A Level Mathematics 9709, Paper 4 Q2 November 2005]

18 The diagram shows the vertical cross-section of a surface. A and B are two points on the cross-section, and A is 5 m higher than B. A particle of mass 0.35 kg passes through A with speed 7 m s^{-1}, moving on the surface towards B.

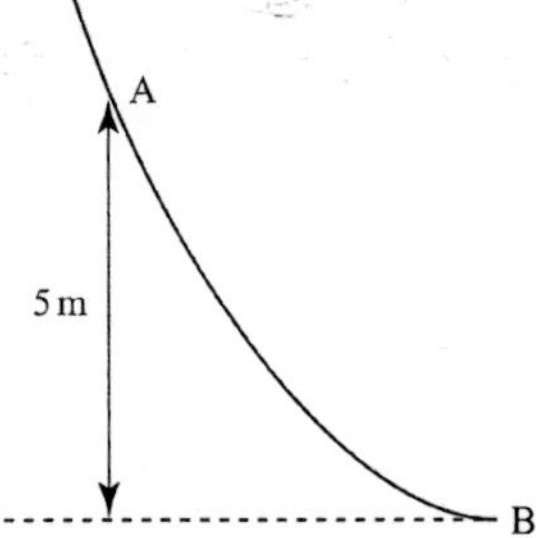

(i) Assuming that there is no resistance to motion, find the speed with which the particle reaches B.

(ii) Assuming instead that there is a resistance to motion, and that the particle reaches B with speed 11 m s^{-1}, find the work done against this resistance as the particle moves from A to B.

[Cambridge AS and A Level Mathematics 9709, Paper 4 Q4 November 2007]

19 A load of mass 160 kg is lifted vertically by a crane, with constant acceleration. The load starts from rest at the point O. After 7 s, it passes through the point A with speed 0.5 m s^{-1}. By considering energy, find the work done by the crane in moving the load from O to A.

[Cambridge AS and A Level Mathematics 9709, Paper 4 Q4 November 2008]

Power

It is claimed that a motorcycle engine can develop a maximum *power* of 26.5 kW at a top *speed* of 165 km h^{-1}. This suggests that power is related to speed and this is indeed the case.

Power is the rate at which work is being done. A powerful car does work at a greater rate than a less powerful one.

You might find it helpful to think in terms of a force, F, acting for a very short time t over a small distance s. Assume F to be constant over this short time.

Power is the rate of working so

$$\begin{aligned} \text{power} &= \frac{\text{work}}{\text{time}} \\ &= \frac{Fs}{t} \\ &= Fv \end{aligned}$$

This gives you the power at an *instant* of time. The result is true whether or not F is constant.

The power of a vehicle moving at speed v under a driving force F is given by Fv.

For a motor vehicle the power is produced by the engine, whereas for a bicycle it is produced by the cyclist. They both make the wheels turn, and the friction between the rotating wheels and the ground produces a forward force on the machine.

The unit of power is the watt (W), named after James Watt. The power produced by a force of 1 N acting on an object that is moving at 1 m s^{-1} is 1 W. Because the watt is such a small unit you will probably use kilowatts more often (1 kW = 1000 W).

EXAMPLE 9.12

A car of mass 1000 kg can produce a maximum power of 45 kW. Its driver wishes to overtake another vehicle. Ignoring air resistance, find the maximum acceleration of the car when it is travelling at

(i) $12\,\text{m}\,\text{s}^{-1}$ **(ii)** $28\,\text{m}\,\text{s}^{-1}$

(these are about $43\,\text{km}\,\text{h}^{-1}$ and $101\,\text{km}\,\text{h}^{-1}$).

SOLUTION

(i) Power = force × velocity

The driving force at $12\,\text{m}\,\text{s}^{-1}$ is F_1 N where

$$45\,000 = F_1 \times 12$$
$$\Rightarrow \quad F_1 = 3750.$$

By Newton's second law $F = ma$

$$\Rightarrow \quad \text{acceleration} = \frac{3750}{1000} = 3.75\,\text{m}\,\text{s}^{-2}.$$

(ii) Now the driving force F_2 is given by

$$45\,000 = F_2 \times 28$$
$$\Rightarrow \quad F_2 = 1607$$
$$\Rightarrow \quad \text{acceleration} = \frac{1607}{1000} = 1.61\,\text{m}\,\text{s}^{-2}.$$

This example shows why it is easier to overtake a slow moving vehicle.

EXAMPLE 9.13

A car of mass 900 kg produces power 45 kW when moving at a constant speed. It experiences a resistance of 1700 N.

(i) What is its speed?

(ii) The car comes to a downhill stretch inclined at 2° to the horizontal. What is its maximum speed downhill if the power and resistance remain unchanged?

SOLUTION

(i) As the car is travelling at a constant speed, there is no resultant force on the car. In this case the forward force of the engine must have the same magnitude as the resistance forces, i.e. 1700 N.

Denoting the speed of the car by $v\,\text{m}\,\text{s}^{-1}$, $P = Fv$ gives

$$v = \frac{P}{F}$$
$$= \frac{45\,000}{1700}$$
$$= 26.5.$$

The speed of the car is $26.5\,\text{m}\,\text{s}^{-1}$ (approximately $95\,\text{km}\,\text{h}^{-1}$).

(ii) The diagram shows the forces acting.

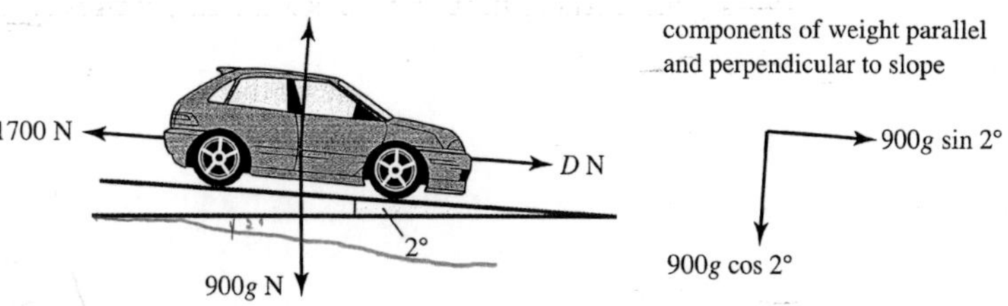

Figure 9.13

At maximum speed there is no acceleration so the resultant force down the slope is zero.

When the driving force is D N

$$D + 900g\sin 2° - 1700 = 0$$

$$\Rightarrow \qquad D = 1386$$

But power is Dv so $\quad 45\,000 = 1386\,v$

$$\Rightarrow \quad v = \frac{45\,000}{1386}$$

The maximum speed is $32.5\,\mathrm{m\,s^{-1}}$ (about $117\,\mathrm{km\,h^{-1}}$).

Historical note

James Watt was born in 1736 in Greenock in Scotland, the son of a house- and ship-builder. As a boy James was frail and he was taught by his mother rather than going to school. This allowed him to spend time in his father's workshop where he developed practical and inventive skills.

As a young man he manufactured mathematical instruments: quadrants, scales, compasses and so on. One day he was repairing a model steam engine for a friend and noticed that its design was very wasteful of steam. He proposed an alternative arrangement, which was to become standard on later steam engines. This was the first of many engineering inventions which made possible the subsequent industrial revolution. James Watt died in 1819, a well known and highly respected man. His name lives on as the S.I. unit for power.

EXERCISE 9C

1 A builder hoists bricks up to the top of the house he is building. Each brick weighs 3.5 kg and the house is 9 m high. In the course of one hour the builder raises 120 bricks from ground level to the top of the house, where they are unloaded by his assistant.

(i) Find the increase in gravitational potential energy of one brick when it is raised in this way.

(ii) Find the total work done by the builder in one hour of raising bricks.

(iii) Find the average power with which he is working.

2 A weightlifter takes 2 seconds to lift 120 kg from the floor to a position 2 m above it, where the weight has to be held stationary.

(i) Calculate the work done by the weightlifter.

(ii) Calculate the average power developed by the weightlifter.

The weight lifter is using the 'clean and jerk' technique. This means that in the first stage of the lift he raises the weight 0.8 m from the floor in 0.5 s. He then holds it stationary for 1 s before lifting it up to the final position in another 0.5 s.

(iii) Find the average power developed by the weightlifter during each of the stages of the lift.

3 A winch is used to pull a crate of mass 180 kg up a rough slope of angle 30° against a frictional force of 450 N. The crate moves at a steady speed, v, of 1.2 m s^{-1}.

(i) Calculate the gravitational potential energy given to the crate during 30 s.

(ii) Calculate the work done against friction during this time.

(iii) Calculate the total work done per second by the winch.

The cable from the winch to the crate runs parallel to the slope.

(iv) Calculate the tension, T, in the cable.

(v) What information is given by $T \times v$?

4 The power output from the engine of a car of mass 50 kg which is travelling along level ground at a constant speed of 33 m s^{-1} is 23 200 W.

(i) Find the total resistance on the car under these conditions.

(ii) You were given one piece of unnecessary information. Which is it?

5 A Kawasaki GPz 305 motorcycle has a maximum power output of 26.5 kW and a top speed of 46 m s^{-1} (about 165 km h^{-1}). Find the force exerted by the motorcycle engine when the motorcycle is travelling at top speed.

6 A crane is raising a load of 500 tonnes at a steady rate of 5 cm s^{-1}. What power is the engine of the crane producing? (Assume that there are no forces from friction or air resistance.)

7 A cyclist, travelling at a constant speed of 8 m s^{-1} along a level road, experiences a total resistance of 70 N.

(i) Find the power which the cyclist is producing.

(ii) Find the work done by the cyclist in 5 minutes under these conditions.

8 A mouse of mass 15 g is stationary 2 m below its hole when it sees a cat. It runs to its hole, arriving 1.5 seconds later with a speed of 3 m s^{-1}.

(i) Show that the acceleration of the mouse is not constant.

(ii) Calculate the average power of the mouse.

9 A train consists of a diesel shunter of mass 100 tonnes pulling a truck of mass 25 tonnes along a level track. The engine is working at a rate of 125 kW. The resistance to motion of the truck and shunter is 50 N per tonne.

(i) Calculate the constant speed of the train.

While travelling at this constant speed, the truck becomes uncoupled. The shunter engine continues to produce the same power.

(ii) Find the acceleration of the shunter immediately after this happens.

(iii) Find the greatest speed the shunter can now reach.

10 A supertanker of mass 4×10^8 kg is steaming at a constant speed of $8\,\mathrm{m\,s^{-1}}$. The resistance force is 2×10^6 N.

(i) What power are the ship's engines producing?

One of the ship's two engines suddenly fails but the other continues to work at the same rate.

(ii) Find the deceleration of the ship immediately after the failure.

The resistance force is directly proportional to the speed of the ship.

(iii) Find the eventual steady speed of the ship under one engine only, assuming that the single engine maintains constant power output.

11 A car of mass 850 kg has a maximum speed of $50\,\mathrm{m\,s^{-1}}$ and a maximum power output of 40 kW. The resistance force, R N at speed $v\,\mathrm{m\,s^{-1}}$ is modelled by

$$R = kv$$

(i) Find the value of k.

(ii) Find the resistance force when the car's speed is $20\,\mathrm{m\,s^{-1}}$.

(iii) Find the power needed to travel at a constant speed of $20\,\mathrm{m\,s^{-1}}$ along a level road.

(iv) Find the maximum acceleration of the car when it is travelling at $20\,\mathrm{m\,s^{-1}}$

(a) along a level road

(b) up a hill at 5° to the horizontal.

12 A car of mass 1 tonne is moving at a constant velocity of 60 km h^{-1} up an inclined road which makes an angle of 6° with the horizontal.

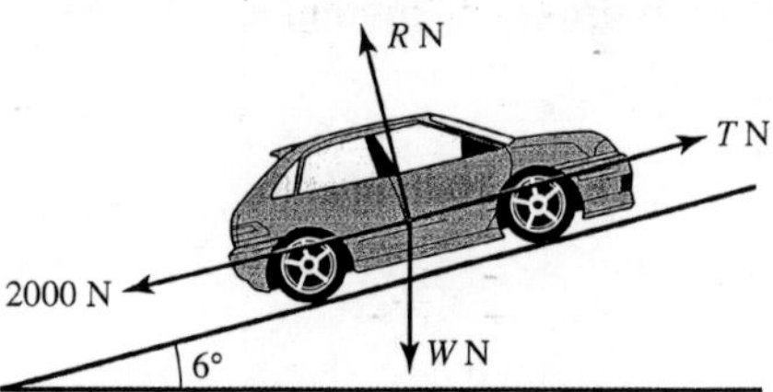

(i) Calculate the weight W of the car and the normal reaction R between the car and the road.

Given that the non-gravitational resistance down the slope is 2000 N, find

(ii) the tractive force T which is propelling the car up the slope

(iii) the rate at which T is doing work.

The engine has a maximum power output of 80 kW.

(iv) Assuming the resistances stay the same as before, calculate the maximum speed of the car up the same slope.

[MEI]

13 A boat of mass 1200 kg is winched a distance 30 m up a flat beach inclined at 10° to the horizontal.

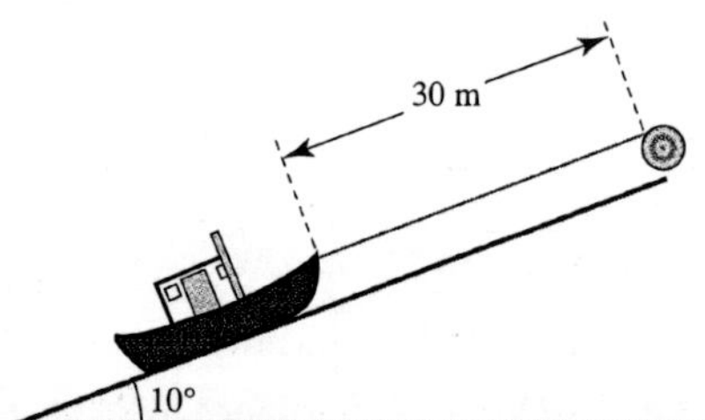

Initially a very approximate model is used in which all resistances are neglected.

(i) Calculate the work done.

(ii) Given that the process takes 2 minutes and that the boat moves at a constant speed, calculate the power of the winch motor.

A better model takes account of the resistance of the beach to the motion. Assuming that the winch motor develops a constant 4.5 kW, the resistance of the beach on the boat is a constant 5 kN and the boat moves at a constant speed,

(iii) calculate how long the winching will take

(iv) show that if the winch cable suddenly broke off at the boat whilst the winching was in progress, the boat would come to rest in about 35 mm.

[MEI]

14 A winch pulls a crate of mass 1500 kg up a slope at 20° to the horizontal. The light wire attached to the winch and the crate is parallel to the slope, as shown in figure **(A)**.

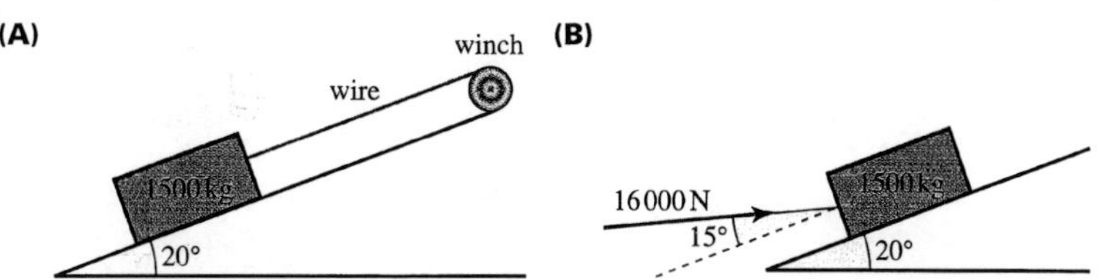

The crate takes 50 seconds to move 25 m up the slope at a constant speed when the power supplied by the winch is 6 kW.

(i) How much work is done by the tension in the wire in the 50 seconds?

(ii) Calculate the resistance to the motion of the crate up the slope.

(iii) Show that the coefficient of friction between the crate and the slope is 0.5 (correct to one decimal place).

The winch breaks down and the crate is then *pushed* up the slope by a mechanical shovel by means of a constant force of 16 000 N inclined at 15° to the slope, as shown in figure **(B)**. You may assume that the crate does not tip up.

(iv) Calculate the distance travelled by the crate up the slope as it speeds up from rest to $2.5\,\text{m}\,\text{s}^{-1}$. (You may assume the coefficient of friction between the crate and the slope is exactly 0.5.)

[MEI, *adapted*]

15 A car of mass 1000 kg moves along a horizontal straight road, passing through points A and B. The power of its engine is constant and equal to 15 000 W. The driving force exerted by the engine is 750 N at A and 500 N at B. Find the speed of the car at A and at B, and hence find the increase in the car's kinetic energy as it moves from A to B.

[Cambridge AS and A Level Mathematics 9709, Paper 41 Q1 November 2009]

16 A car of mass 1150 kg travels up a straight hill inclined at 1.2° to the horizontal. The resistance to motion of the car is 975 N. Find the acceleration of the car at an instant when it is moving with speed $16\,\text{m}\,\text{s}^{-1}$ and the engine is working at a power of 35 kW.

[Cambridge AS and A Level Mathematics 9709, Paper 41 Q1 June 2010]

17 A car of mass 1200 kg travels along a horizontal straight road. The power provided by the car's engine is constant and equal to 20 kW. The resistance to the car's motion is constant and equal to 500 N. The car passes through the points A and B with speeds $10\,\text{m}\,\text{s}^{-1}$ and $25\,\text{m}\,\text{s}^{-1}$ respectively. The car takes 30.5 s to travel from A to B.

(i) Find the acceleration of the car at A.

(ii) By considering work and energy, find the distance AB.

[Cambridge AS and A Level Mathematics 9709, Paper 4 Q7 June 2005]

INVESTIGATION

Crawler lanes

Sometimes on single carriageway roads or even some highways, crawler lanes are introduced for slow-moving, heavily laden lorries going uphill. Investigate how steep a slope can be before a crawler lane is needed.

Data:	Typical power output for a large lorry:	45 kW
	Typical mass of a large laden lorry:	32 tonnes

EXPERIMENT

Energy losses

Figure 9.14

Set up a track like the one above. Release cars or trolleys from different heights and record the heights that they reach on the opposite side. Use your results to formulate a model for the force of resistance acting on them.

KEY POINTS

1. The work done by a constant force F is given by Fs where s is the distance moved in the direction of the force.
2. The kinetic energy (K.E.) of a body of mass m moving with speed v is given by $\frac{1}{2}mv^2$. Kinetic energy is the energy a body possesses on account of its motion.
3. The work–energy principle states that the total work done by all the forces acting on a body is equal to the increase in the kinetic energy of the body.
4. The gravitational potential energy of a body mass m at height h above a given reference level is given by mgh. It is the work done against the force of gravity in raising the body.
5. Mechanical energy (K.E. and P.E.) is conserved when no forces other than gravity do work.
6. Power is the rate of doing work, and is given by Fv.
7. The S.I. unit for energy is the joule and that for power is the watt.

Mechanics 2

M2

10 Motion of a projectile

Swift of foot was Hiawatha;
He could shoot an arrow from him,
And run forward with such fleetness,
That the arrow fell behind him!
Strong of arm was Hiawatha;
He could shoot ten arrows upwards,
Shoot them with such strength and swiftness,
That the last had left the bowstring,
Ere the first to earth had fallen!

The Song of Hiawatha, Longfellow

Look at the water jet in the picture. Every drop of water in a water jet follows its own path which is called its *trajectory*. You can see the same sort of trajectory if you throw a small object across a room. Its path is a parabola. Objects moving through the air like this are called *projectiles*.

Modelling assumptions for projectile motion

The path of a cricket ball looks parabolic, but what about a boomerang? There are modelling assumptions which must be satisfied for the motion to be parabolic. These are

- a projectile is a particle
- it is not powered
- the air has no effect on its motion.

Equations for projectile motion

A projectile moves in two dimensions under the action of only one force, the force of gravity, which is constant and acts vertically downwards. This means that the acceleration of the projectile is $g\,\mathrm{m\,s^{-2}}$ vertically downwards and there is no horizontal acceleration. You can treat the horizontal and vertical motions separately using the equations for constant acceleration.

To illustrate the ideas involved, think of a ball being projected with a speed of $20\,\mathrm{m\,s^{-2}}$ at 60° to the ground as illustrated in figure 10.1. This could be a first model for a football, a chip shot from the rough at golf or a lofted shot at cricket.

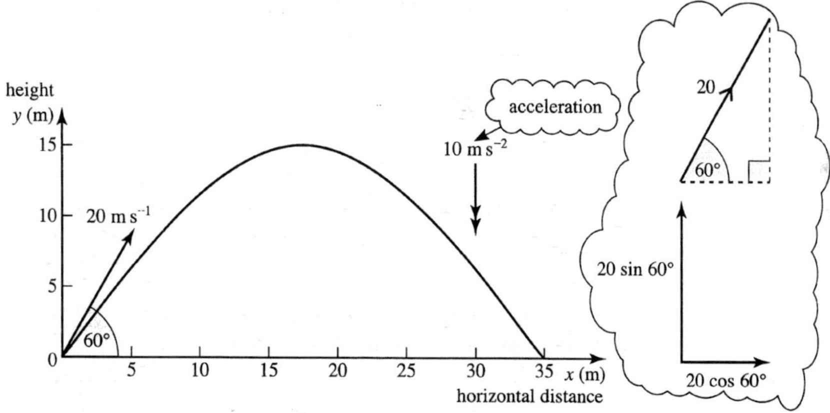

Figure 10.1

Using axes as shown, the components are:

	Horizontal	Vertical
Initial position	0	0
Acceleration	$a_x = 0$	$a_y = -10$
Initial velocity	$u_x = 20\cos 60°$ $= 10$	$u_y = 20\sin 60°$ $= 17.32$

This is negative because the positive y axis is upwards.

as vectors
$\mathbf{a} = \begin{pmatrix} 0 \\ -10 \end{pmatrix}$
$\mathbf{u} = \begin{pmatrix} 20\cos 60° \\ 20\sin 60° \end{pmatrix}$

Using $v = u + at$ in the two directions gives the components of velocity.

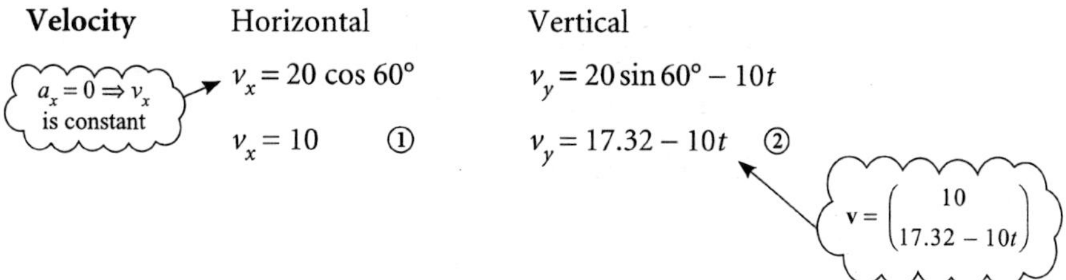

Velocity	Horizontal	Vertical
	$v_x = 20\cos 60°$	$v_y = 20\sin 60° - 10t$
	$v_x = 10$ ①	$v_y = 17.32 - 10t$ ②

Using $s = ut + \frac{1}{2}at^2$ in the two directions gives the components of position.

Position Horizontal Vertical

$x = (20 \cos 60°)t$ $y = (20 \sin 60°)t - 5t^2$

$x = 10t$ ③ $y = 17.32t - 5t^2$ ④

$\mathbf{r} = \begin{pmatrix} 10t \\ 17.32t - 5t^2 \end{pmatrix}$

You can summarise these results in a table.

	Horizontal motion	Vertical motion
initial position	0	0
a	0	−10
u	$u_x = 20 \cos 60° = 10$	$u_y = 20 \sin 60° = 17.32$
v	$v_x = 10$ ①	$v_y = 17.32 - 10t$ ②
r	$x = 10t$ ③	$y = 17.32t - 5t^2$ ④

The four equations ①, ②, ③ and ④ for velocity and position can be used to find several things about the motion of the ball.

What is true at

(i) the top-most point of the path of the ball?

(ii) the point where it is just about to hit the ground?

When you have decided the answer to these questions you have sufficient information to find the greatest height reached by the ball, the time of flight and the total distance travelled horizontally before it hits the ground. This is called the *range* of the ball.

The maximum height

When the ball is at its maximum height, H m, the *vertical* component of its velocity is zero. It still has a horizontal component of 10 m s^{-1} which is constant.

Equation ② gives the vertical component as

$$v_y = 17.32 - 10t$$

At the top: $0 = 17.32 - 10t$

$$t = \frac{17.32}{10} = 1.732$$

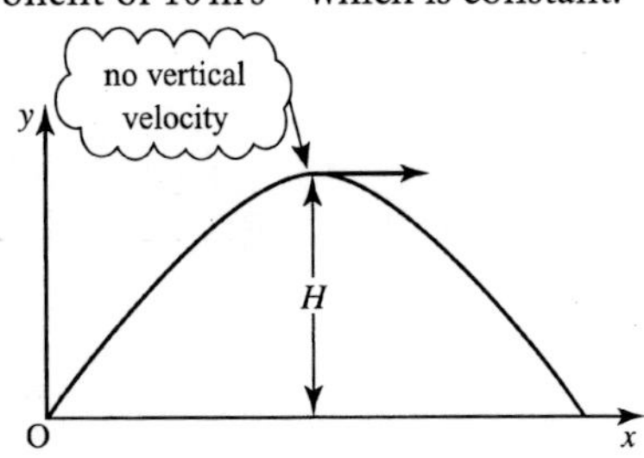

Figure 10.2

To find the maximum height, you now need to find y at this time. Substituting for t in equation ④,

$$y = 17.32t - 5t^2$$
$$y = 17.32 \times 1.732 - 5 \times 1.732^2$$
$$= 15.0$$

The maximum height is 15.0 m.

The time of flight

The flight ends when the ball returns to the ground, that is when $y = 0$. Substituting $y = 0$ in equation ④,

$$y = 17.32t - 5t^2$$
$$17.32t - 5t^2 = 0$$
$$t(17.32 - 5t) = 0$$
$$t = 0 \text{ or } t = 3.46$$

Clearly $t = 0$ is the time when the ball is thrown, so $t = 3.46$ is the time when it lands and the flight time is 3.46 s.

The range

The range, R m, of the ball is the horizontal distance it travels before landing.

R is the value of x when $y = 0$.

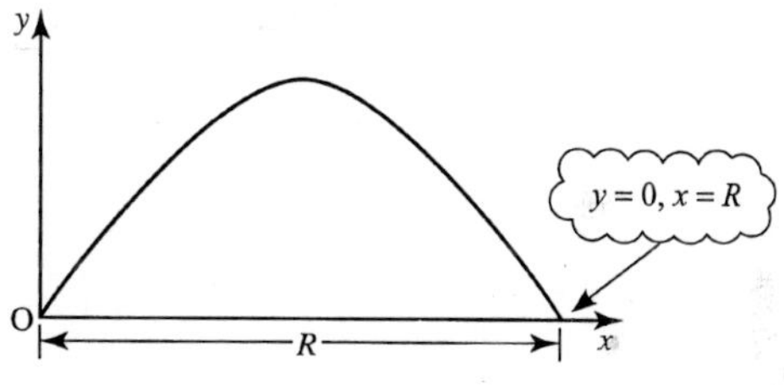

Figure 10.3

R can be found by substituting $t = 3.46$ in equation ③ : $x = 10t$. The range is $10 \times 3.46 = 34.6$ m.

?

1 Notice in this example that the time to maximum height is half the flight time. Is this always the case?

2 Decide which of the following could be modelled as projectiles.

a balloon	a bird	a bullet shot from a gun	a glider
a golf ball	a parachutist	a rocket	a tennis ball

What special conditions would have to apply in particular cases?

EXERCISE 10A

In this exercise take upwards as positive. All the projectiles start at the origin.

1 In each of the following cases you are given the initial velocity of a projectile.

(a) Draw a diagram showing the initial velocity and path.

(b) Write down the horizontal and vertical components of the initial velocity.

(c) Write down equations for the velocity after time t seconds.

(d) Write down equations for the position after time t seconds.

(i) $10\,\text{m s}^{-1}$ at 35° above the horizontal.

(ii) $2\,\text{m s}^{-1}$ horizontally, $5\,\text{m s}^{-1}$ vertically.

(iii) $4\,\text{m s}^{-1}$ horizontally.

(iv) $10\,\text{m s}^{-1}$ at 13° below the horizontal.

(v) $U\,\text{m s}^{-1}$ at angle α above the horizontal.

(vi) $u_0\,\text{m s}^{-1}$ horizontally, $v_0\,\text{m s}^{-1}$ vertically.

2 In each of the following cases find

(a) the time taken for the projectile to reach its highest point

(b) the maximum height.

(i) Initial velocity $5\,\text{m s}^{-1}$ horizontally and $15\,\text{m s}^{-1}$ vertically.

(ii) Initial velocity $10\,\text{m s}^{-1}$ at 30° above the horizontal.

3 In each of the following cases find

(a) the time of flight of the projectile

(b) the horizontal range.

(i) Initial velocity $20\,\text{m s}^{-1}$ horizontally and $20\,\text{m s}^{-1}$ vertically.

(ii) Initial velocity $5\,\text{m s}^{-1}$ at 60° above the horizontal.

Projectile problems

When doing projectile problems, you can treat each direction separately or you can write them both together as vectors. Example 10.1 shows both methods.

EXAMPLE 10.1

A ball is thrown horizontally at $5\,\text{m s}^{-1}$ out of a window 4 m above the ground.

(i) How long does it take to reach the ground?

(ii) How far from the building does it land?

(iii) What is its speed just before it lands and at what angle to the ground is it moving?

SOLUTION

Figure 10.4 shows the path of the ball. It is important to decide at the outset where the origin and axes are. You may choose any axes that are suitable, but you must specify them carefully to avoid making mistakes. Here the origin is taken to be at ground level below the point of projection of the ball and upwards is positive. With these axes, the acceleration is $-g\,\text{m s}^{-2}$.

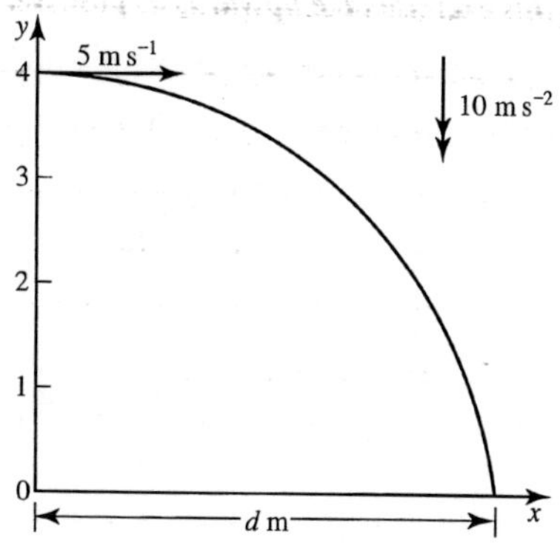

Figure 10.4

Method 1: Resolving into components

(i) *Position*: Using axes as shown and $s = s_0 + ut + \frac{1}{2}at^2$ in the two directions,

Horizontally: $x_0 = 0,\ u_x = 5,\ a_x = 0$

$$x = 5t \qquad ①$$

Vertically: $y_0 = 4,\ u_y = 0,\ a_y = -10$

$$y = 4 - 5t^2 \qquad ②$$

The ball reaches the ground when $y = 0$. Substituting in equation ② gives

$$0 = 4 - 5t^2$$
$$t^2 = \frac{4}{5}$$
$$t = 0.894...$$

The ball hits the ground after 0.894 s (to 3 s.f.).

(ii) When the ball lands $x = d$ so, from equation ①,

$$d = 5t = 5 \times 0.894... = 4.47...$$

The ball lands 4.47 m (to 3 s.f.) from the building.

(iii) *Velocity*: Using $v = u + at$ in the two directions,

Horizontally $v_x = 5 + 0$
Vertically $v_y = 0 - 10t$

To find the speed and direction just before it lands:
The ball lands when $t = 0.894...$ so $v_x = 5$ and $v_y = -8.94...$.
The components of velocity are shown in the diagram.
The speed of the ball is

$$\sqrt{5^2 + 8.94...^2} = 10.25\ \text{m s}^{-1} \text{ (to 4 s.f.)}$$

It hits the ground moving downwards at an angle α to the horizontal where

$$\tan\alpha = \frac{8.94}{5}$$
$$\alpha = 60.8°$$

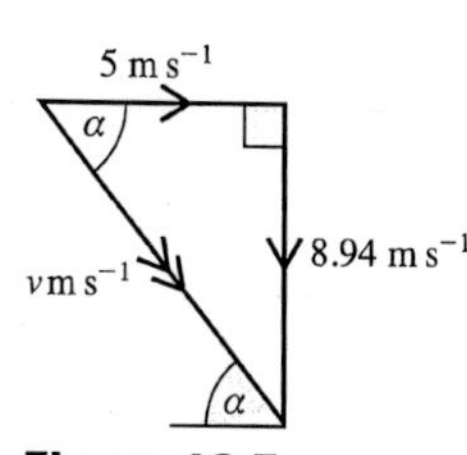

Figure 10.5

Method 2: Using vectors

Using perpendicular vectors in the horizontal (x) and vertical (y) directions, the initial position is $\mathbf{r}_0 = \begin{pmatrix} 0 \\ 4 \end{pmatrix}$ and the ball hits the ground when $\mathbf{r} = \begin{pmatrix} d \\ 0 \end{pmatrix}$. The initial velocity, $\mathbf{u} = \begin{pmatrix} 5 \\ 0 \end{pmatrix}$ and the acceleration $\mathbf{a} = \begin{pmatrix} 0 \\ -10 \end{pmatrix}$.

Using $\quad \mathbf{r} = \mathbf{r}_0 + \mathbf{u}t + \frac{1}{2}\mathbf{a}t^2$

$$\begin{pmatrix} d \\ 0 \end{pmatrix} = \begin{pmatrix} 0 \\ 4 \end{pmatrix} + \begin{pmatrix} 5 \\ 0 \end{pmatrix} t + \frac{1}{2}\begin{pmatrix} 0 \\ -10 \end{pmatrix} t^2$$

$$d = 5t \qquad ①$$

and

$$0 = 4 - 5t^2 \qquad ②$$

(i) Equation ② gives $t = 0.894$ and substituting this into ① gives **(ii)** $d = 4.47$.

(iii) The speed and direction of motion are the magnitude and direction of the velocity of the ball. Using

$$\mathbf{v} = \mathbf{u} + \mathbf{a}t$$

$$\begin{pmatrix} v_x \\ v_y \end{pmatrix} = \begin{pmatrix} 5 \\ 0 \end{pmatrix} + \begin{pmatrix} 0 \\ -10 \end{pmatrix} t$$

So when $t = 0.894$, $\begin{pmatrix} v_x \\ v_y \end{pmatrix} = \begin{pmatrix} 5 \\ -8.94 \end{pmatrix}$

You can find the speed and angle as before.

Notice that in both methods the time forms a link between the motions in the two directions. You can often find the time from one equation and then substitute it in another to find out more information.

Representing projectile motion by vectors

The diagram shows a possible path for a marble which is thrown across a room from the moment it leaves the hand until just before it hits the floor.

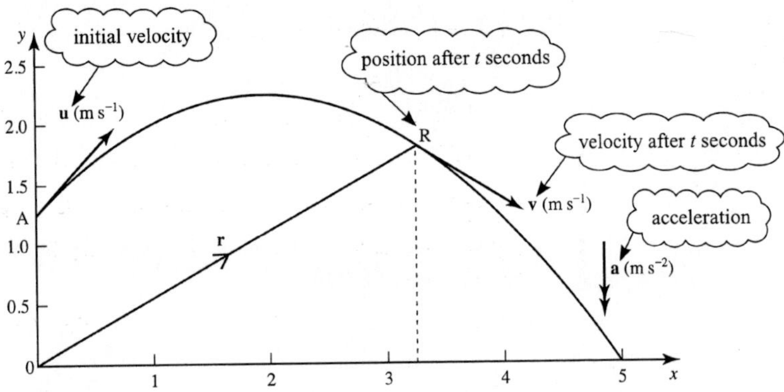

Figure 10.6

The vector $\mathbf{r} = \vec{OR}$ is the position vector of the marble after a time t seconds and the vector $\mathbf{v}$ represents its velocity in $\mathrm{m\,s^{-1}}$ at that instant of time (to a different scale).

Notice that the graph shows the trajectory of the marble. It is its path through space, not a position–time graph.

You can use equations for constant acceleration in vector form to describe the motion as in Example 10.1, Method 2.

velocity $\mathbf{v} = \mathbf{u} + \mathbf{a}t$

displacement $\mathbf{r} - \mathbf{r}_0 = \mathbf{u}t + \frac{1}{2}\mathbf{a}t^2$ so $\mathbf{r} = \mathbf{r}_0 + \mathbf{u}t + \frac{1}{2}\mathbf{a}t^2$

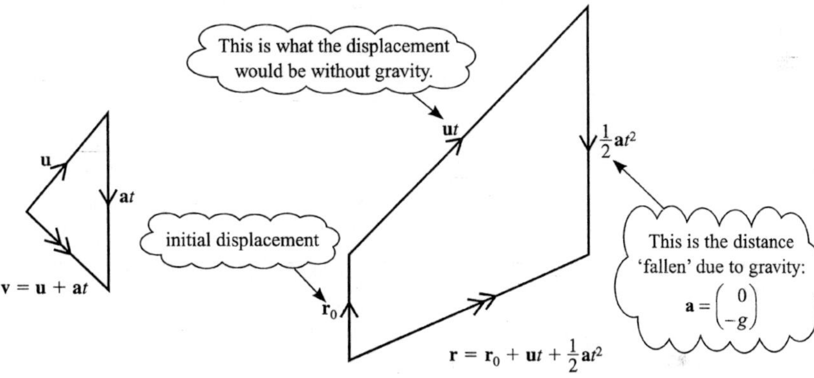

Figure 10.7

Always check whether or not the projectile starts at the origin. The change in position is the vector $\mathbf{r} - \mathbf{r}_0$. This is the equivalent of $s - s_0$ in one dimension.

EXERCISE 10B

In this exercise take upwards as positive.

1 In each of the following cases

(a) draw a diagram showing the initial velocity and path

(b) write the velocity after time t s in vector form

(c) write the position after time t s in vector form.

(i) Initial position (0, 10 m); initial velocity $4\,\mathrm{m\,s^{-1}}$ horizontally.

(ii) Initial position (0, 7 m); initial velocity $10\,\mathrm{m\,s^{-1}}$ at 35° above the horizontal.

(iii) Initial position (0, 20 m); initial velocity $10\,\mathrm{m\,s^{-1}}$ at 13° below the horizontal.

(iv) Initial position O; initial velocity $\begin{pmatrix} 7 \\ 24 \end{pmatrix}\mathrm{m\,s^{-1}}$.

(v) Initial position (a, b) m; initial velocity $\begin{pmatrix} u_0 \\ v_0 \end{pmatrix}\mathrm{m\,s^{-1}}$.

2 In each the following cases find

(a) the time taken for the projectile to reach its highest point

(b) the maximum height above the origin.

(i) Initial position (0, 15 m); velocity $5\,\text{m s}^{-1}$ horizontally and $14.7\,\text{m s}^{-1}$ vertically.

(ii) Initial position (0, 10 m); initial velocity $\begin{pmatrix}5\\3\end{pmatrix}\,\text{m s}^{-1}$.

3 Find the horizontal range for these projectiles which start from the origin.

(i) Initial velocity $\begin{pmatrix}2\\7\end{pmatrix}\,\text{m s}^{-1}$.

(ii) Initial velocity $\begin{pmatrix}7\\2\end{pmatrix}\,\text{m s}^{-1}$.

(iii) Sketch the paths of these two projectiles using the same axes.

Further examples

EXAMPLE 10.2

In this question neglect air resistance.

In an attempt to raise money for a charity, participants are sponsored to kick a ball over some vans. The vans are each 2 m high and 1.8 m wide and stand on horizontal ground. One participant kicks the ball at an initial speed of $22\,\text{m s}^{-1}$ inclined at 30° to the horizontal.

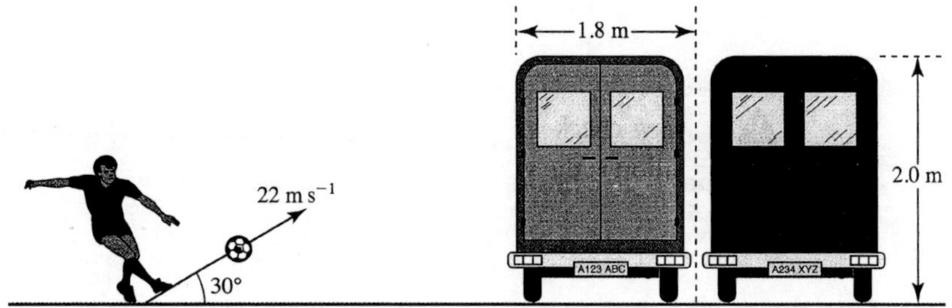

Figure 10.8

(i) What are the initial values of the vertical and horizontal components of velocity?

(ii) Show that while in flight the vertical height y metres at time t seconds satisfies the equation $y = 11t - 5t^2$.
Calculate at what times the ball is at least 2 m above the ground.

The ball should pass over as many vans as possible.

(ii) Deduce that the ball should be placed about 3.8 m from the first van and find how many vans the ball will clear.

(iv) What is the greatest vertical distance between the ball and the top of the vans?

[MEI]

SOLUTION

(i) *Initial velocity*

horizontally: $22\cos 30° = 19.05\,\mathrm{m\,s^{-1}}$

vertically: $22\sin 30° = 11\,\mathrm{m\,s^{-1}}$

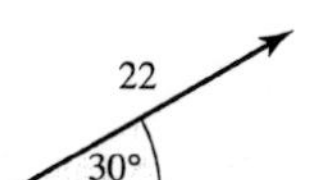

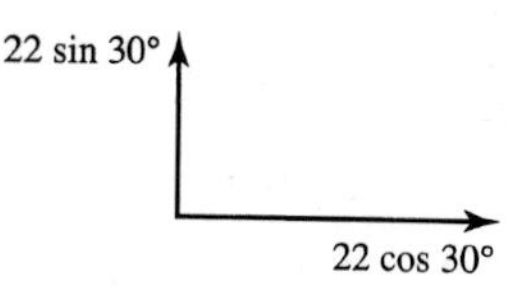

Figure 10.9

(ii) *When the ball is above 2 m*

Using axes as shown and $s = ut + \frac{1}{2}at^2$ vertically

$$\Rightarrow \quad y = 11t - 5t^2$$

The ball is 2 m above the ground when $y = 2$, then

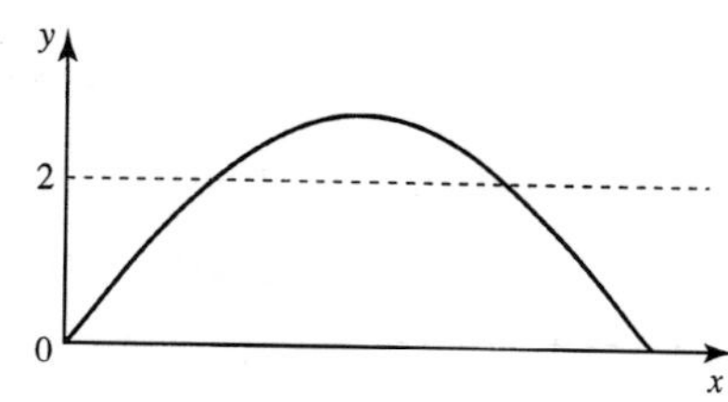

Figure 10.10

$$2 = 11t - 5t^2$$
$$5t^2 - 11t + 2 = 0$$
$$(5t - 1)(t - 2) = 0$$
$$t = 0.2 \text{ or } 2$$

$a = -10$ because the positive direction is upwards.

The ball is at least 2 m above the ground when $0.2 \leqslant t \leqslant 2$.

(iii) *How many vans?*

Horizontally, $s = ut + \frac{1}{2}at^2$ with $a = 0$

$$\Rightarrow \quad x = 19.05t$$

When $t = 0.2$, $\quad x = 3.81$ (at A)

when $t = 2$, $\quad x = 38.1$ (at B)

To clear as many vans as possible, the ball should be placed about 3.8 m in front of the first van.

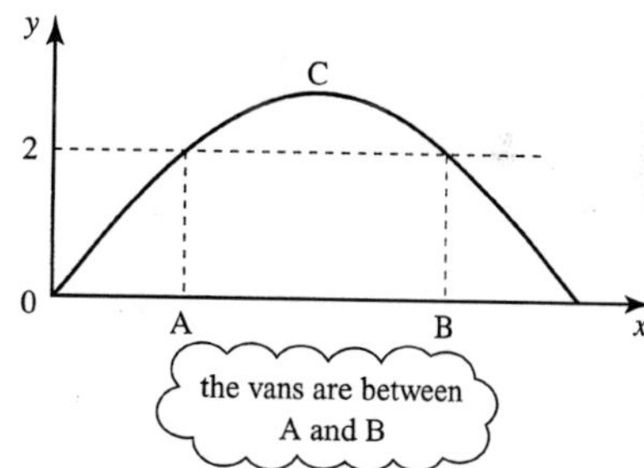

Figure 10.11

$$AB = 38.1 - 3.81\,\mathrm{m} = 34.29\,\mathrm{m}$$
$$\frac{34.29}{1.8} = 19.05$$

The maximum possible number of vans is 19.

(iv) *Maximum height*

At the top (C), vertical velocity = 0, so using $v = u + at$ vertically

$$\Rightarrow \quad 0 = 11 - 10t$$
$$t = 1.1$$

Substituting in $y = 11t - 5t^2$, maximum height is

$$11 \times 1.1 - 5 \times 1.1^2 = 6.05\,\mathrm{m}$$

The ball clears the tops of the vans by about 4 m.

EXAMPLE 10.3 *In this question use 9.8 m s⁻² for g.*

Sharon is diving into a swimming pool. During her flight she may be modelled as a particle. Her initial velocity is $1.8\,\text{m s}^{-1}$ at angle 30° above the horizontal and initial position 3.1 m above the water. Air resistance may be neglected.

(i) Find the greatest height above the water that Sharon reaches during her dive.

(ii) Show that the time t, in seconds, that it takes Sharon to reach the water is given by $4.9t^2 - 0.9t - 3.1 = 0$ and solve the equation to find t.
Explain the significance of the other root of the equation.

Just as Sharon is diving a small boy jumps into the swimming pool. He hits the water at a point in line with the diving board and 1.5 m from its end.

(iii) Is there an accident?

SOLUTION

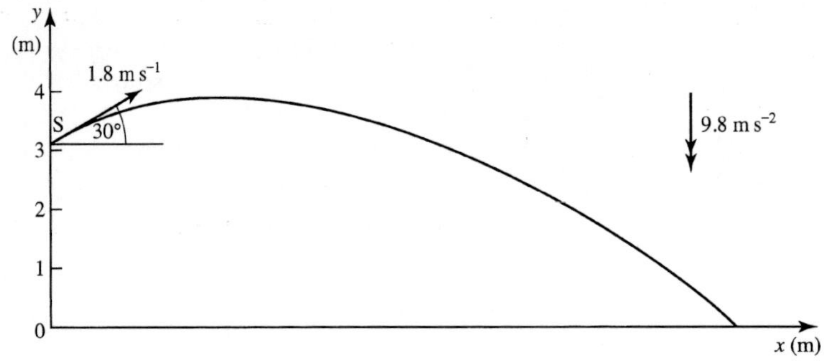

Figure 10.12

Referring to the axes shown:

	Horizontal motion		Vertical motion	
initial position	0		3.1	
a	0		−9.8	
u	$u_x = 1.8\cos 30° = 1.56$		$u_y = 1.8\sin 30° = 0.9$	
v	$v_x = 1.56$	①	$v_y = 0.9 - 9.8t$	②
r	$x = 1.56t$	③	$y = 3.1 + 0.9t - 4.9t^2$	④

(i) At the top $v_y = 0$ $0 = 0.9 - 9.8t \Rightarrow t = 0.092$ from ②

When $t = 0.092$ $y = 3.1 + 0.9 \times 0.092 - 4.9 \times 0.092^2 = 3.14$ from ④

Sharon's greatest height above the water is 3.14 m.

(ii) Sharon reaches the water when $y = 0$

$$0 = 3.1 + 0.9t - 4.9t^2 \qquad \text{from ④}$$

$$4.9t^2 - 0.9t - 3.1 = 0$$

$$t = \frac{0.9 \pm \sqrt{0.9^2 + 4 \times 4.9 \times 3.1}}{9.8}$$

$$t = -0.71 \text{ or } 0.89$$

Sharon hits the water after 0.89 s. The negative value of t gives the point on the parabola at water level to the left of the point (S) where Sharon dives.

(iii) At time t the horizontal distance from the diving board,

$$x = 1.56t \qquad \text{from ③}$$

When Sharon hits the water

$$x = 1.56 \times 0.89 = 1.39$$

Assuming that the particles representing Sharon and the boy are located at their centres of mass, the difference of 11 cm between 1.39 m and 1.5 m is not sufficient to prevent an accident.

Note

When the point S is taken as the origin in the above example, the initial position is (0, 0) and $y = 0.9t - 4.9t^2$. In this case, Sharon hits the water when $y = -3.1$. This gives the same equation for t.

EXAMPLE 10.4

A boy kicks a small ball from the floor of a gymnasium with an initial velocity of $12\,\text{m s}^{-1}$ inclined at an angle α to the horizontal. Air resistance may be neglected.

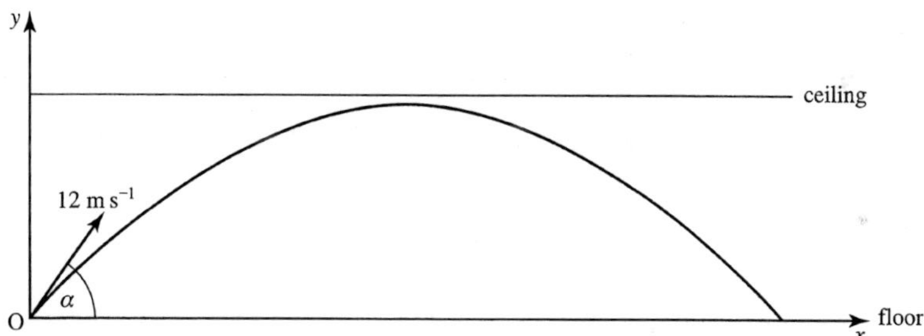

Figure 10.13

(i) Write down expressions in terms of α for the vertical speed of the ball and vertical height of the ball after t seconds.

The ball just fails to touch the ceiling which is 4 m high. The highest point of the motion of the ball is reached after T seconds.

(ii) Use one of your expressions to show that $6\sin\alpha = 5T$ and the other to form a second equation involving $\sin\alpha$ and T.

(iii) Eliminate $\sin\alpha$ from your two equations to show that T has a value of about 0.89.

(iv) Find the horizontal range of the ball when kicked at $12\,\text{m s}^{-1}$ from the floor of the gymnasium so that it just misses the ceiling.

[MEI]

SOLUTION

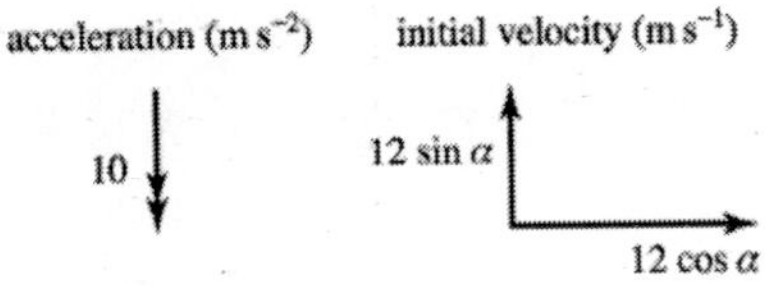

acceleration ($m\,s^{-2}$)　initial velocity ($m\,s^{-1}$)

10　　12 sin α　　12 cos α

Figure 10.14

(i) *Vertical components*

speed $v_y = 12\sin\alpha - 10t$ ①

height $y = (12\sin\alpha)t - 5t^2$ ②

(ii) *Time to highest point*

At the top $v_y = 0$ and $t = T$, so equation ① gives

$$12\sin\alpha - 10T = 0$$
$$12\sin\alpha = 10T$$
$$6\sin\alpha = 5T \qquad ③$$

When $t = T$, $y = 4$, so from ②

$$4 = (12\sin\alpha)T - 5T^2 \qquad ④$$

(iii) Substituting for $6\sin\alpha$ from ③ into ④ gives

$$4 = 2 \times 5T \times T - 5T^2$$
$$4 = 5T^2$$
$$T = \sqrt{0.8} = 0.89 \text{ (to 2 d.p.)}$$

(iv) *Range*

The path is symmetrical so the time of flight is $2T$ seconds.

Horizontally $a = 0$ and $u_x = 12\cos\alpha$

$$\Rightarrow \quad x = (12\cos\alpha)t$$

The range is $12\cos\alpha \times 2T = 21.47\cos\alpha$ m.

From ③ $\quad 6\sin\alpha = 5T = 4.47$

$$\alpha = 48.19°$$

The range is $21.47\cos 48.19° = 14.3$ m (to 3 s.f.).

? Two marbles start simultaneously from the same height. One (P) is dropped and the other (Q) is projected horizontally. Which reaches the ground first?

EXERCISE 10C

1 A ball is thrown from a point at ground level with velocity $20\,m\,s^{-1}$ at 30° to the horizontal. The ground is level and horizontal and you should ignore air resistance.

(i) Find the horizontal and vertical components of the ball's initial velocity.

(ii) Find the horizontal and vertical components of the ball's acceleration.

(iii) Find the horizontal distance travelled by the ball before its first bounce.

(iv) Find how long the ball takes to reach maximum height.

(v) Find the maximum height reached by the ball.

2 In this question use $9.8\,\text{m}\,\text{s}^{-2}$ for g.

Nick hits a golf ball with initial velocity $50\,\text{m}\,\text{s}^{-1}$ at 35° to the horizontal.

(i) Find the horizontal and vertical components of the ball's initial velocity.

(ii) Specify suitable axes and calculate the position of the ball at one second intervals for the first six seconds of its flight.

(iii) Draw a graph of the path of the ball (its trajectory) and use it to estimate

(a) the maximum height of the ball

(b) the horizontal distance the ball travels before bouncing.

(iv) Calculate the maximum height the ball reaches and the horizontal distance it travels before bouncing. Compare your answers with the estimates you found from your graph.

(v) State the modelling assumptions you made in answering this question.

3 Clare scoops a hockey ball off the ground, giving it an initial velocity of $19\,\text{m}\,\text{s}^{-1}$ at 25° to the horizontal.

(i) Find the horizontal and vertical components of the ball's initial velocity.

(ii) Find the time that elapses before the ball hits the ground.

(iii) Find the horizontal distance the ball travels before hitting the ground.

(iv) Find how long it takes for the ball to reach maximum height.

(v) Find the maximum height reached.

(vi) A member of the opposing team is standing 20 m away from Clare in the direction of the ball's flight. How high is the ball when it passes her? Can she stop the ball?

4 A footballer is standing 30 m in front of the goal. He kicks the ball towards the goal with velocity $18\,\text{m}\,\text{s}^{-1}$ and angle 55° to the horizontal. The height of the goal's crossbar is 2.5 m. Air resistance and spin may be ignored.

(i) Find the horizontal and vertical components of the ball's initial velocity.

(ii) Find the time it takes for the ball to cross the goal-line.

(iii) Does the ball bounce in front of the goal, go straight into the goal or go over the crossbar?

(iv) In fact the goalkeeper is standing 5 m in front of the goal and will stop the ball if its height is less than 2.8 m when it reaches him. Does the goalkeeper stop the ball?

5 A plane is flying at a speed of $300\,\text{m}\,\text{s}^{-1}$ and maintaining an altitude of 10 000 m when a bolt becomes detached. Ignoring air resistance, find

(i) the time that the bolt takes to reach the ground

(ii) the horizontal distance between the point where the bolt leaves the plane and the point where it hits the ground

(iii) the speed of the bolt when it hits the ground

(iv) the angle to the horizontal at which the bolt hits the ground.

6 Reena is learning to serve in tennis. She hits the ball from a height of 2 m. For her serve to be legal it must pass over the net which is 12 m away from her and 0.91 m high, and it must land within 6.4 m of the net. Make the following modelling assumptions to answer the questions.

- She hits the ball horizontally.
- Air resistance may be ignored.
- The ball may be treated as a particle.
- The ball does not spin.
- She hits the ball straight down the middle of the court.

(i) How long does the ball take to fall to the level of the top of the net?

(ii) How long does the ball take from being hit to first reaching the ground?

(iii) What is the lowest speed with which Reena must hit the ball to clear the net?

(iv) What is the greatest speed with which she may hit it if it is to land within 6.4 m of the net?

7 A stunt motorcycle rider attempts to jump over a gorge 50 m wide. He uses a ramp at 25° to the horizontal for his take-off and has a speed of $30\,\mathrm{m\,s^{-1}}$ at this time.

(i) Assuming that air resistance is negligible, find out whether the rider crosses the gorge successfully.

The stunt man actually believes that in any jump the effect of air resistance is to reduce his distance by 40%.

(ii) Calculate his minimum safe take-off speed for this jump.

8 A catapult projects a small pellet at speed $20\,\mathrm{m\,s^{-1}}$ and can be directed at any angle to the horizontal.

(i) Find the range of the catapult when the angle of projection is

(a) 30° **(b)** 40° **(c)** 45° **(d)** 50° **(e)** 60°.

(ii) Show algebraically that the range is the same when the angle of projection is α as it is when the angle is $90° - \alpha$.

The catapult is angled with the intention that the pellet should hit a point on the ground 36 m away.

(iii) Verify that one appropriate angle of projection would be 32.1° and write down another suitable angle.

In fact the angle of projection from the catapult is liable to error.

(iv) Find the distance by which the pellet misses the target in each of the cases in **(iii)** when the angle of projection is subject to an error of +0.5°. Which angle should you use for greater accuracy?

9 A cricketer hits the ball on the half-volley, that is when the ball is at ground level. The ball leaves the ground at an angle of 30° to the horizontal and travels towards a fielder standing on the boundary 60 m away.

(i) Find the initial speed of the ball if it hits the ground for the first time at the fielder's feet.

(ii) Find the initial speed of the ball if it is at a height of 3.2 m (well outside the fielder's reach) when it passes over the fielder's head.

In fact the fielder is able to catch the ball without moving provided that its height, h m, when it reaches him satisfies the inequality $0.25 \leqslant h \leqslant 2.1$.

(iii) Find a corresponding range of values for u, the initial speed of the ball.

10 A horizontal tunnel has a height of 3 m. A ball is thrown inside the tunnel with an initial speed of $18\,\mathrm{m\,s^{-1}}$. What is the greatest horizontal distance that the ball can travel before it bounces for the first time?

11 Use $g = 9.8\,\mathrm{m\,s^{-2}}$ in this question.

The picture shows Romeo trying to attract Juliet's attention without her nurse, who is in a downstairs room, noticing. He stands 10 m from the house and lobs a small pebble at her bedroom window. Romeo throws the pebble from a height of 1 m with a speed of $11.5\,\mathrm{m\,s^{-1}}$ at an angle of 60° to the horizontal.

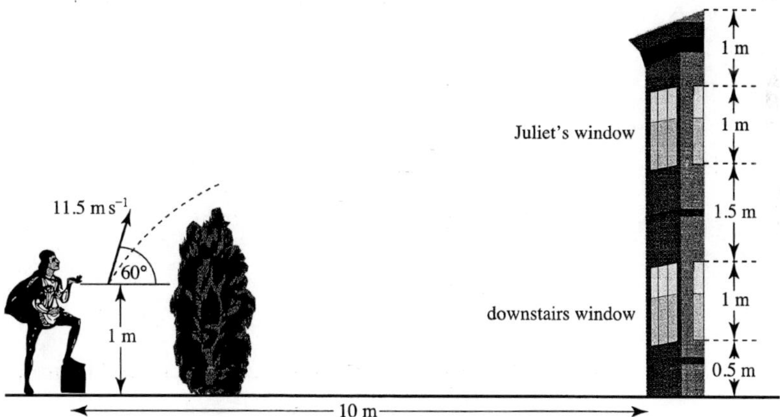

(i) How long does the pebble take to reach the house?

(ii) Does the pebble hit Juliet's window, the wall of the house or the downstairs room window?

(iii) What is the speed of the pebble when it hits the house? [MEI]

12 A firework is buried so that its top is at ground level and it projects sparks all at a speed of $8\,\text{m s}^{-1}$. Air resistance may be neglected.

(i) Calculate the height reached by a spark projected vertically and explain why no spark can reach a height greater than this.

(ii) For a spark projected at 30° to the horizontal over horizontal ground, show that its height in metres t seconds after projection is $4t - 5t^2$ and hence calculate the distance it lands from the firework.

(iii) For what angle of projection will a spark reach a maximum height of 2 m?

[MEI]

13 A stone is projected from a point O on horizontal ground with speed $V\,\text{m s}^{-1}$ at an angle θ above the horizontal, where $\sin\theta = \frac{3}{5}$. The stone is at its highest point when it has travelled a horizontal distance of 19.2 m.

(i) Find the value of V.

After passing through its highest point the stone strikes a vertical wall at a point 4 m above the ground.

(ii) Find the horizontal distance between O and the wall.

At the instant when the stone hits the wall the horizontal component of the stone's velocity is halved in magnitude and reversed in direction. The vertical component of the stone's velocity does not change as a result of the stone hitting the wall.

(iii) Find the distance from the wall of the point where the stone reaches the ground.

[Cambridge AS and A Level Mathematics 9709, Paper 5 Q7 June 2006]

14 A particle A is released from rest at time $t = 0$, at a point P which is 7 m above horizontal ground. At the same instant as A is released, a particle B is projected from a point O on the ground. The horizontal distance of O from P is 24 m. Particle B moves in the vertical plane containing O and P, with initial speed $V\,\text{m s}^{-1}$ and initial direction making an angle of θ above the horizontal (see diagram).

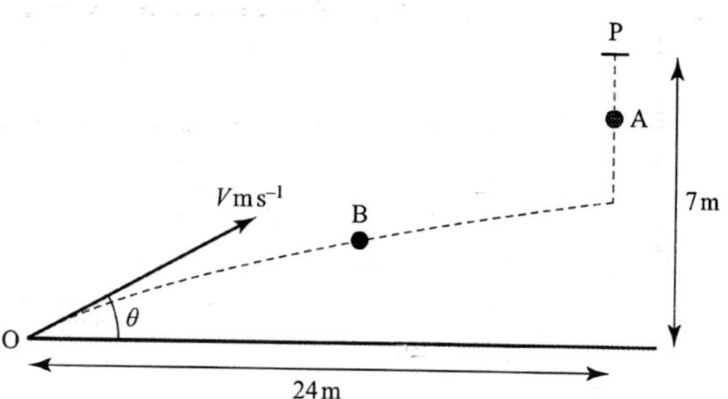

Write down

(i) an expression for the height of A above the ground at time t s,

(ii) an expression in terms of V, θ and t for

(a) the horizontal distance of B from O,

(b) the height of B above the ground.

At time $t = T$ the particles A and B collide at a point above the ground.

(iii) Show that $\tan\theta = \frac{7}{24}$ and that $VT = 25$.

(iv) Deduce that $7V^2 > 3125$.

[Cambridge AS and A Level Mathematics 9709, Paper 5 Q7 June 2005]

15 A particle P is released from rest at a point A which is 7 m above horizontal ground. At the same instant that P is released a particle Q is projected from a point O on the ground. The horizontal distance of O from A is 24 m. Particle Q moves in the vertical plane containing O and A, with initial speed $50\,\text{m s}^{-1}$ and initial direction making an angle θ above the horizontal, where $\tan\theta = \frac{7}{24}$ (see diagram). Show that the particles collide.

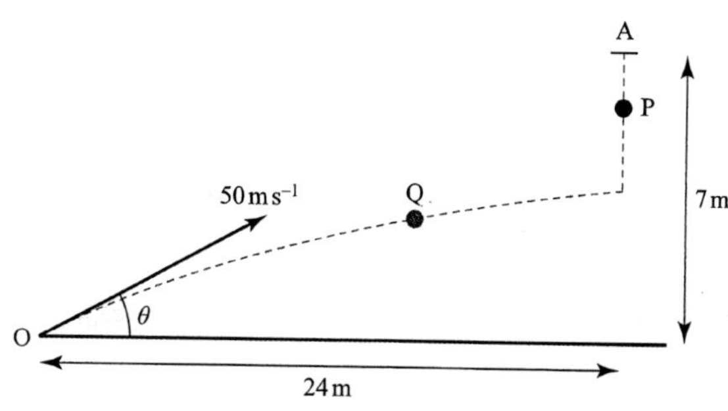

[Cambridge AS and A Level Mathematics 9709, Paper 52 Q3 November 2009]

The path of a projectile

Look at the equations

$$x = 20t$$
$$y = 6 + 30t - 5t^2$$

They represent the path of a projectile.

? What is the initial velocity of the projectile? What is its initial position? What value of g is assumed?

These equations give x and y in terms of a third variable t. (They are called *parametric equations* and t is the *parameter.*)

You can find the *cartesian equation* connecting x and y directly by eliminating t as follows:

$$x = 20t \Rightarrow t = \frac{x}{20}$$

so $\quad y = 6 + 30t - 5t^2$

can be written as $\quad y = 6 + 30 \times \frac{x}{20} - 5 \times \left(\frac{x}{20}\right)^2$

$$y = 6 + 1.5x - \frac{x^2}{80}$$ ← This is the cartesian equation.

EXERCISE 10D

1 Find the cartesian equation of the path of these projectiles by eliminating the parameter t.

(i) $x = 4t \qquad y = 5t^2$

(ii) $x = 5t \qquad y = 6 + 2t - 5t^2$

(iii) $x = 2 - t \qquad y = 3t - 5t^2$

(iv) $x = 1 + 5t \qquad y = 8 + 10t - 5t^2$

(v) $x = ut \qquad y = 2ut - \frac{1}{2}gt^2$

2 A particle is projected with initial velocity $50\,\text{m}\,\text{s}^{-1}$ at an angle of $36.9°$ to the horizontal. The point of projection is taken to be the origin, with the x axis horizontal and the y axis vertical in the plane of the particle's motion.

(i) Show that at time t s, the height of the particle in metres is given by

$$y = 30t - 5t^2$$

and write down the corresponding expression for x.

(ii) Eliminate t between your equations for x and y to show that

$$y = \frac{3x}{4} - \frac{x^2}{320}.$$

(iii) Plot the graph of y against x.

(iv) Mark on your graph the points corresponding to the position of the particle after 1, 2, 3, 4, ... seconds.

3 A golfer hits a ball with initial velocity $50\,\text{m}\,\text{s}^{-1}$ at an angle α to the horizontal where $\sin\alpha = 0.6$.

(i) Find the equation of its trajectory, assuming that air resistance may be neglected. The flight of the ball is recorded on film and its position vector, from the point where it was hit, is calculated. The unit vectors **i** and **j** are horizontal and vertical in the plane of the ball's motion. The results (to the nearest 0.5 m) are as shown in the table.

Time (s)	0	1	2	3	4	5	6
Position (m)	$\begin{pmatrix}0\\0\end{pmatrix}$	$\begin{pmatrix}39.5\\24.5\end{pmatrix}$	$\begin{pmatrix}78\\39\end{pmatrix}$	$\begin{pmatrix}116.5\\44\end{pmatrix}$	$\begin{pmatrix}152\\39\end{pmatrix}$	$\begin{pmatrix}187.5\\24.5\end{pmatrix}$	$\begin{pmatrix}222\\0\end{pmatrix}$

(ii) On the same piece of graph paper draw the trajectory you found in part **(i)** and that found from analysing the film. Compare the two graphs and suggest a reason for any differences.

(iii) It is suggested that the horizontal component of the resistance to the motion of the golf ball is almost constant. Are the figures consistent with this?

General equations

The work done in this chapter can now be repeated for the general case using algebra. Assume a particle is projected from the origin with speed u at an angle α to the horizontal and that the only force acting on the particle is the force due to gravity. The x and y axes are horizontal and vertical through the origin, O, in the plane of motion of the particle.

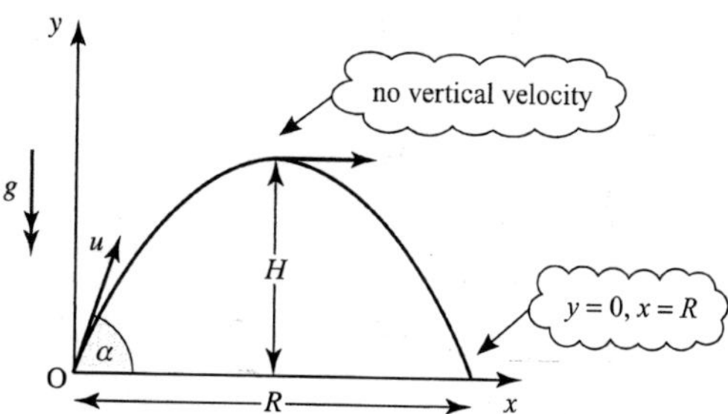

Figure 10.15

The components of velocity and position

	Horizontal motion		Vertical motion	
Initial position	0		0	
a	0		$-g$	
u	$u_x = u\cos\alpha$		$u_y = u\sin\alpha$	
v	$v_x = u\cos\alpha$	①	$v_y = u\sin\alpha - gt$	②
r	$x = ut\cos\alpha$	③	$y = ut\sin\alpha - \frac{1}{2}gt^2$	④

$ut\cos\alpha$ is preferable to $u\cos\alpha t$ because this could mean $u\cos(\alpha t)$ which is incorrect.

The maximum height

At its greatest height, the vertical component of velocity is zero.

From equation ②

$$u \sin\alpha - gt = 0$$

$$t = \frac{u\sin\alpha}{g}$$

Substitute in equation ④ to obtain the height of the projectile:

$$y = u \times \frac{u\sin\alpha}{g} \times \sin\alpha - \tfrac{1}{2}g \times \frac{(u\sin\alpha)^2}{g^2}$$

$$= \frac{u^2\sin^2\alpha}{g} - \frac{u^2\sin^2\alpha}{2g}$$

The greatest height is

$$H = \frac{u^2\sin^2\alpha}{2g}$$

The time of flight

When the projectile hits the ground, $y = 0$.

From equation ④ $\quad y = ut\sin\alpha - \tfrac{1}{2}gt^2$

$$0 = ut\sin\alpha - \tfrac{1}{2}gt^2$$

$$0 = t(u\sin\alpha - \tfrac{1}{2}gt)$$

The solution $t = 0$ is at the start of the motion.

The time of flight is $\quad t = \dfrac{2u\sin\alpha}{g}$

The range

The range of the projectile is the value of x when $t = \dfrac{2u\sin\alpha}{g}$

From equation ④: $\quad x = ut\cos\alpha$

$$\Rightarrow \quad R = u \times \frac{2u\sin\alpha}{g} \times \cos\alpha$$

$$R = \frac{2u^2\sin\alpha\cos\alpha}{g}$$

It can be shown that $2\sin\alpha\cos\alpha = \sin 2\alpha$, so the range can be expressed as

$$R = \frac{u^2\sin 2\alpha}{g}$$

The range is a maximum when $\sin 2\alpha = 1$, that is when $2\alpha = 90°$ or $\alpha = 45°$. The maximum possible horizontal range for projectiles with initial speed u is

$$R_{max} = \frac{u^2}{g}.$$

The equation of the path

From equation ③ $t = \dfrac{x}{u\cos\alpha}$

Substitute into equation ④ to give

$$y = u \times \frac{x}{u\cos\alpha} \times \sin\alpha - \tfrac{1}{2}g \times \frac{x^2}{(u\cos\alpha)^2}$$

$$y = x\frac{\sin\alpha}{\cos\alpha} - \frac{gx^2}{2u^2\cos^2\alpha}$$

So the equation of the trajectory is

$$y = x\tan\alpha - \frac{gx^2}{2u^2\cos^2\alpha}.$$

In *Pure Mathematics 2* you learn that $\dfrac{1}{\cos^2\alpha} = 1 + \tan^2\alpha$ so

$$y = x\tan\alpha - \frac{gx^2}{2u^2}(1 + \tan^2\alpha)$$

It is important that you understand the methods used to derive these formulae and don't rely on learning the results by heart. They are only true when the given assumptions apply and the variables are as defined in figure 10.15.

What are the assumptions on which this work is based?

EXERCISE 10E

In this exercise use the modelling assumptions that air resistance can be ignored and the ground is horizontal.

1 A projectile is launched from the origin with an initial velocity $30\,\mathrm{m\,s^{-1}}$ at an angle of 45° to the horizontal.

(i) Write down the position of the projectile after time t.

(ii) Show that the equation of the path is the parabola $y = x - 0.011x^2$.

(iii) Find y when $x = 10$.

(iv) Find x when $y = 20$.

2 Jack throws a cricket ball at a wicket 0.7 m high with velocity $10\,\mathrm{m\,s^{-1}}$ at 14° above the horizontal. The ball leaves his hand 1.5 m above the origin.

(i) Show that the equation of the path is the parabola $y = 1.5 + 0.25x - 0.053x^2$.

(ii) How far from the wicket is he standing if the ball just hits the top?

3 In this question, take $g = 9.8\,\text{m}\,\text{s}^{-2}$.

While practising his tennis serve, Matthew hits the ball from a height of 2.5 m with a velocity of magnitude $25\,\text{m}\,\text{s}^{-1}$ at an angle of 5° above the horizontal as shown in the diagram.

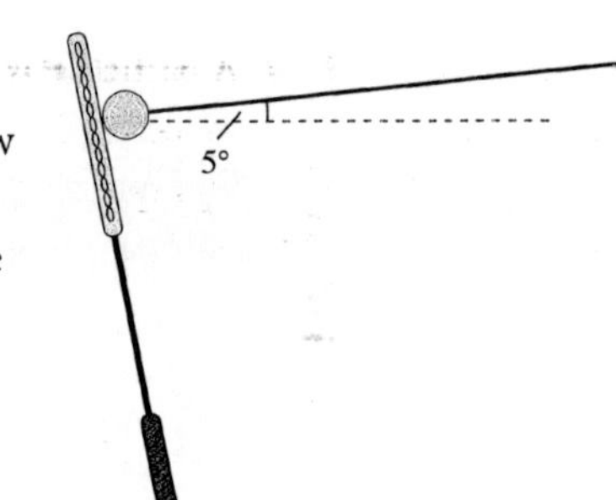

(i) Show that while in flight
$y = 2.5 + 0.087x - 0.0079x^2$.

(ii) Find the horizontal distance from the serving point to the spot where the ball lands.

(iii) Determine whether the ball would clear the net, which is 1 m high and 12 m from the serving position in the horizontal direction.

4 Ching is playing volleyball. She hits the ball with initial speed $u\,\text{m}\,\text{s}^{-1}$ from a height of 1m at an angle of 35° to the horizontal.

(i) Define a suitable origin and x and y axes and find the equation of the trajectory of the ball in terms of x, y and u.

The rules of the game require the ball to pass over the net, which is at height 2 m, and land inside the court on the other side, which is of length 5 m. Ching hits the ball straight along the court and is 3 m from the net when she does so.

(ii) Find the minimum value of u for the ball to pass over the net.

(iii) Find the maximum value of u for the ball to land inside the court.

5 A particle is projected from horizontal ground with speed $u\,\text{m}\,\text{s}^{-1}$ at an angle of $\theta°$ above the horizontal. The greatest height reached by the particle is 10 m and the particle hits the ground at a distance of 40 m from the point of projection. In either order,

(i) find the values of u and θ,

(ii) find the equation of the trajectory, in the form $y = ax - bx^2$, where xm and ym are the horizontal and vertical displacements of the particle from the point of projection.

[Cambridge AS and A Level Mathematics 9709, Paper 5 Q4 November 2005]

6 A particle is projected from a point O at an angle of 35° above the horizontal. At time Ts later the particle passes through a point A whose horizontal and vertically upward displacements from O are 8 m and 3 m respectively.

(i) By using the equation of the particle's trajectory, or otherwise, find (in either order) the speed of projection of the particle from O and the value of T.

(ii) Find the angle between the direction of motion of the particle at A and the horizontal.

[Cambridge AS and A Level Mathematics 9709, Paper 5 Q6 November 2007]

7 A particle P is projected from a point O on horizontal ground with speed $V \text{ m s}^{-1}$ and direction 60° upwards from the horizontal. At time t s later the horizontal and vertical displacements of P from O are x m and y m respectively.

(i) Write down expressions for x and y in terms of V and t and hence show that the equation of the trajectory of P is

$$y = (\sqrt{3})x - \frac{20x^2}{V^2}.$$

P passes through the point A at which $x = 70$ and $y = 10$. Find

(ii) the value of V,

(iii) the direction of motion of P at the instant it passes through A.

[Cambridge AS and A Level Mathematics 9709, Paper 5 Q7 November 2008]

8 A particle is projected from a point O on horizontal ground. The velocity of projection has magnitude 20 m s^{-1} and direction upwards at an angle θ to the horizontal. The particle passes through the point which is 7 m above the ground and 16 m horizontally from O, and hits the ground at the point A.

(i) Using the equation of the particle's trajectory and the identity $\sec^2\theta = 1 + \tan^2\theta$, show that the possible values of $\tan\theta$ are $\frac{3}{4}$ and $\frac{17}{4}$.

(ii) Find the distance OA for each of the two possible values of $\tan\theta$.

(iii) Sketch in the same diagram the two possible trajectories.

[Cambridge AS and A Level Mathematics 9709, Paper 51 Q5 June 2010]

EXPERIMENT

The diagram shows how a wet ball projected on to a sloping table can be used to simulate a projectile.

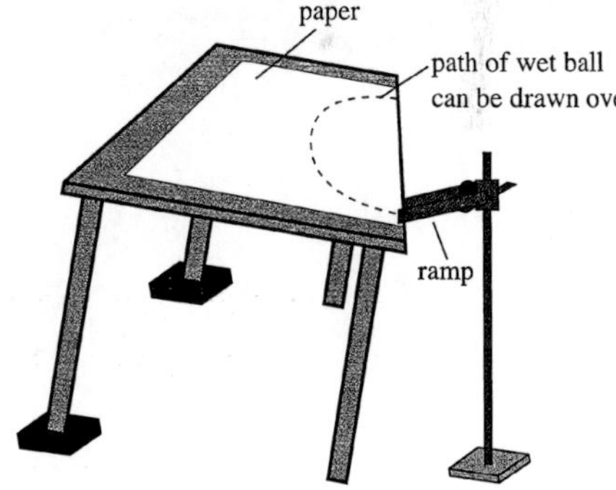

Figure 10.16

1 Ignoring rotation and friction, what is the ball's acceleration?

2 Does the mass of the ball affect the motion?

Set up the apparatus so that you can move the ramp to make different angles of projection with the same speed.

3 Can the same range be achieved using two different angles?

4 What angle gives the maximum range?

5 What is the shape of the curve containing all possible paths with the same initial speed?

INVESTIGATION

Fireworks

A firework sends out sparks from ground level with the same speed, 20 m s^{-1}, in all directions. A spark starts at an angle α to the horizontal. Investigate the accessible points for this speed by plotting the trajectory for different values of α.

Using a graphic calculator or other graph plotter investigate the shape of the curve which forms the outer limit for all possible sparks with trajectories which lie in a vertical plane.

Show that the trajectory of a spark is given by $y = x\tan\alpha - \frac{1}{80}x^2(1 + \tan^2\alpha)$.

KEY POINTS

1 Modelling assumptions for projectile motion with acceleration due to gravity:

- a projectile is a particle
- it is not powered
- the air has no effect on its motion.

2 Projectile motion is usually considered in terms of horizontal and vertical components.

When the initial position is O

Angle is projection $= \alpha$

Initial velocity, $\mathbf{u} = \begin{pmatrix} u\cos\alpha \\ u\sin\alpha \end{pmatrix}$

Acceleration, $\mathbf{g} = \begin{pmatrix} 0 \\ -g \end{pmatrix}$

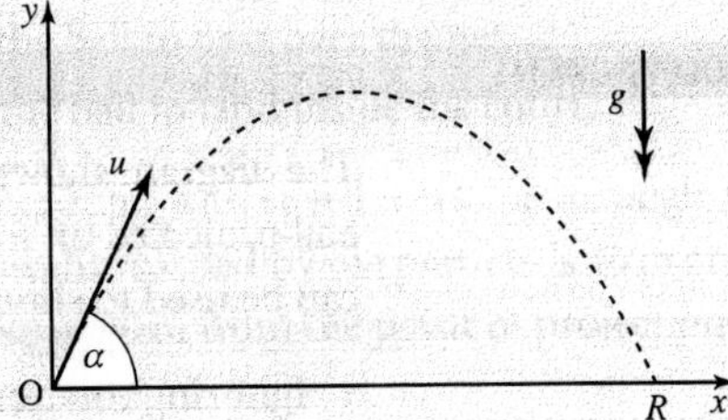

- At time t, velocity, $\mathbf{v} = \mathbf{u} + \mathbf{a}t$ $\qquad \begin{pmatrix} v_x \\ v_y \end{pmatrix} = \begin{pmatrix} u\cos\alpha \\ u\sin\alpha \end{pmatrix} + \begin{pmatrix} 0 \\ -g \end{pmatrix}t$

$$v_x = u\cos\alpha \qquad ①$$

$$v_y = u\sin\alpha - gt \qquad ②$$

- Displacement, $\mathbf{r} = \mathbf{u}t + \frac{1}{2}\mathbf{a}t^2$ $\qquad \begin{pmatrix} x \\ y \end{pmatrix} = \begin{pmatrix} u\cos\alpha \\ u\sin\alpha \end{pmatrix}t + \frac{1}{2}\begin{pmatrix} 0 \\ -g \end{pmatrix}t^2$

$$x = ut\cos\alpha \qquad ③$$

$$y = ut\sin\alpha - \tfrac{1}{2}gt^2$$

3 At a maximum height $v_y = 0$.

4 $y = 0$ when the projectile lands.

5 The time to hit the ground is twice the time to maximum height.

6 When the point of projection is (x_0, y_0) rather than $(0, 0)$

$$\mathbf{r} = \mathbf{r}_0 + \mathbf{u}t + \frac{1}{2}\mathbf{a}t^2 \qquad \begin{pmatrix} x \\ y \end{pmatrix} = \begin{pmatrix} x_0 \\ y_0 \end{pmatrix} + \begin{pmatrix} u\cos\alpha \\ u\sin\alpha \end{pmatrix} t + \frac{1}{2}\begin{pmatrix} 0 \\ -g \end{pmatrix} t^2$$

7 The equation of the trajectory of a projectile is

$$y = x\tan\alpha - \frac{gx^2}{2u^2\cos^2\alpha}$$

11 Moments of forces

Give me a firm place to stand and I will move the earth.

Archimedes

Figure 11.1

The illustration shows a swing bridge over a canal. It can be raised to allow barges and boats to pass. It is operated by hand, even though it is very heavy. How is this possible?

The bridge depends on the turning effects or *moments* of forces. To understand these you might find it helpful to look at a simpler situation.

Two children sit on a simple see-saw, made of a plank balanced on a fulcrum as in figure 11.2. Will the see-saw balance?

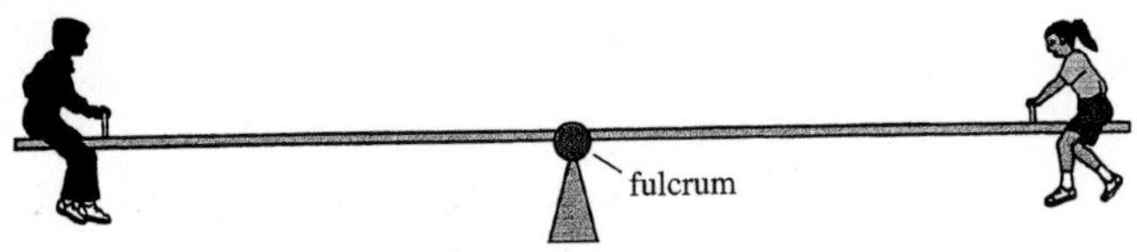

Figure 11.2

If both children have the same mass and sit the same distance from the fulcrum, then you expect the see-saw to balance.

Now consider possible changes to this situation:

(i) If one child is heavier than the other then you expect the heavier one to go down;

(ii) If one child moves nearer the centre you expect that child to go up.

You can see that both the weights of the children and their distances from the fulcrum are important.

What about this case? One child has mass 35 kg and sits 1.6 m from the fulcrum and the other has mass 40 kg and sits on the opposite side 1.4 m from the fulcrum (see figure 11.3).

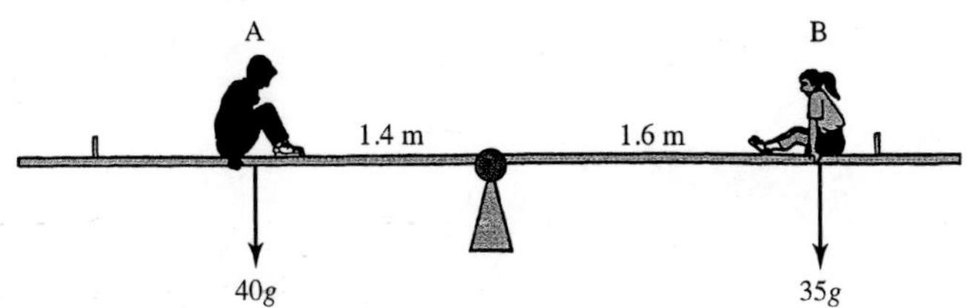

Figure 11.3

Taking the products of their weights and their distances from the fulcrum, gives

A: $40g \times 1.4 = 56g$

B: $35g \times 1.6 = 56g$

So you might expect the see-saw to balance and this indeed is what would happen.

Rigid bodies

Until now the particle model has provided a reasonable basis for the analysis of the situations you have met. In examples like the see-saw however, where turning is important, this model is inadequate because the forces do not all act through the same point.

In such cases you need the *rigid body model* in which an object, or *body*, is recognised as having size and shape, but is assumed not to be deformed when forces act on it.

Suppose that you push a tray lying on a smooth table with one finger so that the force acts parallel to one edge and through the centre of mass (figure 11.4).

Figure 11.4

The particle model is adequate here: the tray travels in a straight line in the direction of the applied force.

If you push the tray equally hard with two fingers as in figure 11.5, symmetrically either side of the centre of mass, the particle model is still adequate.

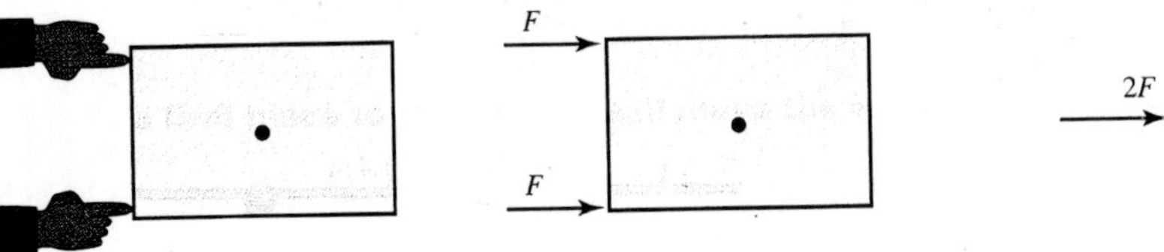

Figure 11.5

However, if the two forces are not equal or are not symmetrically placed, or as in figure 11.6 are in different directions, the particle model cannot be used.

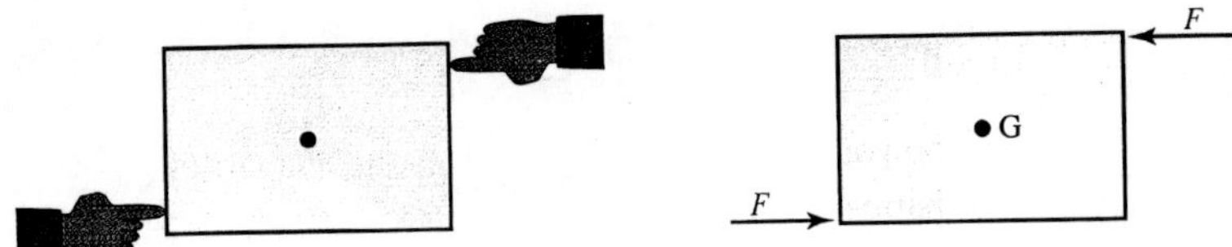

Figure 11.6

The resultant force is now zero, since the individual forces are equal in magnitude but opposite in direction. What happens to the tray? Experience tells us that it starts to rotate about G. How fast it starts to rotate depends, among other things, on the magnitude of the forces and the width of the tray. The rigid body model allows you to analyse the situation.

Moments

In the example of the see-saw we looked at the product of each force and its distance from a fixed point. This product is called the *moment* of the force about the point.

The see-saw balances because the moments of the forces on either side of the fulcrum are the same magnitude and in opposite directions. One would tend to make the see-saw turn clockwise, the other anticlockwise. By contrast, the moments about G of the forces on the tray in the last situation do not balance. They both tend to turn it anticlockwise, so rotation occurs.

Conventions and units

The moment of a force F about a point O is defined by

$$\text{moment} = Fd$$

where d is the perpendicular distance from the point O to the line of action of the force (figure 11.7).

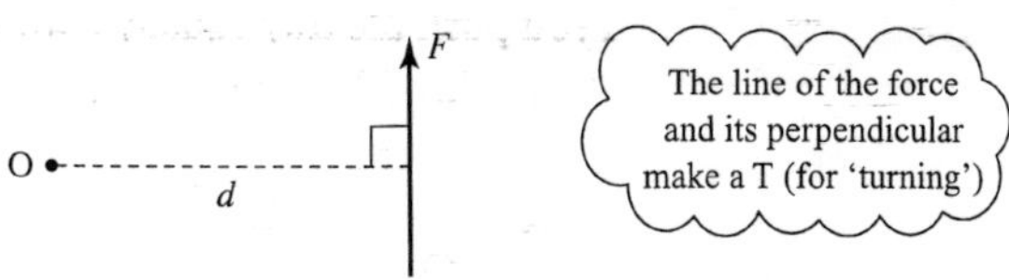

Figure 11.7

In two dimensions, the sense of a moment is described as either positive (anticlockwise) or negative (clockwise) as shown in figure 11.8.

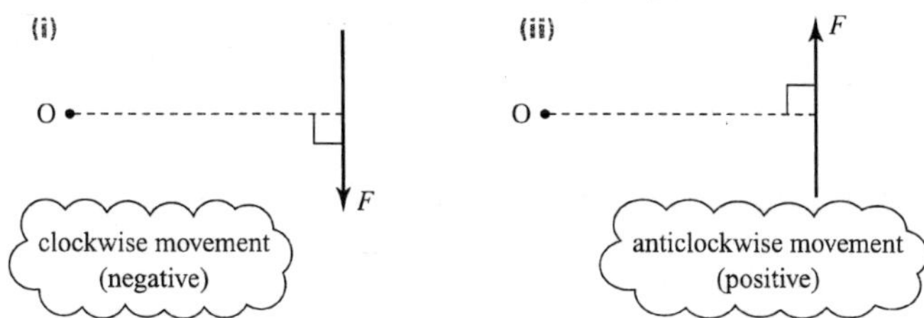

Figure 11.8

If you imagine putting a pin at O and pushing along the line of *F*, your page would turn clockwise for (i) and anticlockwise for (ii).

In the S.I. system the unit for moment is the newton metre (Nm), because a moment is the product of a force, the unit of which is the newton, and distance, the unit of which is the metre.

Remember that moments are always taken about a point and you must always specify what that point is. A force acting through the point will have no moment about that point because in that case $d = 0$.

? Figure 11.9 shows two tools for undoing wheel nuts on a car. Discuss the advantages and disadvantages of each.

(i) **(ii)**

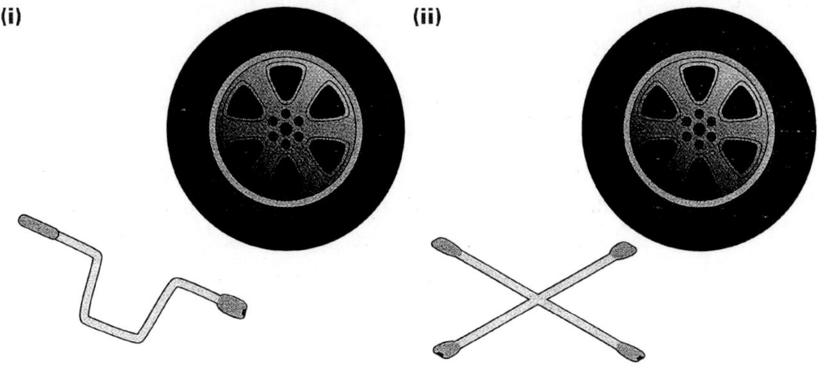

Figure 11.9

When using the spider wrench (the tool with two 'arms'), you apply equal and opposite forces either side of the nut. These produce moments in the same direction. One advantage of this method is that there is no resultant force and hence no tendency for the nut to snap off.

Couples

Whenever two forces of the same magnitude act in opposite directions along different lines, they have a zero resultant force, but do have a turning effect. In fact the moment will be Fd about any point, where d is the perpendicular distance between the forces. This is demonstrated in figure 11.10.

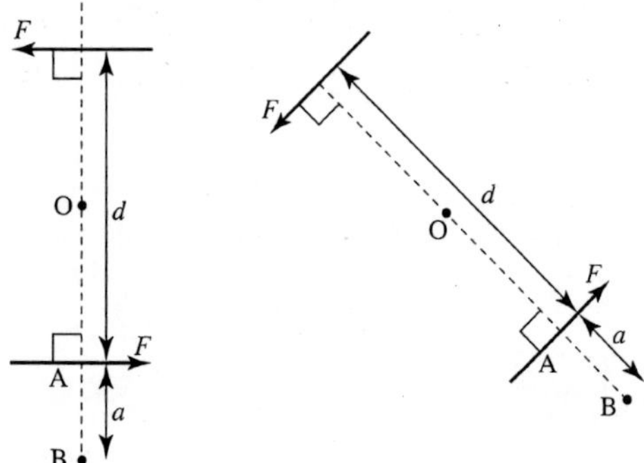

Figure 11.10

In each of these situations:

Moment about O $\quad F\frac{d}{2} + F\frac{d}{2} = Fd$ ← anticlockwise is positive

Moment about A $\quad 0 + Fd = Fd$

Moment about B $\quad -aF + (a + d)\,F = Fd$

Any set of forces like these with a zero resultant but a non-zero total moment is known as a *couple*. The effect of a couple on a rigid body is to cause rotation.

Equilibrium revisited

In Chapter 3 we said that an object is in equilibrium if the resultant force on the object is zero. This definition is adequate provided all the forces act through the same point on the object. However, we are now concerned with forces acting at different points, and in this situation even if the forces balance there may be a resultant moment.

Figure 11.11 shows a tray on a smooth surface being pushed equally hard at opposite corners.

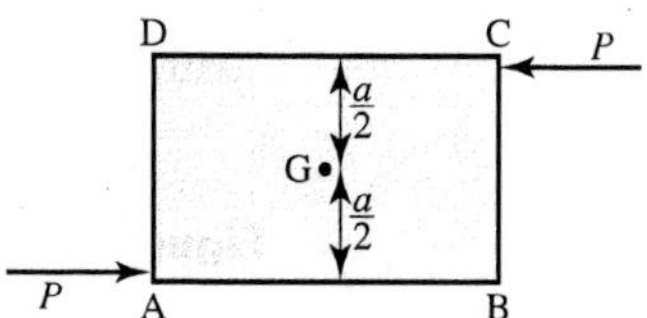

Figure 11.11

The resultant force on the tray is clearly zero, but the resultant moment about its centre point, G, is

$$P \times \frac{a}{2} + P \times \frac{a}{2} = Pa.$$

The tray will start to rotate about its centre and so it is clearly not in equilibrium.

Note

You could have taken moments about any of the corners, A, B, C or D, or any other point in the plane of the paper and the answer would have been the same, *Pa* anticlockwise.

So we now tighten our mathematical definition of equilibrium to include moments. For an object to remain at rest (or moving at constant velocity) when a system of forces is applied, both the resultant force and the total moment must be zero.

To check that an object is in equilibrium under the action of a system of forces, you need to check two things:

(i) that the resultant force is zero;
(ii) that the resultant moment about any point is zero. (You only need to check one point.)

EXAMPLE 11.1

Two children are playing with a door. Kerry tries to open it by pulling on the handle with a force of 50 N at right angles to the plane of the door, at a distance 0.8 m from the hinges. Peter pushes at a point 0.6 m from the hinges, also at right angles to the door and with sufficient force just to stop Kerry opening it.

(i) What is the moment of Kerry's force about the hinges?
(ii) With what force does Peter push?
(iii) Describe the resultant force on the hinges.

Figure 11.12

SOLUTION

Looking down from above, the line of the hinges becomes a point, H. The door opens clockwise. Anticlockwise is taken to be positive.

(i)

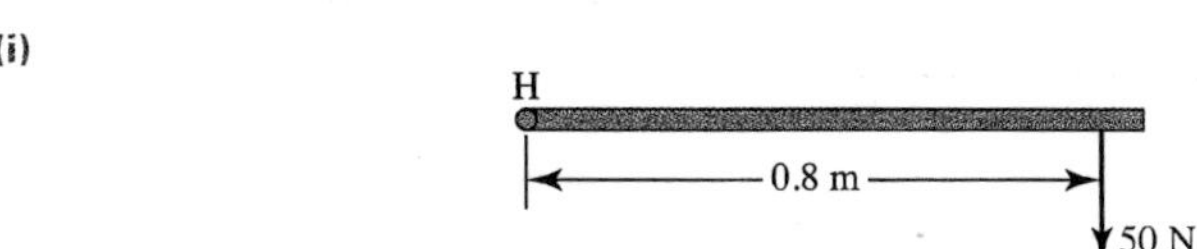

Figure 11.13

Kerry's moment about H $= -50 \times 0.8$

$= -40\,\text{Nm}$

The moment of Kerry's force about the hinges is $-40\,\text{Nm}$.
(Note that it is a clockwise moment and so negative.)

(ii)

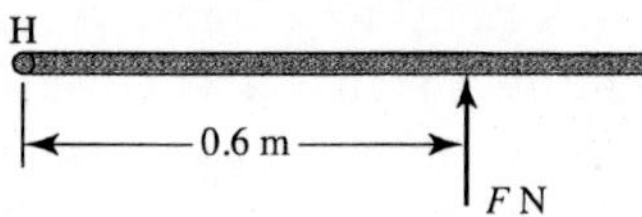

Figure 11.14

Peter's moment about $H = +F \times 0.6$

Since the door is in equilibrium, the total moment on it must be zero, so

$$\begin{aligned} F \times 0.6 - 40 &= 0 \\ F &= \frac{40}{0.6} \\ &= 66.7 \text{ (to 3 s.f.)} \end{aligned}$$

Peter pushes with a force of 66.7 N.

(iii) Since the door is in equilibrium the overall resultant force on it must be zero.

All the forces are at right angles to the door, as shown in the diagram.

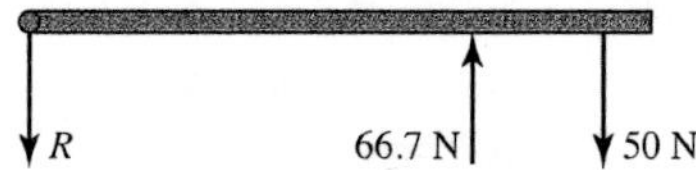

Figure 11.15

Resolve perpendicular to door:

$$\begin{aligned} R + 50 &= 66.7 \\ R &= 16.7 \text{ (to 3 s.f.)} \end{aligned}$$

The total reaction at the hinges is a force of 16.7 N in the same direction as Kerry is pulling.

Note

The reaction force at a hinge may act in any direction, according to the forces elsewhere in the system. A hinge can be visualised in cross section as shown in figure 11.16. If the hinge is well oiled, and the friction between the inner and outer parts is negligible, the hinge cannot exert any moment. In this situation the door is said to be 'freely hinged'.

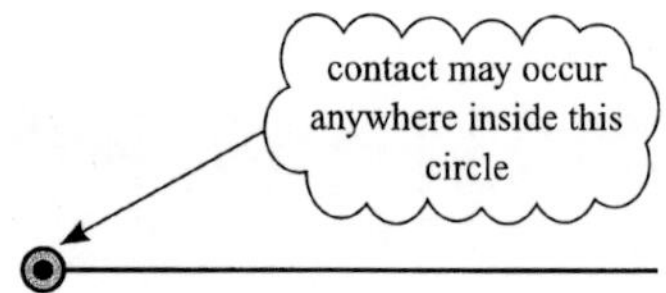

Figure 11.16

EXAMPLE 11.2

The diagram shows a man of weight 600 N standing on a footbridge that consists of a uniform wooden plank just over 2 m long of weight 200 N. Find the reaction forces exerted on each end of the plank.

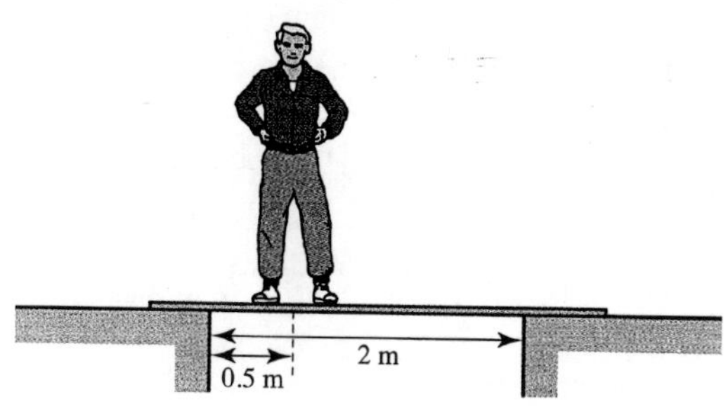

Figure 11.17

SOLUTION

The diagram shows the forces acting on the plank.

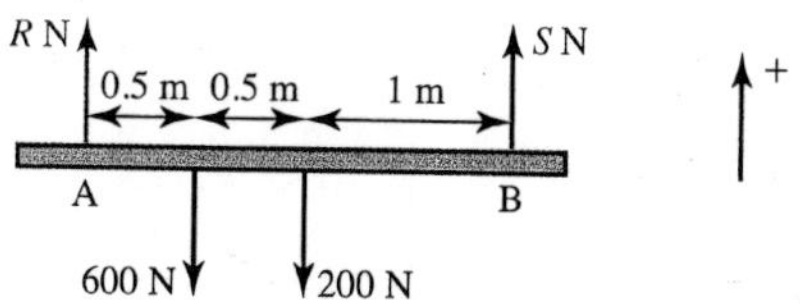

Figure 11.18

For equilibrium both the resultant force and the total moment must be zero.

As all the forces act vertically we have

$$R + S - 800 = 0 \qquad ①$$

Taking moments about the point A gives

$$(\curvearrowleft) \qquad R \times 0 - 600 \times 0.5 - 200 \times 1 + S \times 2 = 0 \qquad ②$$

From equation ② $S = 250$ and so equation ① gives $R = 550$.

The reaction forces are 250 N at A and 550 N at B.

Notes

1. You cannot solve this problem without taking moments.
2. You can take moments about any point and can, for example, show that by taking moments about B you get the same answer.
3. The whole weight of the plank is being considered to act at its centre.
4. When a force acts through the point about which moments are being taken, its moment about that point is zero.

Levers

A lever can be used to lift or move a heavy object using a relatively small force. Levers depend on moments for their action.

Two common lever configurations are shown below. In both cases a load W is being lifted by an applied force F, using a lever of length l. The calculations assume equilibrium.

Case 1

The fulcrum is at one end of the lever, figure 11.19.

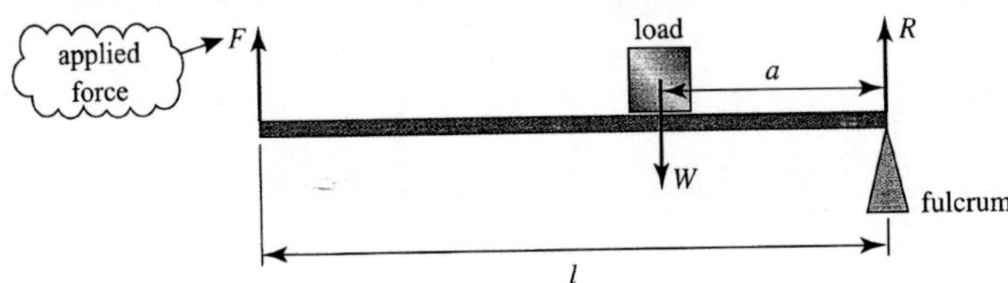

Figure 11.19

Taking moments about the fulcrum:

$$(\curvearrowright) \quad F \times l - W \times a = 0$$

$$F = W \times \frac{a}{l}$$

Since a is much smaller than l, the applied force F is much smaller than the load W.

Case 2

The fulcrum is within the lever, figure 11.20.

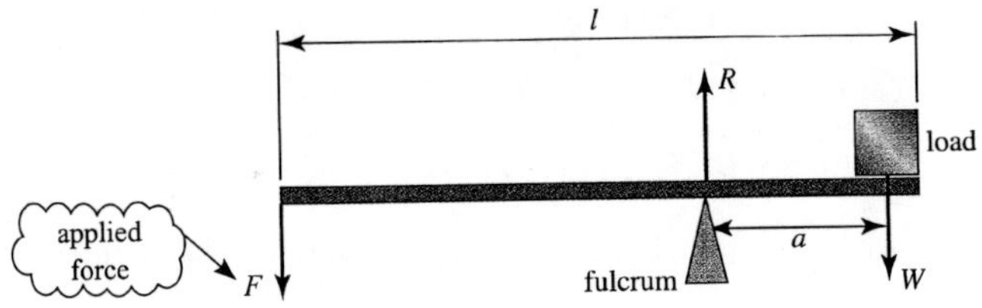

Figure 11.20

Taking moments about the fulcrum:

$$(\curvearrowleft) \quad F \times (l - a) - W \times a = 0$$

$$F = W \times \frac{a}{l - a}$$

Provided that the fulcrum is nearer the end with the load, the applied force is less than the load.

These examples also indicate how to find a single force equivalent to two parallel forces. The force equivalent to F and W should be equal and opposite to R and with the same line of action.

Describe the single force equivalent to P and Q in each of these cases.

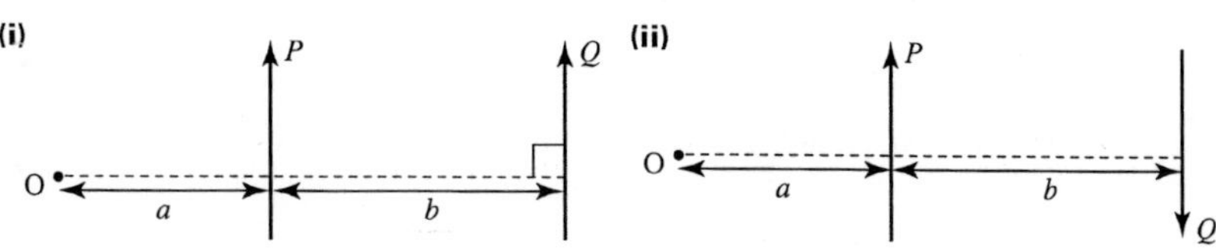

Figure 11.21

In each case state its magnitude and line of action.

How do you use moments to open a screw-top jar?
Why is it an advantage to press hard when it is stiff?

EXERCISE 11A

1 In each of the situations shown below, find the moment of the force about the point and state whether it is positive (anticlockwise) or negative (clockwise).

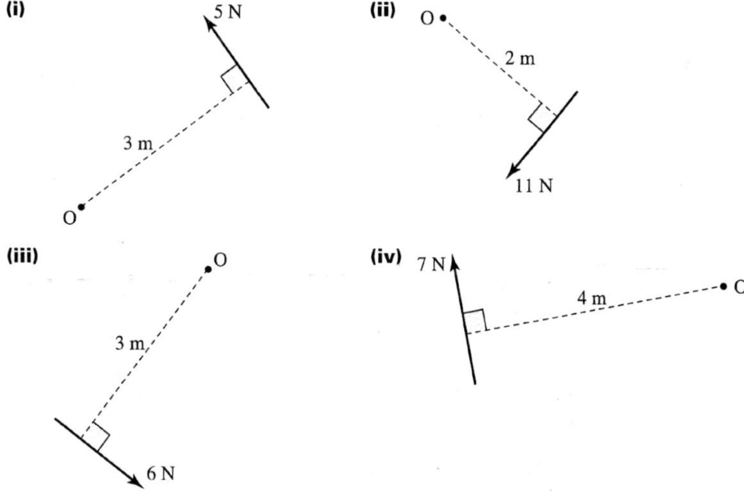

2 The situations below involve several forces acting on each object. For each one, find the total moment.

(i)

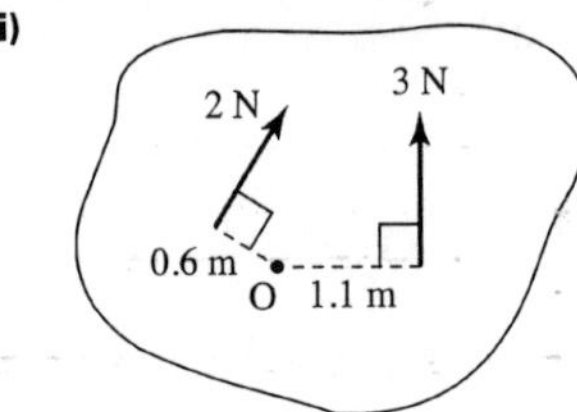

(ii)

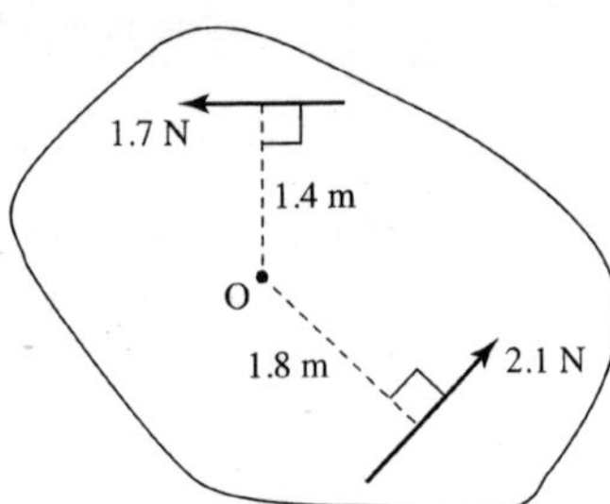

(iii)

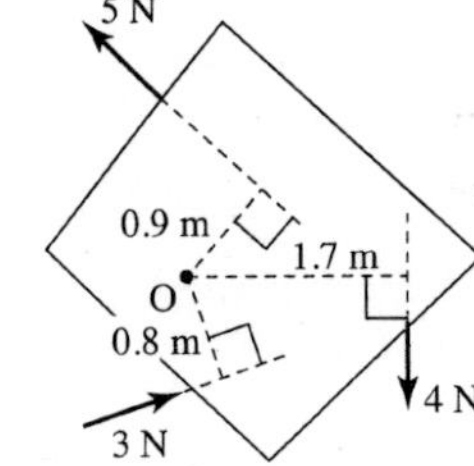

(iv)

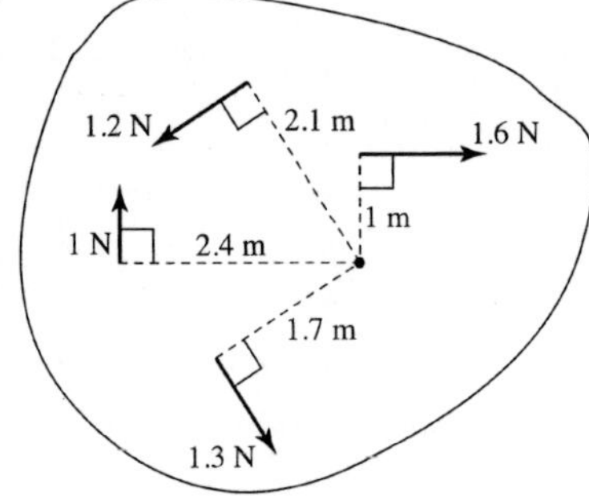

3 A uniform horizontal bar of mass 5 kg has length 30 cm and rests on two vertical supports, 10 cm and 22 cm from its left-hand end. Find the magnitude of the reaction force at each of the supports.

4 The diagram shows a motorcycle of mass 250 kg, and its rider whose mass is 80 kg. The centre of mass of the motorcycle lies on a vertical line midway between its wheels. When the rider is on the motorcycle, his centre of mass is 1 m behind the front wheel.

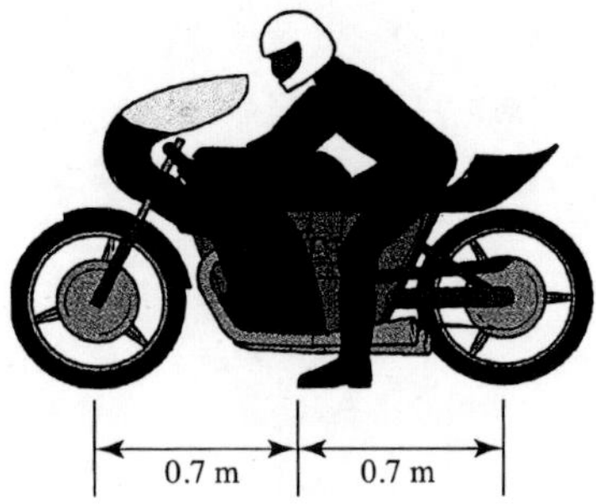

Find the vertical reaction forces acting through the front and rear wheels when

(i) the rider is not on the motorcycle

(ii) the rider is on the motorcycle.

5 Find the reaction forces on the hi-fi shelf shown below. The shelf itself has weight 25 N and its centre of mass is midway between A and D.

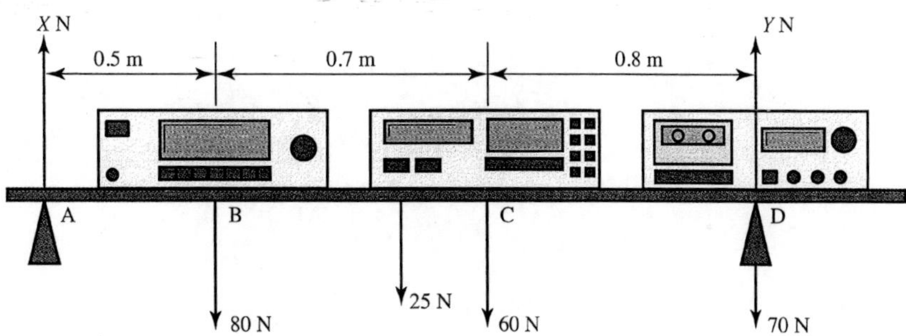

6 Karen and Jane are trying to find the positions of their centres of mass. They place a uniform board of mass 8 kg symmetrically on two bathroom scales whose centres are 2 m apart. When Karen lies flat on the board, Jane notes that scale A reads 37 kg and scale B reads 26 kg.

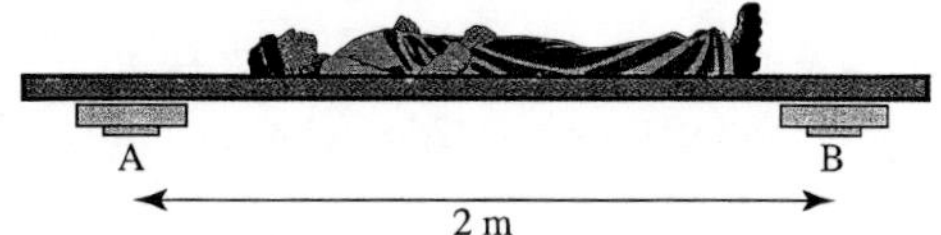

(i) Draw a diagram showing the forces acting on Karen and the board and calculate Karen's mass.

(ii) How far from the centre of scale A is her centre of mass?

7 The diagram shows two people, an adult and a child, sitting on a uniform bench of mass 40 kg; their positions are as shown. The mass of the child is 50 kg, that of the adult is 85 kg.

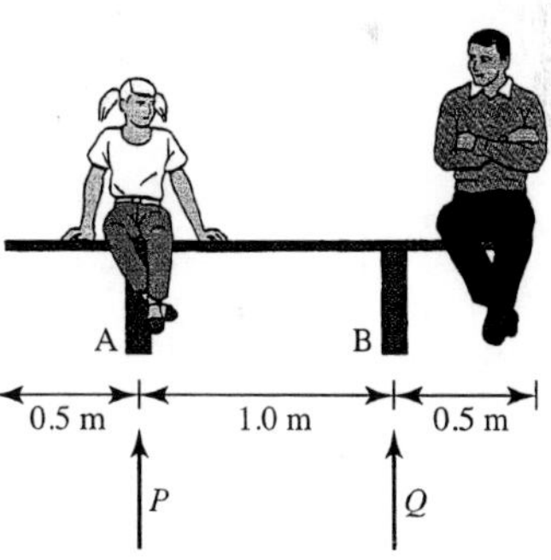

(i) Find the reaction forces, P and Q (in N), from the ground on the two supports of the bench.

(ii) The child now moves to the mid-point of the bench. What are the new values of P and Q?

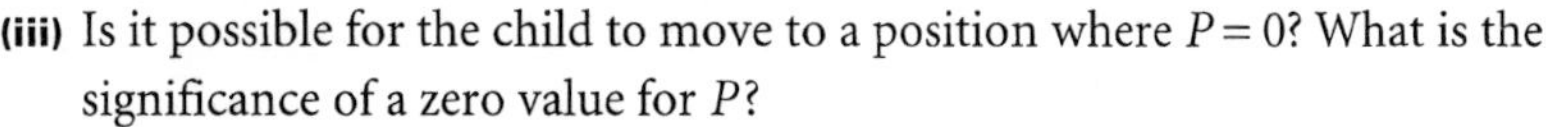

(iii) Is it possible for the child to move to a position where $P = 0$? What is the significance of a zero value for P?

(iv) What happens if the child leaves the bench?

8 The diagram shows a diving board which some children have made. It consists of a uniform plank of mass 20 kg and length 3 m, with 1 m of its length projecting out over a pool. They have put a boulder of mass 25 kg on the end over the land; and there is a support at the water's edge.

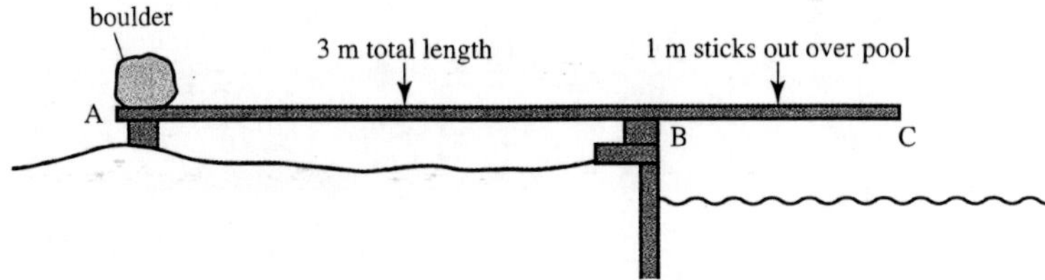

(i) Find the forces at the two supports when nobody is using the diving board.

(ii) A child of mass 50 kg is standing on the end of the diving board over the pool. What are the forces at the two supports?

(iii) Some older children arrive and take over the diving board. One of these is a heavy boy of mass 90 kg. What is the reaction at A if the board begins to tip over?

(iv) How far can the boy walk from B before the board tips over?

9 A lorry of mass 5000 kg is driven across a Bailey bridge of mass 20 tonnes. The bridge is a roadway of length 10 m which is supported at both ends.

(i) Find expressions for the reaction forces at each end of the bridge in terms of the distance x in metres travelled by the lorry from the start of the bridge.

(ii) From what point of the lorry is the distance x measured?

Two identical lorries cross the bridge at the same speed, starting at the same instant, from opposite directions.

(iii) How do the reaction forces of the supports on the bridge vary as the lorries cross the bridge?

10 A simple suspension bridge across a narrow river consists of a uniform beam, 4 m long and of mass 60 kg, supported by vertical cables attached at a distance 0.75 m from each end of the beam.

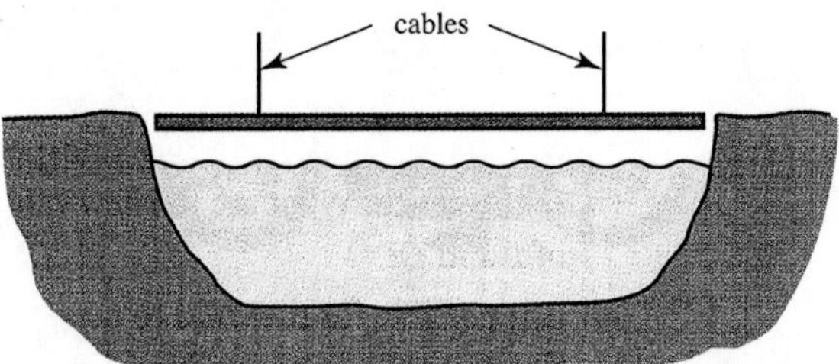

(i) Find the tension in each cable when a boy of mass 50 kg stands 1 m from the end of the bridge.

(ii) Can a couple walking hand-in-hand cross the bridge safely, without it tipping, if their combined mass is 115 kg?

(iii) What is the mass of a person standing on the end of the bridge when the tension in one cable is four times that in the other cable?

EXPERIMENT

Set up the apparatus shown in figure 11.22 below and experiment with two or more weights in different positions.

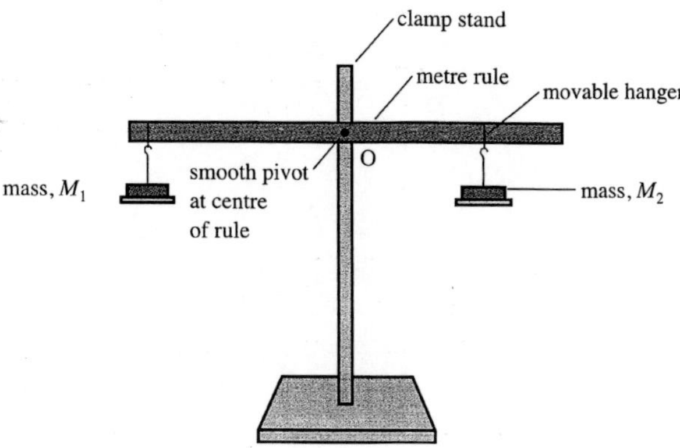

Figure 11.22

Record your results in a table showing weights, distances from O and moments about O.

Two masses are suspended from the rule in such a way that the rule balances in a horizontal position. What happens when the rule is then moved to an inclined position and released?

Now attach a pulley as in figure 11.23. Start with equal weights and measure d and l. Then try different weights and pulley positions.

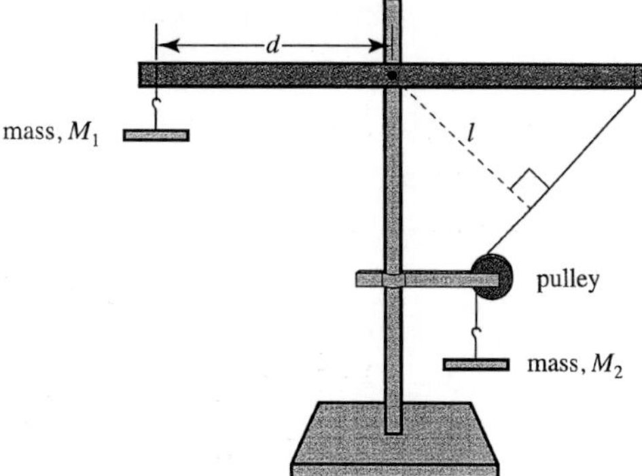

Figure 11.23

The moment of a force which acts at an angle

From the experiment you will have seen that the moment of a force about the pivot depends on the *perpendicular distance* from the pivot to the line of the force.

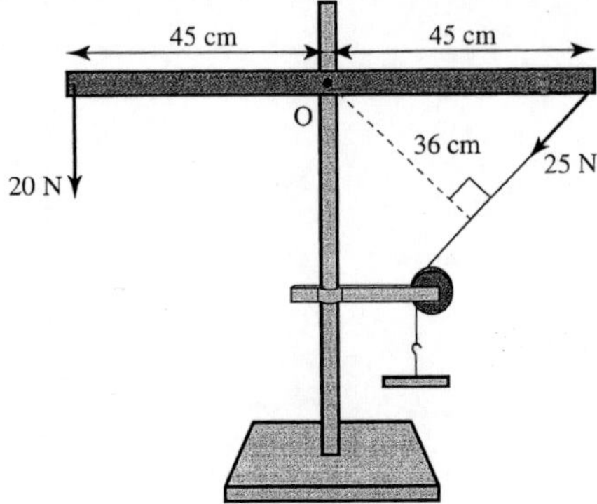

Figure 11.24

In figure 11.24, where the system remains at rest, the moment about O of the 20 N force is $20 \times 0.45 = 9$ Nm. The moment about O of the 25 N force is $-25 \times 0.36 = -9$ Nm. The system is in equilibrium even though unequal forces act at equal distances from the pivot.

The magnitude of the moment of the force F about O in figure 11.25 is given by

$$F \times l = Fd \sin \alpha$$

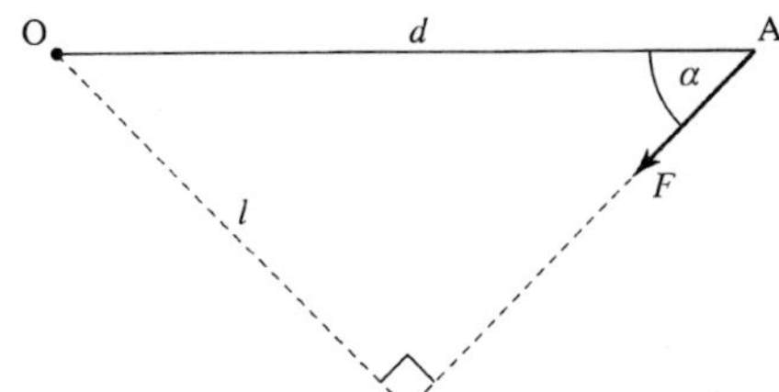

Figure 11.25

Alternatively the moment can be found by noting that the force F can be resolved into components $F \cos \alpha$ parallel to AO and $F \sin \alpha$ perpendicular to AO, both acting through A (figure 11.26). The moment of each component can be found and then summed to give the total moment.

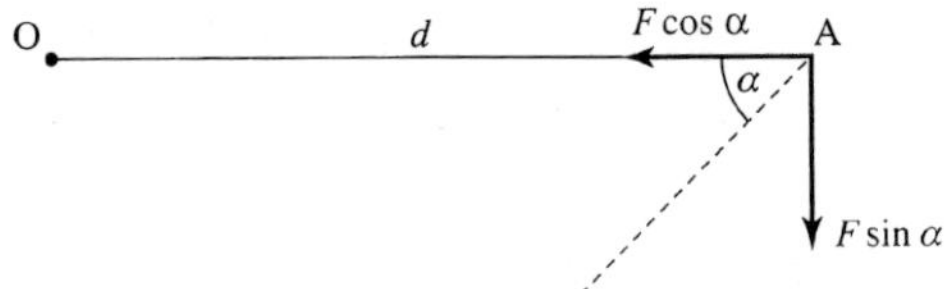

Figure 11.26

The moment of the component along AO is zero because it acts through O. The magnitude of the moment of the perpendicular component is $F \sin \alpha \times d$ so the total moment is $Fd \sin \alpha$, as expected.

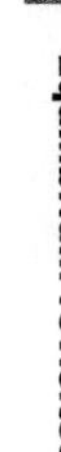

M2 11

EXAMPLE 11.3 A force of 40 N is exerted on a rod as shown. Find the moment of the force about the point marked O.

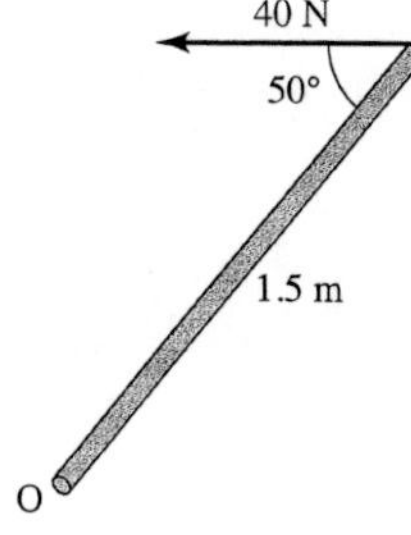

Figure 11.27

SOLUTION

In order to calculate the moment, the perpendicular distance between O and the line of action of the force must be found. This is shown on the diagram.

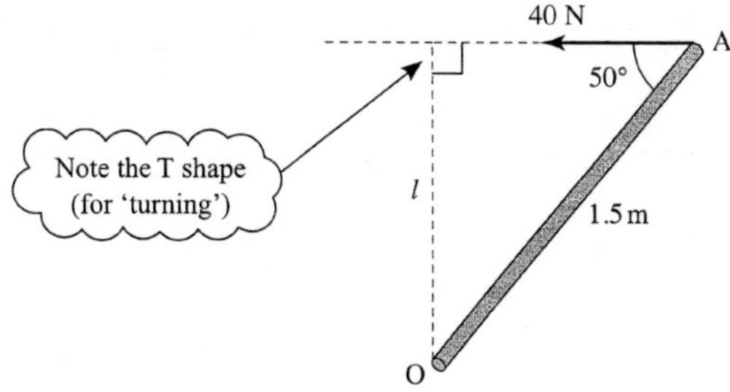

Figure 11.28

Here $l = 1.5 \times \sin 50°$.

So the moment about O is

$$F \times l = 40 \times (1.5 \times \sin 50°)$$
$$= 46.0\,\text{Nm}.$$

Alternatively you can resolve the 40 N force into components as in Figure 11.29.

The component of the force parallel to AO is 40 cos 50° N. The component perpendicular to AO is 40 sin 50° (or 40 cos 40°) N.

So the moment about O is

$$40 \sin 50° \times 1.5 = 60 \sin 50°$$
$$= 46.0\,\text{Nm as before.}$$

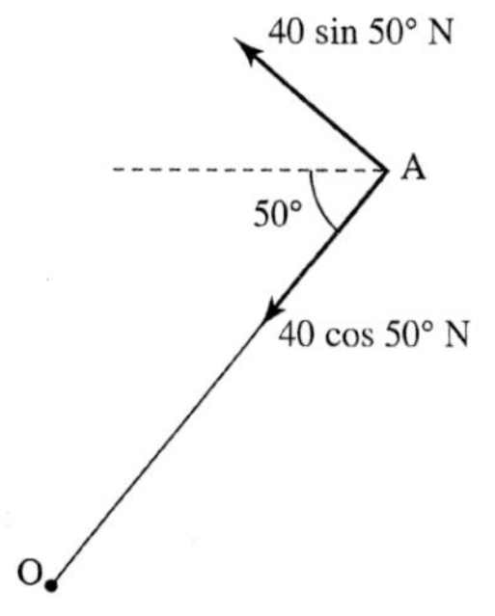

Figure 11.29

EXAMPLE 11.4

A sign is attached to a light rod of length 1 m which is freely hinged to the wall and supported in a vertical plane by a light string as in the diagram. The sign is assumed to be a uniform rectangle of mass 10 kg. The angle of the string to the horizontal is 25°.

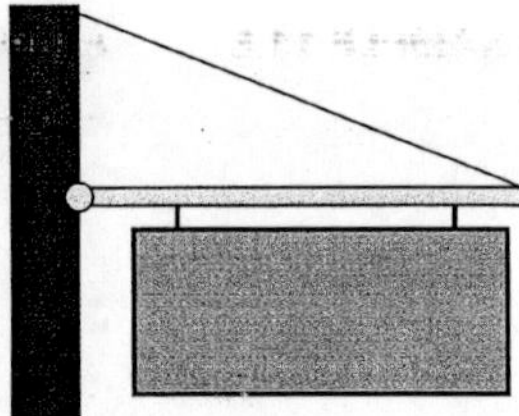

Figure 11.30

(i) Find the tension in the string.

(ii) Find the magnitude and direction of the reaction force of the hinge on the sign.

SOLUTION

(i) The diagram shows the forces acting on the rod, where R_H and R_V are the magnitudes of the horizontal and vertical components of the reaction **R** on the rod at the wall.

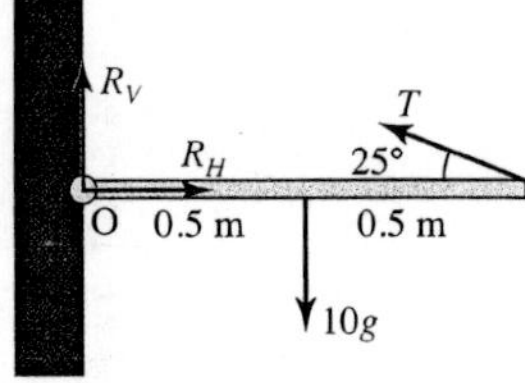

Figure 11.31

Taking moments about O:

$$0 \times R_V + 0 \times R_H - 10g \times 0.5 + T \sin 25° \times 1 = 0$$

$$\Rightarrow \qquad T \sin 25° = 5g$$

$$T = 118 \text{ (to 3 s.f.)}$$

The tension is 118 N.

(ii) You can resolve to find the reaction at the wall.

Horizontally:

$$R_H = T \cos 25°$$

$$\Rightarrow \quad R_H = 107$$

Vertically:

$$R_V + T \sin 25° = 10g$$

$$\Rightarrow \quad R_V = 10g - 5g = 50$$

$$R = \sqrt{107^2 + 50^2}$$

$$= 118$$

$$\tan \theta = \frac{50}{107}$$

$$\theta = 25° \text{ (to the nearest degree)}$$

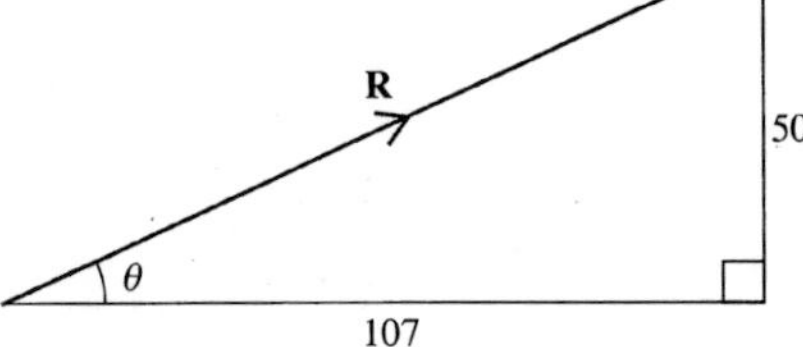

Figure 11.32

The reaction at the hinge has magnitude 118 N and acts at 25° above the horizontal.

❓ Is it by chance that R and T have the same magnitude and act at the same angle to the horizontal?

EXAMPLE 11.5 A uniform ladder is standing on rough ground and leaning against a smooth wall at an angle of 60° to the ground. The ladder has length 4 m and mass 15 kg. Find the normal reaction forces at the wall and ground and the friction force at the ground.

SOLUTION

The diagram shows the forces acting on the ladder. The forces are in newtons.

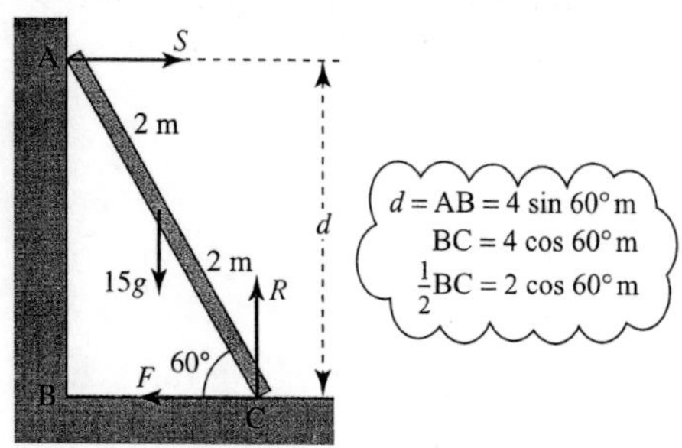

Figure 11.33

The diagram shows that there are three unknown forces S, R and F so we need three equations from which to find them. If the ladder remains at rest (in equilibrium) then the resultant force is zero and the resultant moment is zero. These two conditions provide the three necessary equations.

Equilibrium of horizontal components: $S - F = 0$ ①

Equilibrium of vertical components: $R - 15g = 0$ ②

Moments about the foot of the ladder:

$$R \times 0 + F \times 0 + 15g \times 2\cos 60° - S \times 4\sin 60° = 0$$

$$\Rightarrow \quad 150 - 4S\sin 60° = 0 \qquad ③$$

$$\Rightarrow \quad S = \frac{150}{4\sin 60°} = 43.3 \text{ (to 3 s.f.)}$$

From ① $F = S = 43.3$

From ② $R = 150$

The force at the wall is 43.3 N, those at the ground are 43.3 N horizontally and 150 N vertically.

EXERCISE 11B

1 Find the moment about O of each of the forces illustrated below.

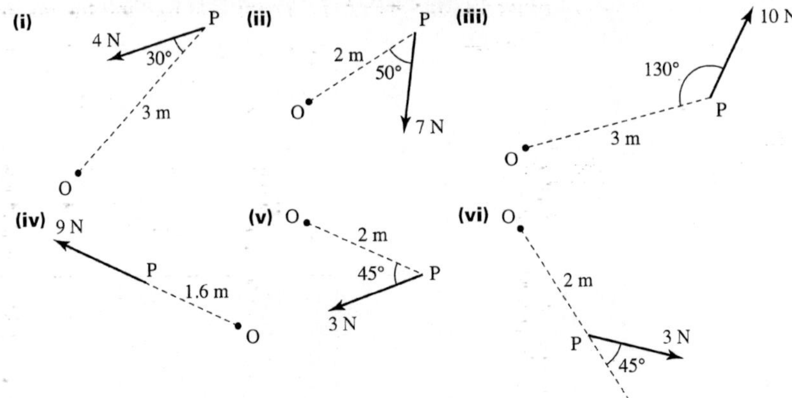

2 The diagram shows three children pushing a playground roundabout. Hannah and David want it to go one way but Rabina wants it to go the other way. Who wins?

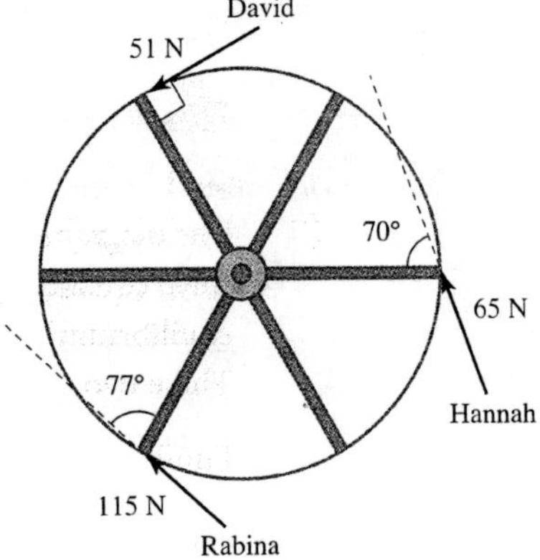

3 The operating instructions for a small crane specify that when the jib is at an angle of 25° above the horizontal, the maximum safe load for the crane is 5000 kg. Assuming that this maximum load is determined by the maximum moment that the pivot can support, what is the maximum safe load when the angle between the jib and the horizontal is

(i) 40° **(ii)** an angle θ?

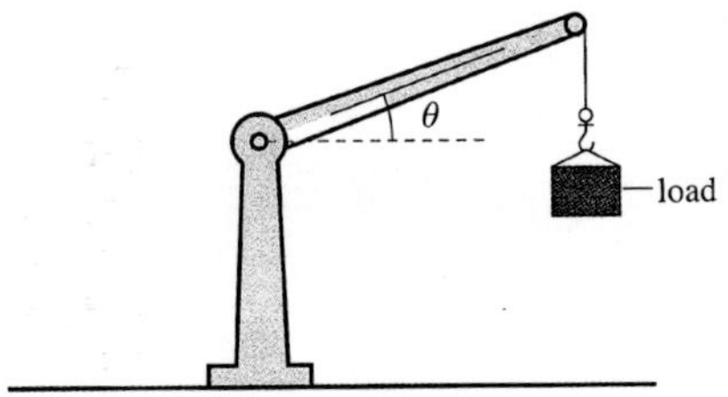

4 In each of these diagrams, a uniform beam of mass 5 kg and length 4 m, freely hinged at one end, A, is in equilibrium. Find the magnitude of the force T in each case.

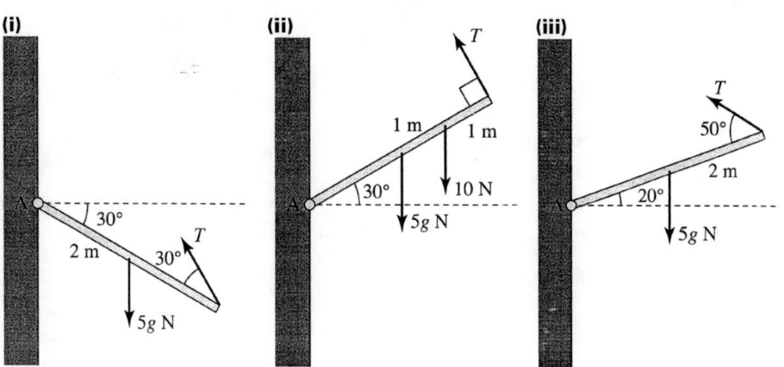

5 The diagram shows a uniform rectangular sign ABCD, 3 m × 2 m, of weight 20 N. It is freely hinged at A and supported by the string CE, which makes an angle of 30° with the horizontal. The tension in the string is T (in N).

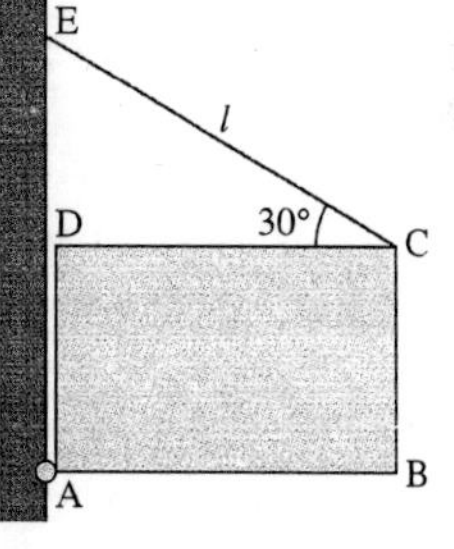

(i) Resolve the tension T into horizontal and vertical components.

(ii) Hence show that the moment of the tension in the string about A is given by

$$2T\cos 30° + 3T\sin 30°.$$

(iii) Write down the moment of the sign's weight about A.

(iv) Hence show that $T = 9.28$.

(v) Hence find the horizontal and vertical components of the reaction on the sign at the hinge, A.

You can also find the moment of the tension in the string about A as $d \times T$, where d is the length of AF as shown in the diagram.

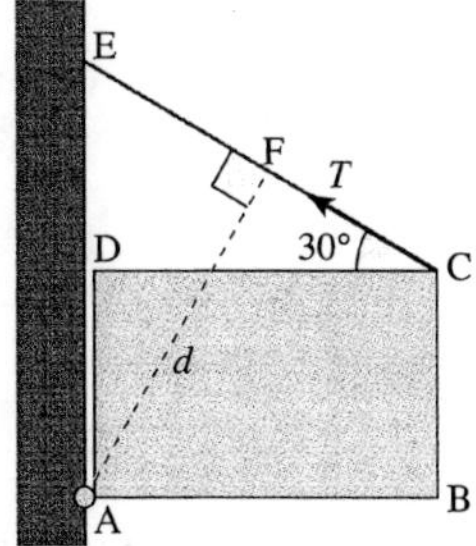

(vi) Find **(a)** the angle ACD **(b)** the length d.

(vii) Show that you get the same value for T when it is calculated in this way.

6 The diagram shows a simple crane. The weight of the jib (AB) may be ignored. The crane is in equilibrium in the position shown.

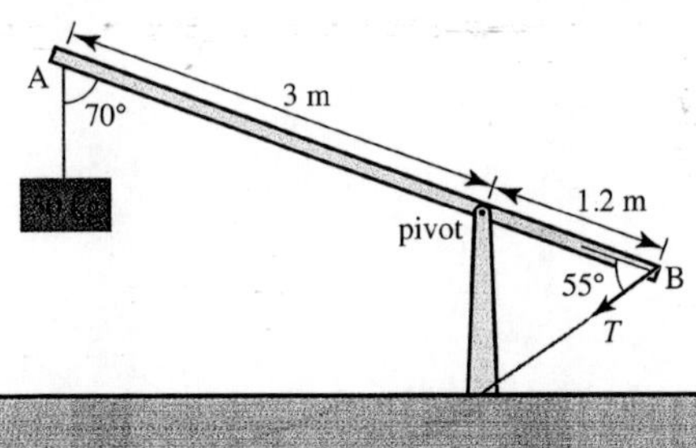

(i) By taking moments about the pivot, find the magnitude of the tension T (in N).

(ii) Find the reaction of the pivot on the jib in the form of components parallel and perpendicular to the jib.

(iii) Show that the total moment about the end A of the forces acting on the jib is zero.

(iv) What would happen if

(a) the rope holding the 50 kg mass snapped?

(b) the rope with tension T snapped?

7 A uniform plank, AB, of mass 50 kg and length 6 m is in equilibrium leaning against a smooth wall at an angle of 60° to the horizontal. The lower end, A, is on rough horizontal ground.

(i) Draw a diagram showing all the forces acting on the plank.

(ii) Write down the total moment about A of all the forces acting on the plank.

(iii) Find the normal reaction of the wall on the plank at point B.

(iv) Find the frictional force on the foot of the plank. What can you deduce about the coefficient of friction between the ground and the plank?

(v) Show that the total moment about B of all the forces acting on the plank is zero.

8 A uniform ladder of mass 20 kg and length $2l$ rests in equilibrium with its upper end against a smooth vertical wall and its lower end on a rough horizontal floor. The coefficient of friction between the ladder and the floor is μ. The normal reaction at the wall is S, the frictional force at the ground is F and the normal reaction at the ground is R. The ladder makes an angle α with the horizontal.

(i) Draw a diagram showing the forces acting on the ladder.

For each of the cases, **(a)** $\alpha = 60°$, **(b)** $\alpha = 45°$

(ii) find the magnitudes of S, F and R

(iii) find the least possible value of μ.

9 The diagram shows a car's hand brake. The force F is exerted by the hand in operating the brake, and this creates a tension T in the brake cable. The hand brake is freely pivoted at point B and is assumed to be light.

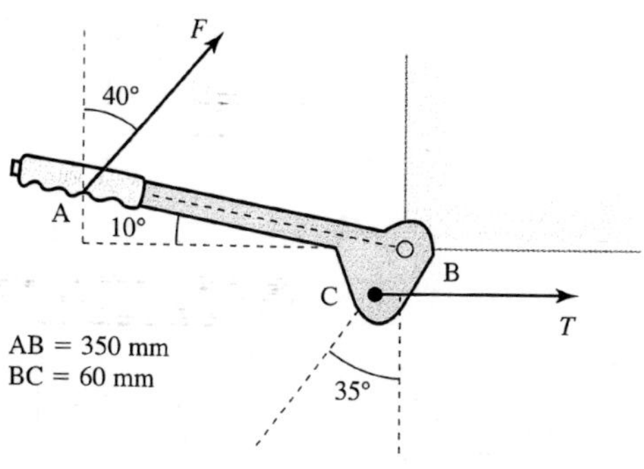

(i) Draw a diagram showing all the forces acting on the hand brake.

(ii) What is the required magnitude of force F if the tension in the brake cable is to be 1000 N?

(iii) A child applies the hand brake with a force of 10 N. What is the tension in the brake cable?

10 The diagram shows four tugs manoeuvring a ship. A and C are pushing it, B and D are pulling it.

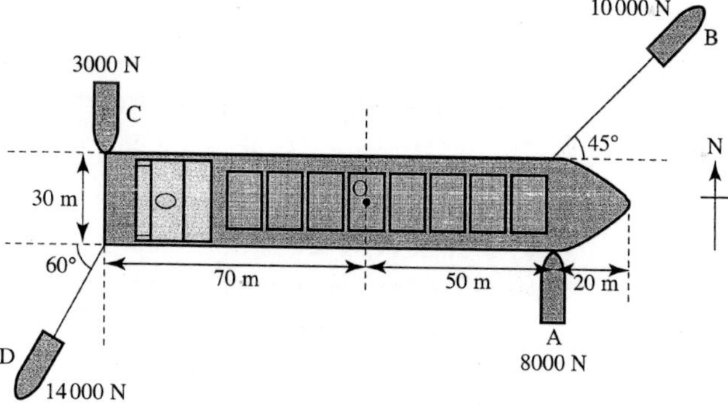

(i) Show that the resultant force on the ship is less than 100 N.

(ii) Find the overall turning moment on the ship about its centre point, O.

A breeze starts to blow from the south, causing a total force of 2000 N to act uniformly along the length of the ship, at right angles to it.

(iii) How (assuming B and D continue to apply the same forces) can tugs A and C counteract the sideways force on the ship by altering the forces with which they are pushing, while maintaining the same overall moment about the centre of the ship?

11 The boom of a fishing boat may be used as a simple crane. The boom AB is uniform, 8 m long and has a mass of 30 kg. It is freely hinged at the end A.

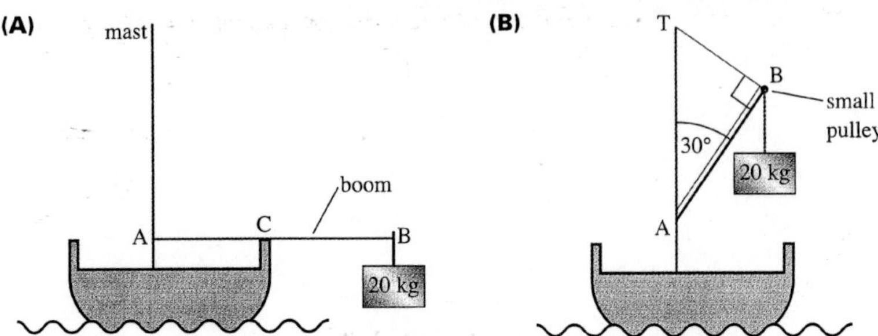

In figure (A), the boom shown is in equilibrium supported at C by the boat's rail, where the length AC is 3.5 m. The boom is horizontal and has a load of mass 20 kg suspended from the end B.

(i) Draw a diagram showing all the forces acting on the boom AB.

(ii) Find the force exerted on the boom by the rail at C.

(iii) Calculate the magnitude and direction of the force acting on the boom at A.

It is more usual to use the boom in a position such as the one shown in figure (B). AT is vertical and the boom is held in equilibrium by the rope section TB, which is perpendicular to it. Angle TAB = 30°. A load of mass 20 kg is supported by a rope passing over a small, smooth pulley at B. The rope then runs parallel to the boom to a fixing point at A.

(iv) Find the tension in the rope section TB when the load is stationary.

[MEI, *part*]

12 Jules is cleaning windows. Her ladder is uniform and stands on rough ground at an angle of 60° to the horizontal and with the top end resting on the edge of a smooth window sill. The ladder has mass 12 kg and length 2.8 m and Jules has mass 56 kg.

(i) Draw a diagram to show the forces on the ladder when nobody is standing on it. Show that the reaction at the sill is then $3g$ N.

(ii) Find the friction and normal reaction forces at the foot of the ladder.

Jules needs to be sure that the ladder will not slip however high she climbs.

(iii) Find the least possible value of μ for the ladder to be safe at 60° to the horizontal.

(iv) The value of μ is in fact 0.4. How far up the ladder can Jules stand before it begins to slip?

13 Overhead cables for a tramway are supported by uniform, rigid, horizontal beams of weight 1500 N and length 5 m. Each beam, AB, is freely pivoted at one end A and supports two cables which may be modelled by vertical loads, each of 1000 N, one 1.5 m from A and the other at 1 m from B.

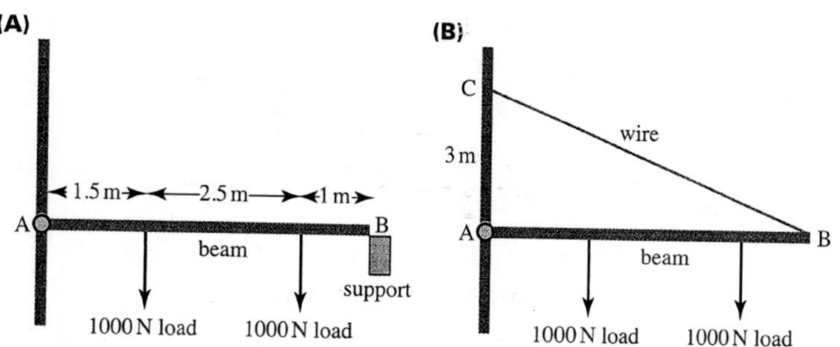

In one situation, the beam is held in equilibrium by resting on a small horizontal support at B, as shown in figure **(A)**.

(i) Draw a diagram showing all the forces acting on the beam AB. Show that the vertical force acting on the beam at B is 1850 N.

In another situation, the beam is supported by a wire, *instead of the support at* B. The wire is light, attached at one end to the beam at B and at the other to the point C which is 3 m vertically above A, as shown in figure **(B)**.

(ii) Calculate the tension in the wire.

(iii) Find the magnitude and direction of the force on the beam at A.

[MEI]

14 A uniform beam AB has length 2 m and mass 10 kg. The beam is hinged at A to a fixed point on a vertical wall, and is held in a fixed position by a light inextensible string of length 2.4 m. One end of the string is attached to the beam at a point 0.7 m from A. The other end of the string is attached to the wall at a point vertically above the hinge. The string is at right angles to AB. The beam carries a load of weight 300 N at B (see diagram).

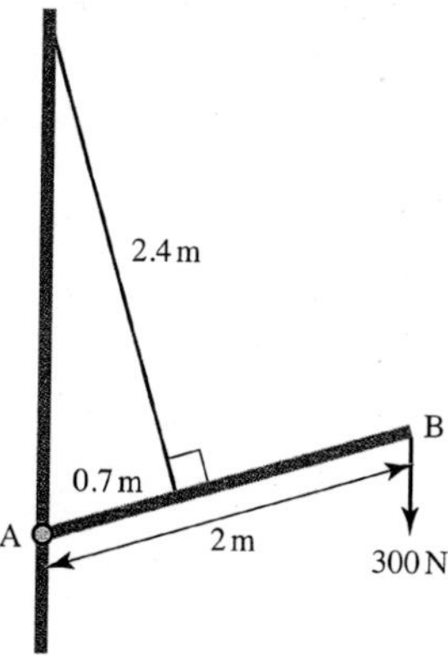

(i) Find the tension in the string.

The components of the force exerted by the hinge on the beam are XN horizontally away from the wall and YN vertically downwards.

(ii) Find the values of X and Y.

[Cambridge AS and A Level Mathematics 9709, Paper 5 Q3 November 2007]

INVESTIGATION

Toolbox

Which tools in a typical toolbox or kitchen drawer depend upon moments for their successful operation?

KEY POINTS

1 The moment of a force F about a point O is given by the product Fd where d is the perpendicular distance from O to the line of action of the force.

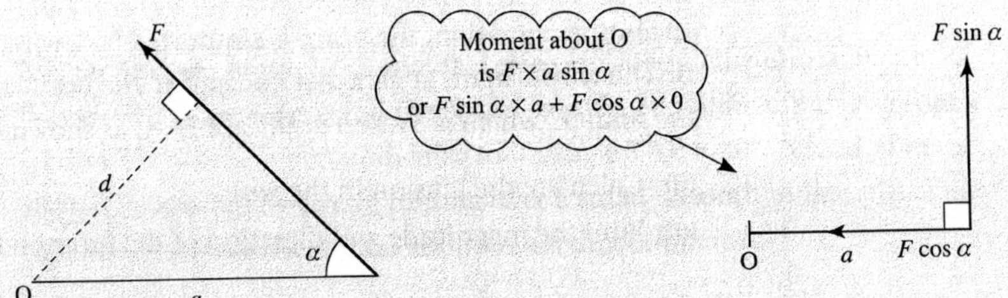

2 The S.I. unit for moment is the newton metre (Nm).

3 Anticlockwise moments are usually called positive, clockwise negative.

4 If a body is in equilibrium the sum of the moments of the forces acting on it, about any point, is zero.

5 Two parallel forces P and Q ($P > Q$) are equivalent to a single force $P + Q$ when P and Q are in the same direction and $P - Q$ when they are in opposite directions. The line of action of the equivalent force is found by taking moments.

Centre of mass

Let man then contemplate the whole of nature in her full and grand mystery ... It is an infinite sphere, the centre of which is everywhere, the circumference nowhere.

Blaise Pascal

Figure 12.1, which is drawn to scale, shows a mobile suspended from the point P. The horizontal rods and the strings are light but the geometrically shaped pieces are made of uniform heavy card. Does the mobile balance? If it does, what can you say about the position of its centre of mass?

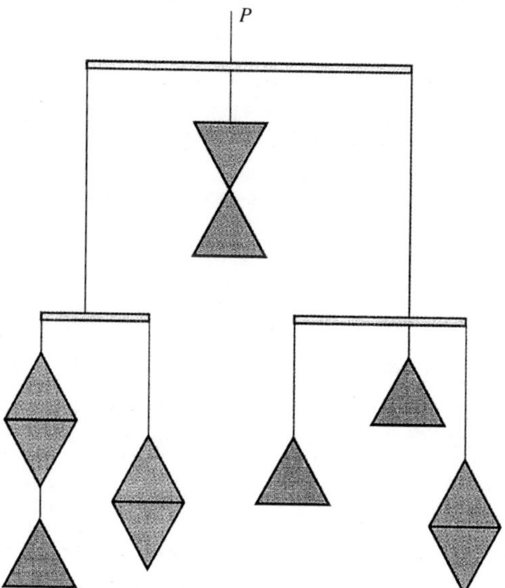

Figure 12.1

You have met the concept of centre of mass in the context of two general models.

- *The particle model*
  The centre of mass is the single point at which the whole mass of the body may be taken to be situated.
- *The rigid body model*
  The centre of mass is the balance point of a body with size and shape.

The following examples show how to calculate the position of the centre of mass of a body.

EXAMPLE 12.1

An object consists of three point masses 8 kg, 5 kg and 4 kg attached to a rigid light rod as shown.

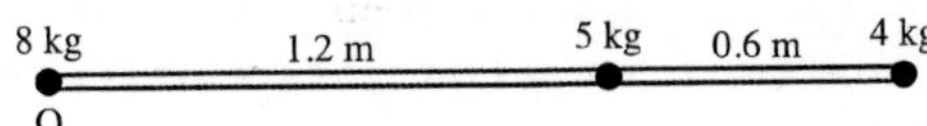

Figure 12.2

Calculate the distance of the centre of mass of the object from end O. (Ignore the mass of the rod.)

SOLUTION

Suppose the centre of mass C is $\bar{x}$ m from O. If a pivot were at this position the rod would balance.

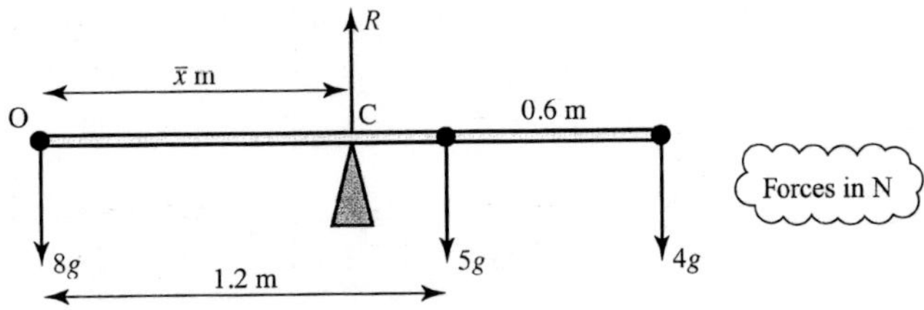

Figure 12.3

For equilibrium $\quad R = 8g + 5g + 4g = 17g$

Taking moments of the forces about O gives:

$$\begin{aligned} \text{Total clockwise moment} &= (8g \times 0) + (5g \times 1.2) + (4g \times 1.8) \\ &= 13.2g \text{ Nm} \\ \text{Total anticlockwise moment} &= R\bar{x} \\ &= 17g\bar{x} \text{ Nm} \end{aligned}$$

The overall moment must be zero for the rod to be in balance, so

$$17g\bar{x} - 13.2g = 0$$

$$\Rightarrow \quad 17\bar{x} = 13.2$$

$$\Rightarrow \quad \bar{x} = \frac{13.2}{17} = 0.776 \text{ (to 3 s.f.)}$$

The centre of mass is 0.776 m from the end O of the rod.

Note that although g was included in the calculation, it cancelled out. The answer depends only on the masses and their distances from the origin and not on the value of g. This leads to the following definition for the position of the centre of mass.

Definition

Consider a set of n point masses m_1, m_2, ..., m_n attached to a rigid light rod (whose mass is neglected) at positions x_1, x_2, ..., x_n from one end O. The situation is shown in figure 12.4.

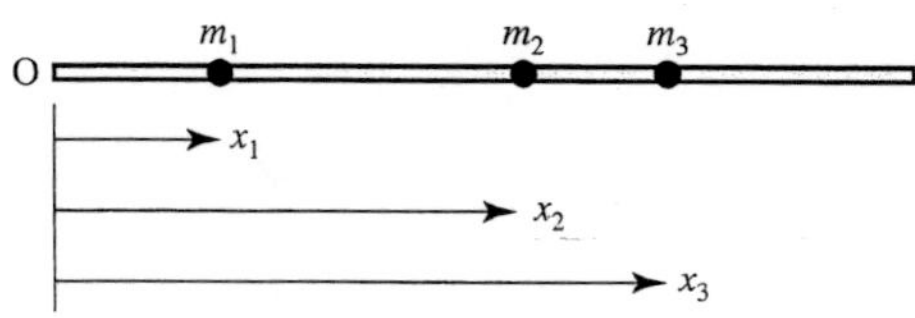

Figure 12.4

The position, $\bar{x}$, of the centre of mass relative to O, is defined by the equation:

- moment of whole mass at centre of mass = sum of moments of individual masses

$$(m_1 + m_2 + m_3 + ...)\bar{x} = m_1x_1 + m_2x_2 + m_3x_3 + ...$$

or

$$M\bar{x} = \sum_{i=1}^{n} m_i x_i$$

where M is the total mass (or $\sum m_i$).

The symbol Σ (sigma) means 'the sum of'.

EXAMPLE 12.2

A uniform rod of length 2 m has mass 5 kg. Masses of 4 kg and 6 kg are fixed at each end of the rod. Find the centre of mass of the rod.

SOLUTION

Since the rod is uniform, it can be treated as a point mass at its centre. Figure 12.5 illustrates this situation.

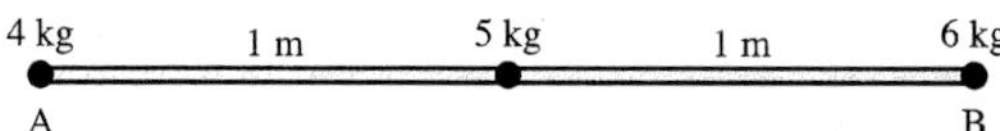

Figure 12.5

Taking the end A as origin,

$$M\bar{x} = \Sigma m_i x_i$$

$$(4 + 5 + 6)\bar{x} = 4 \times 0 + 5 \times 1 + 6 \times 2$$

$$15\bar{x} = 17$$

$$\bar{x} = \frac{17}{15} = 1\frac{2}{15}$$

So the centre of mass is 1.133 m from the 4 kg point mass.

? Check that the rod would balance about a pivot $1\frac{2}{15}$ m from A.

EXAMPLE 12.3

A rod AB of mass 1.1 kg and length 1.2 m has its centre of mass 0.48 m from the end A. What mass should be attached to the end B to ensure that the centre of mass is at the mid-point of the rod?

SOLUTION

Let the extra mass be m kg.

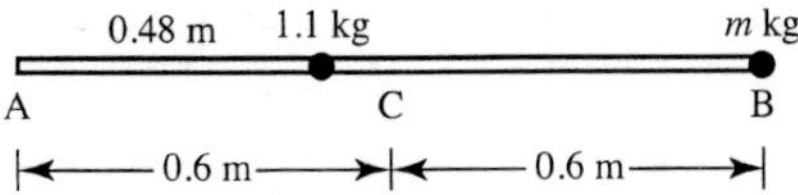

Figure 12.6

Method 1

Refer to the mid-point, C, as origin, so $\bar{x} = 0$. Then

$$(1.1 + m) \times 0 = 1.1 \times (-0.12) + m \times 0.6$$

$$\Rightarrow \qquad 0.6m = 1.1 \times 0.12$$

$$\Rightarrow \qquad m = 0.22.$$

The 1.1 mass has negative x referred to C

A mass of 220 grams should be attached to B.

Method 2

Refer to the end A, as origin, so $\bar{x} = 0.6$. Then

$$(1.1 + m) \times 0.6 = 1.1 \times 0.48 + m \times 1.2$$

$$\Rightarrow \qquad 0.66 + 0.6\,m = 0.528 + 1.2m$$

$$\Rightarrow \qquad 0.132 = 0.6m$$

$$m = 0.22 \text{ as before.}$$

Composite bodies

The position of the centre of mass of a composite body such as a cricket bat, tennis racquet or golf club is important to sports people who like to feel its balance. If the body is symmetric then the centre of mass will lie on the axis of symmetry. The next example shows how to model a composite body as a system of point masses so that the methods of the previous section can be used to find the centre of mass.

EXAMPLE 12.4

A squash racquet of mass 200 g and total length 70 cm consists of a handle of mass 150 g whose centre of mass is 20 cm from the end, and a frame of mass 50 g, whose centre of mass is 55 cm from the end.

Find the distance of the centre of mass from the end of the handle.

SOLUTION

Figure 12.7 shows the squash racquet and its dimensions.

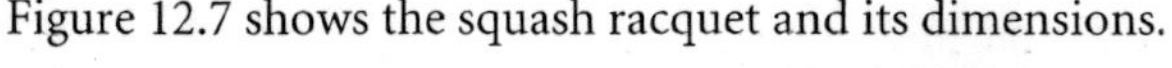

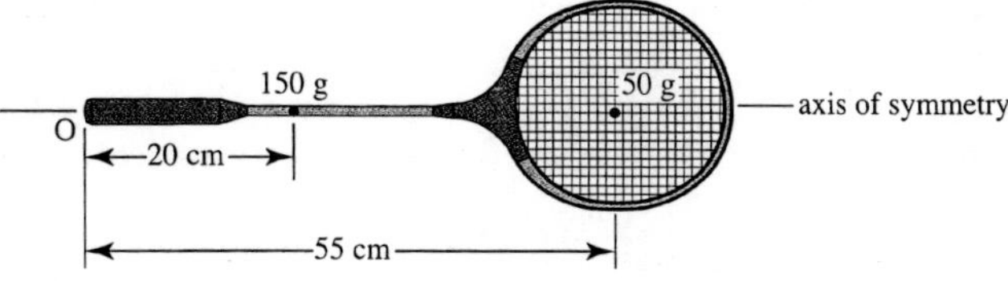

Figure 12.7

The centre of mass lies on the axis of symmetry. Model the handle as a point mass of 0.15 kg a distance 0.2 m from O and the frame as a point mass of 0.05 kg a distance 0.55 m from the end O.

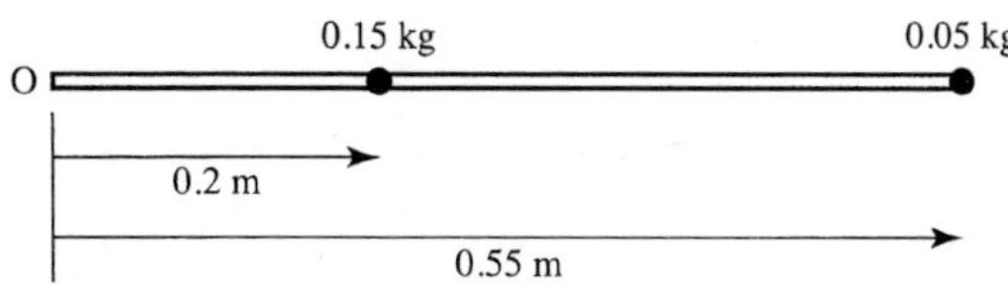

Figure 12.8

The distance, $\bar{x}$, of the centre of mass from O is given by

$$(0.15 + 0.05)\,\bar{x} = (0.15 \times 0.2) + (0.05 \times 0.55)$$
$$\bar{x} = 0.2875.$$

The centre of mass of the squash racquet is 28.75 cm from the end of the handle.

EXERCISE 12A

1 The diagrams show point masses attached to rigid light rods. In each case calculate the position of the centre of mass relative to the point O.

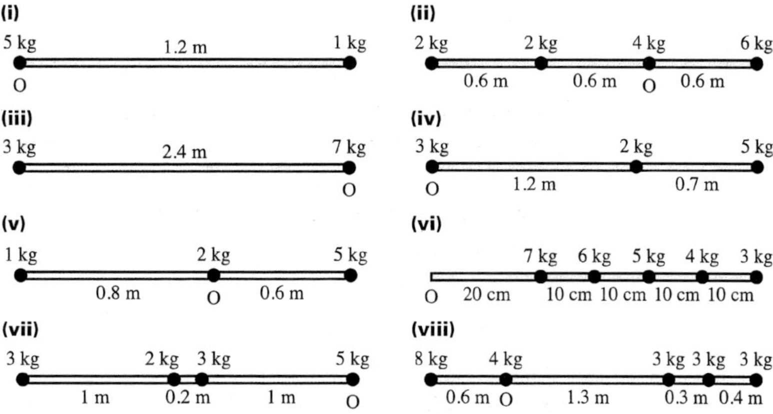

2 A see-saw consists of a uniform plank 4 m long of mass 10 kg. Calculate the centre of mass when two children, of masses 20 kg and 25 kg, sit, one on each end.

3 A weightlifter's bar in a competition has mass 10 kg and length 1 m. By mistake, 50 kg is placed on one end and 60 kg on the other end. How far is the centre of mass of the bar from the centre of the bar itself?

4 The masses of the earth and the moon are 5.98×10^{24} kg and 7.38×10^{22} kg, and the distance between their centres is 3.84×10^5 km. How far from the centre of the earth is the centre of mass of the earth–moon system?

5 A crossing warden carries a sign which consists of a uniform rod of length 1.5 m, and mass 1 kg, on top of which is a circular disc of radius 0.25 m and mass 0.2 kg. Find the distance of the centre of mass from the free end of the stick.

6 A rod has length 2 m and mass 3 kg. The centre of mass should be in the middle but due to a fault in the manufacturing process it is not. This error is corrected by placing a 200 g mass 5 cm from the centre of the rod. Where is the centre of mass of the rod itself?

7 A child's toy consists of four uniform discs, all made out of the same material. They each have thickness 2 cm and their radii are 6 cm, 5 cm, 4 cm and 3 cm. They are placed symmetrically on top of each other to form a tower. How high is the centre of mass of the tower?

8 A standard lamp consists of a uniform heavy metal base of thickness 4 cm, attached to which is a uniform metal rod of length 1.75 m and mass 0.25 kg.

What is the minimum mass for the base if the centre of mass of the lamp is no more than 12 cm from the ground?

9 A uniform scaffold pole of length 5 m has brackets bolted to it as shown in the diagram below. The mass of each bracket is 1 kg.

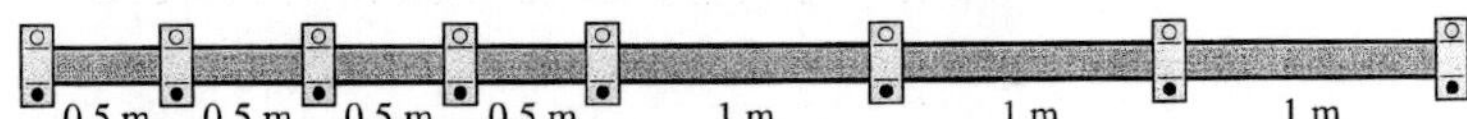

The centre of mass is 2.44 m from the left-hand end. What is the mass of the pole?

10 An object of mass m_1 is placed at one end of a light rod of length l. An object of mass m_2 is placed at the other end. Find the position of the centre of mass.

11 The diagram illustrates a mobile tower crane. It consists of the main vertical section (mass M tonnes), housing the engine, winding gear and controls, and the boom. The centre of mass of the main section is on its centre line. The boom, which has negligible mass, supports the load (L tonnes) and the counterweight (C tonnes). The main section stands on supports at P and Q, distance $2d$ m apart. The counterweight is held at a fixed distance a m from the centre line of the main section and the load at a variable distance l m.

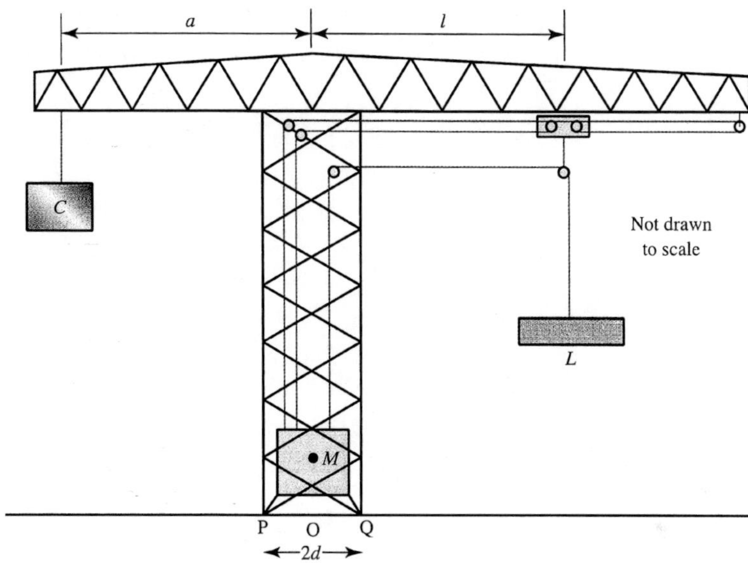

(i) In the case when $C = 3$, $M = 10$, $L = 7$, $a = 8$, $d = 2$ and $l = 13$, find the horizontal position of the centre of mass and say what happens to the crane.

(ii) Show that for these values of C, M, a, d and l the crane will not fall over when it has no load, and find the maximum safe load that it can carry.

(iii) Formulate two inequalities in terms of C, M, L, a, d and l that must hold if the crane is to be safe loaded or unloaded.

(iv) Find, in terms of M, a, d and l, the maximum load that the crane can carry.

Centre of mass for two- and three-dimensional bodies

The techniques developed for finding the centre of mass using moments can be extended into two and three dimensions.

If a two-dimensional body consists of a set of n point masses $m_1, m_2, \ldots, m_n$ located at positions $(x_1, y_1), (x_2, y_2), \ldots, (x_n, y_n)$ as in figure 12.9 (overleaf) then the position of the centre of mass of the body $(\bar{x}, \bar{y})$ is given by

$$M\bar{x} = \Sigma m_i x_i \quad \text{and} \quad M\bar{y} = \Sigma m_i y_i$$

where $M\ (= \Sigma m_i)$ is the total mass of the body.

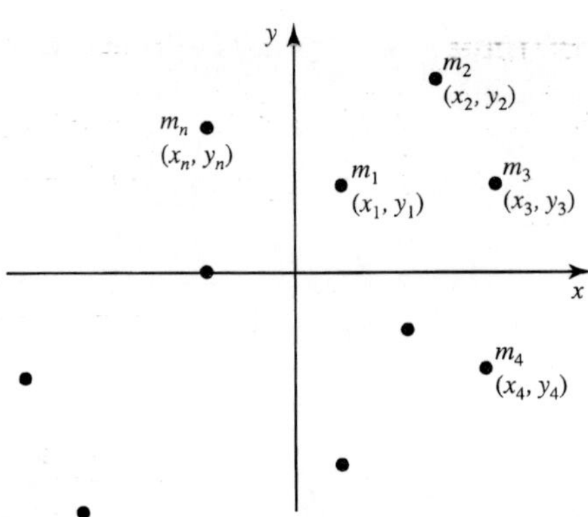

Figure 12.9

In three dimensions, the z co-ordinates are also included; to find $\bar{z}$ use

$$M\bar{z} = \Sigma m_i z_i$$

The centre of mass of any composite body in two or three dimensions can be found by replacing each component by a point mass at its centre of mass.

EXAMPLE 12.5

Joanna makes herself a pendant in the shape of a letter J made up of rectangular shapes as shown in figure 12.10.

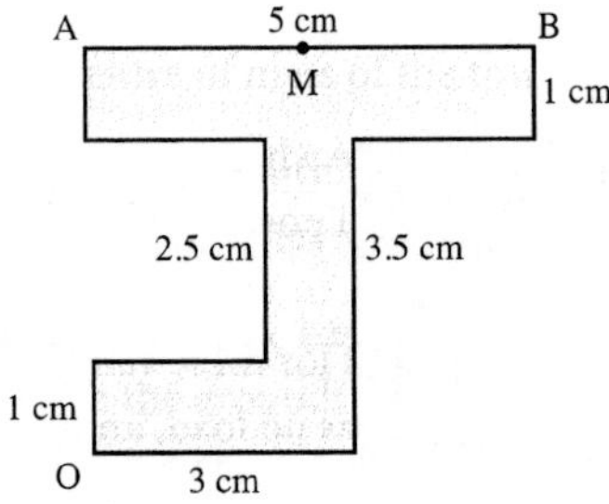

Figure 12.10

(i) Find the position of the centre of mass of the pendant.

(ii) Find the angle that AB makes with the horizontal if she hangs the pendant from a point, M, in the middle of AB.

She wishes to hang the pendant so that AB is horizontal.

(iii) How far along AB should she place the ring that the suspending chain will pass through?

SOLUTION

(i) The first step is to split the pendant into three rectangles. The centre of mass of each of these is at its middle, as shown in figure 12.11.

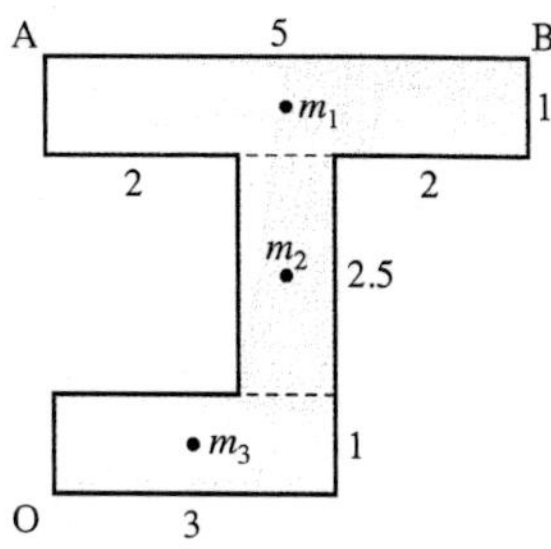

Figure 12.11

You can model the pendant as three point masses m_1, m_2 and m_3, which are proportional to the areas of the rectangular shapes. Since the areas are $5\,\text{cm}^2$, $2.5\,\text{cm}^2$ and $3\,\text{cm}^2$, the masses, in suitable units, are 5, 2.5 and 3, and the total mass is $5 + 2.5 + 3 = 10.5$ (in the same units).

The table below gives the mass and position of m_1, m_2 and m_3.

Mass	m_1	m_2	m_3	M
Mass units	5	2.5	3	10.5
Position of centre of mass x y	2.5 4	2.5 2.25	1.5 0.5	$\bar{x}$ $\bar{y}$

Now it is possible to find $\bar{x}$:

$$M\bar{x} = \Sigma m_i x_i$$
$$10.5\bar{x} = 5 \times 2.5 + 2.5 \times 2.5 + 3 \times 1.5$$
$$\bar{x} = \frac{23.25}{10.5} = 2.2\,\text{cm}$$

Similarly for $\bar{y}$:

$$M\bar{y} = \Sigma m_i y_i$$
$$10.5\bar{y} = 5 \times 4 + 2.5 \times 2.25 + 3 \times 0.5$$
$$\bar{y} = \frac{27.125}{10.5} = 2.6\,\text{cm}$$

The centre of mass is at (2.2, 2.6).

(ii) When the pendant is suspended from M, the centre of mass, G, is vertically below M, as shown in figure 12.12 (overleaf).

The pendant hangs like the first diagram but you might find it easier to draw your own diagram like the second.

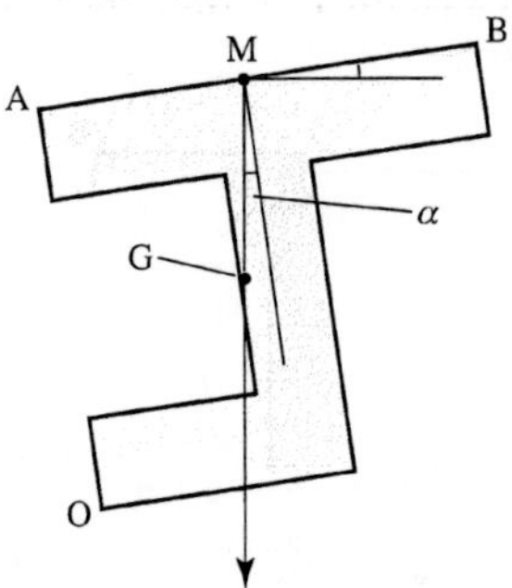

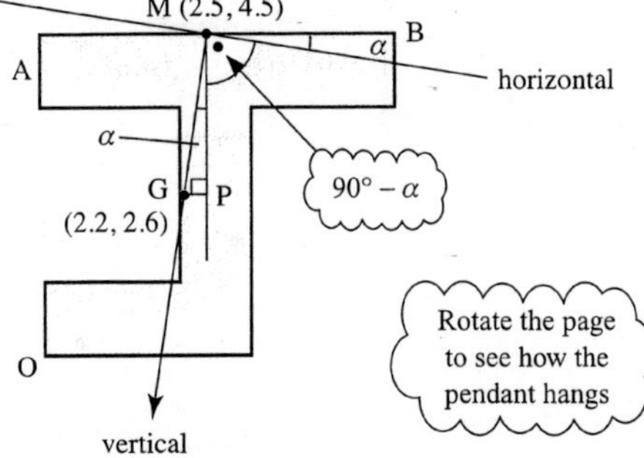

Figure 12.12

$$\mathrm{GP} = 2.5 - 2.2 = 0.3$$

$$\mathrm{MP} = 4.5 - 2.6 = 1.9$$

$$\therefore \quad \tan\alpha = \frac{0.3}{1.9} \Rightarrow \alpha = 9°$$

AB makes an angle of 9° with the horizontal (or 8.5° working with unrounded figures).

(iii) For AB to be horizontal the point of suspension must be directly above the centre of mass, and so it is 2.2 cm from A.

EXAMPLE 12.6

Find the centre of mass of a body consisting of a square plate of mass 3 kg and side length 2 m, with small objects of mass 1 kg, 2 kg, 4 kg and 5 kg at the corners of the square.

SOLUTION

Figure 12.13 shows the square plate, with the origin taken at the corner at which the 1 kg mass is located. The mass of the plate is represented by a 3 kg point mass at its centre.

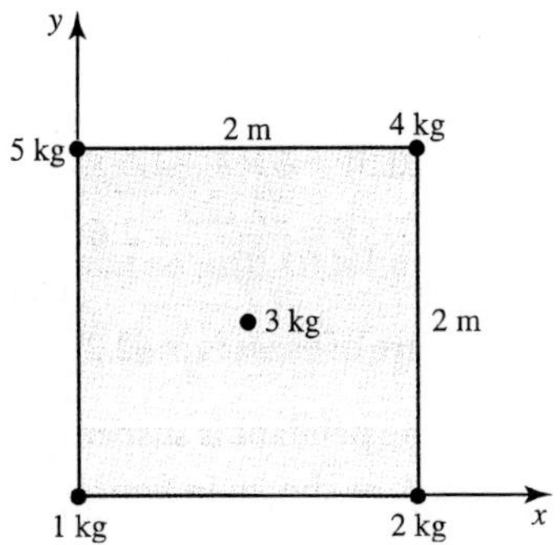

Figure 12.13

In this example the total mass M (in kilograms) is $1 + 2 + 4 + 5 + 3 = 15$.

The two formulae for $\bar{x}$ and $\bar{y}$ can be combined into one using column vector notation:

$$\begin{pmatrix} M\bar{x} \\ M\bar{y} \end{pmatrix} = \begin{pmatrix} \Sigma m_i x_i \\ \Sigma m_i y_i \end{pmatrix}$$

which is equivalent to

$$M\begin{pmatrix} \bar{x} \\ \bar{y} \end{pmatrix} = \Sigma m_i \begin{pmatrix} x_i \\ y_i \end{pmatrix}$$

Substituting our values for M and m_i and x_i and y_i:

$$15\begin{pmatrix} \bar{x} \\ \bar{y} \end{pmatrix} = 1\begin{pmatrix} 0 \\ 0 \end{pmatrix} + 2\begin{pmatrix} 2 \\ 0 \end{pmatrix} + 4\begin{pmatrix} 2 \\ 2 \end{pmatrix} + 5\begin{pmatrix} 0 \\ 2 \end{pmatrix} + 3\begin{pmatrix} 1 \\ 1 \end{pmatrix}$$

$$15\begin{pmatrix} \bar{x} \\ \bar{y} \end{pmatrix} = \begin{pmatrix} 15 \\ 21 \end{pmatrix}$$

$$\begin{pmatrix} \bar{x} \\ \bar{y} \end{pmatrix} = \begin{pmatrix} 1 \\ 1.4 \end{pmatrix}$$

The centre of mass is at the point (1, 1.4).

EXAMPLE 12.7

A metal disc of radius 15 cm has a hole of radius 5 cm cut in it as shown in figure 12.14. Find the centre of mass of the disc.

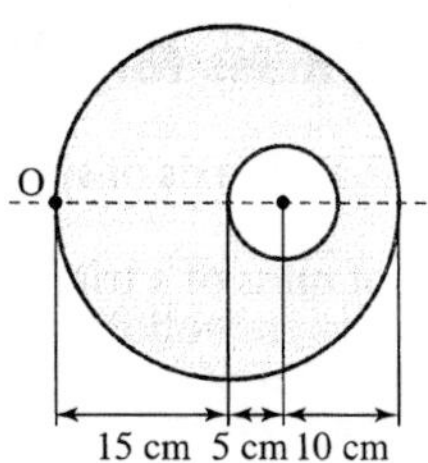

Figure 12.14

SOLUTION

Think of the original uncut disc as a composite body made up of the final body and a disc to fit into the hole. Since the material is uniform the mass of each part is proportional to its area.

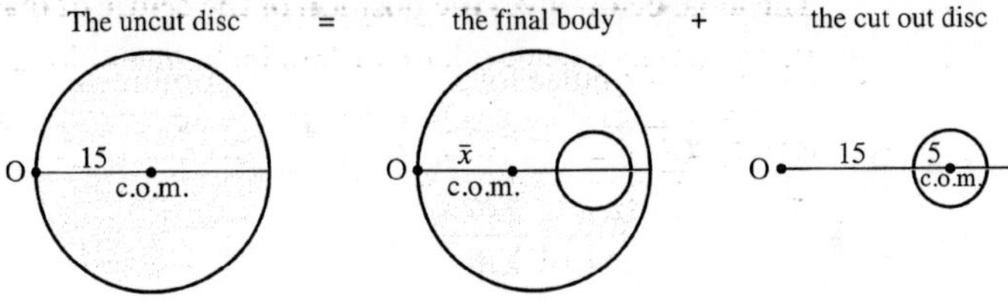

Figure 12.15

	Uncut disc	Final body	Cut out disc
Area	$15^2\pi = 225\pi$	$15^2\pi - 5^2\pi = 200\pi$	$5^2\pi = 25\pi$
Distance from O to centre of mass	15 cm	$\bar{x}$ cm	20 cm

Taking moments about O:

$$225\pi \times 15 = 200\pi \times \bar{x} + 25\pi \times 20$$

Divide by π

$$\Rightarrow \qquad \bar{x} = \frac{225 \times 15 - 25 \times 20}{200}$$

$$= 14.375$$

The centre of mass is 14.4 cm from O, that is 0.6 cm to the left of the centre of the disc.

Centres of mass for different shapes

If an object has an axis of symmetry, then the centre of mass lies on it.

The centre of mass of a triangular lamina lies on the intersection of the medians.

The triangle in figure 12.16 is divided up into thin strips parallel to the side AB.

The median of a triangle joins a vertex to the mid-point of the opposite side.

The centre of mass of each strip lies in the middle of the strip, at the points $C_1, C_2, C_3, \ldots$.

When these points are joined they form the median of the triangle drawn from C.

Similarly, the centre of mass also lies on the medians from B and from C. Therefore, the centre of mass lies at the intersection of the three medians; this is the *centroid* of the triangle. This point is $\frac{2}{3}$ of the distance along the median from the vertex.

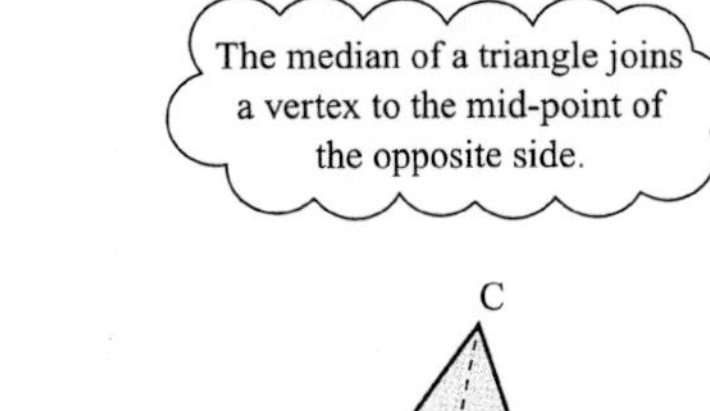

Figure 12.16

The table below gives the position of the centre of mass of some uniform objects that you may encounter, or wish to include within models of composite bodies.

Body	Position of centre of mass	Diagram
Triangular lamina	$\frac{2}{3}$ along the median from vertex	
Solid cone or pyramid	$\frac{3}{4}h$ from vertex	
Solid hemisphere	$\frac{3}{8}r$ from centre	
Hemispherical shell	$\frac{1}{2}r$ from centre	
Circular sector of radius r and angle 2α radians	$\frac{2r\sin\alpha}{3\alpha}$ from centre	
Circular arc of radius r and angle 2α radians	$\frac{r\sin\alpha}{\alpha}$ from centre	

EXERCISE 12B

1 Find the centre of mass of the following sets of point masses.

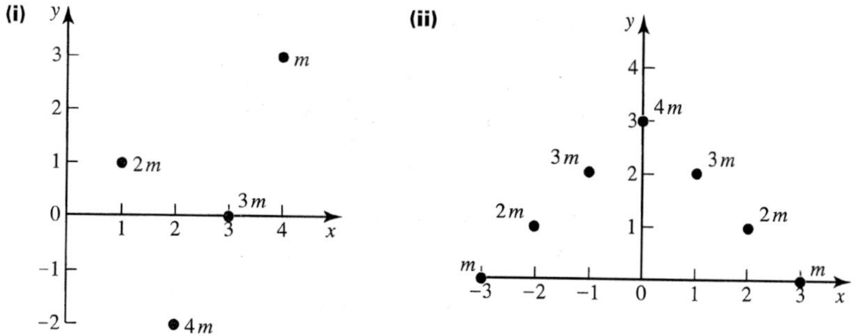

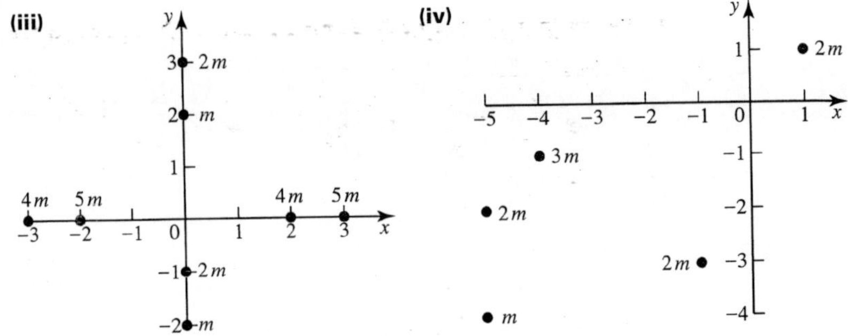

2 Masses of 1, 2, 3 and 4 grams are placed at the corners A, B, C and D of a square piece of uniform cardboard of side 10 cm and mass 5 g. Find the position of the centre of mass relative to axes through AB and AD.

3 As part of an illuminated display, letters are produced by mounting bulbs in holders 30 cm apart on light wire frames. The combined mass of a bulb and its holder is 200 g. Find the position of the centre of mass for each of the letters shown below, in terms of its horizontal and vertical displacement from the bottom left-hand corner of the letter.

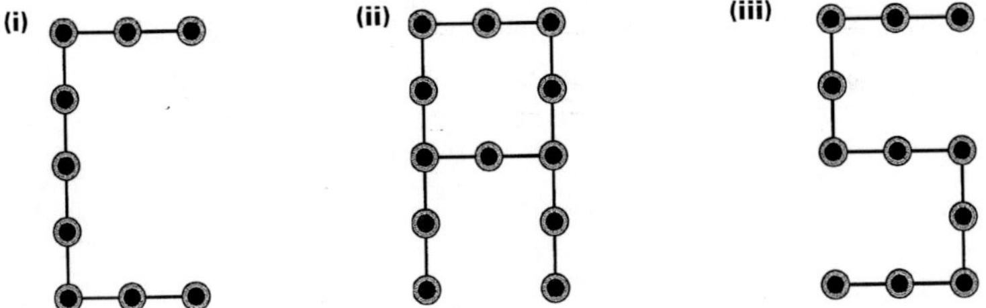

4 Four people of masses 60 kg, 65 kg, 62 kg and 75 kg sit on the four seats of the fairground ride shown below. The seats and the connecting arms are light. Find the radius of the circle described by the centre of mass when the ride rotates about O.

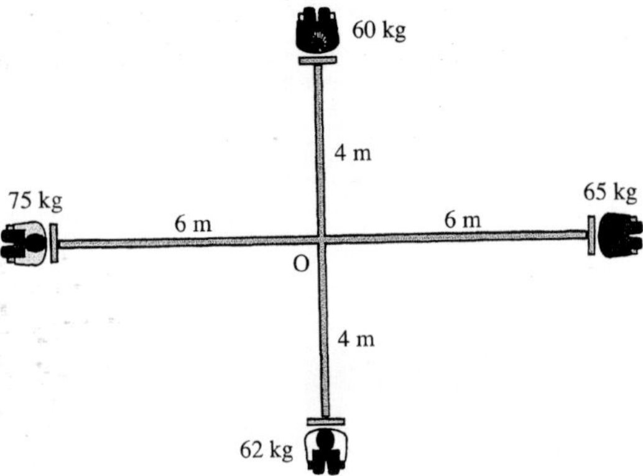

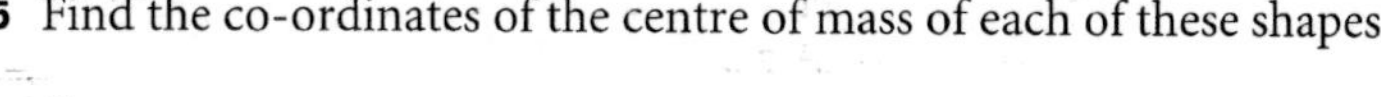

5 Find the co-ordinates of the centre of mass of each of these shapes.

(i)

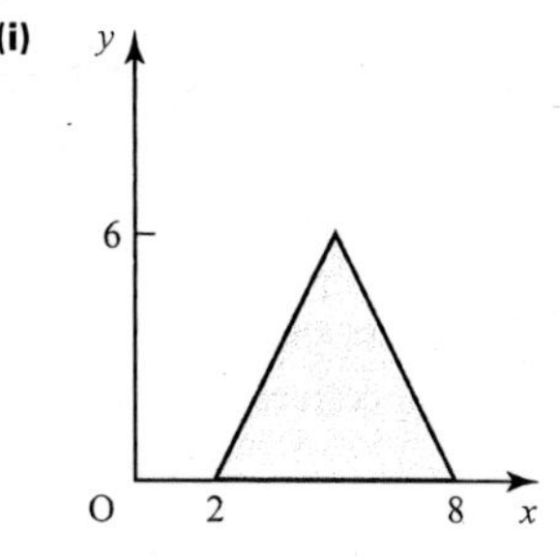

(ii)

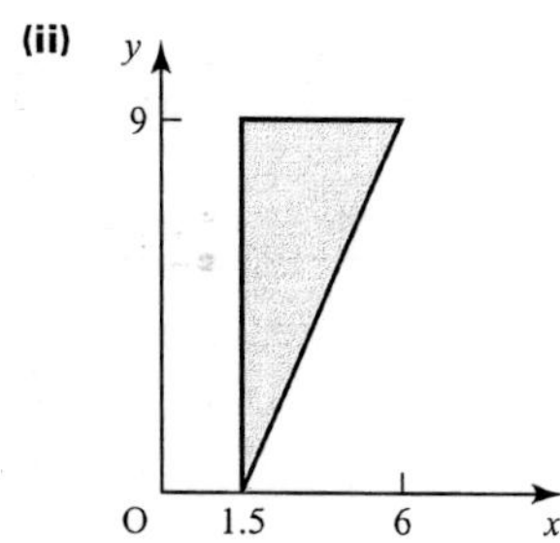

(iii)

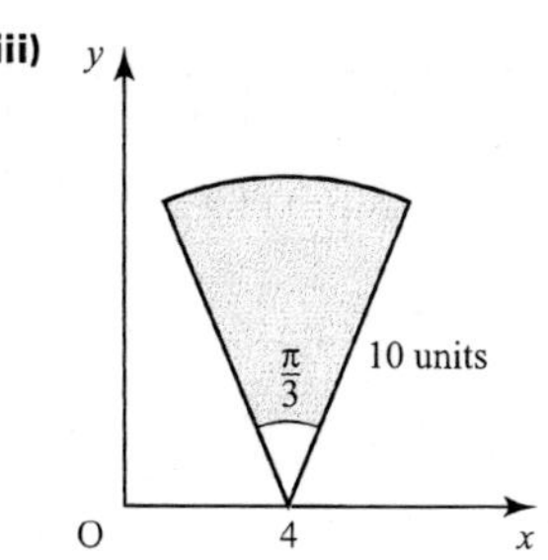

(iv)

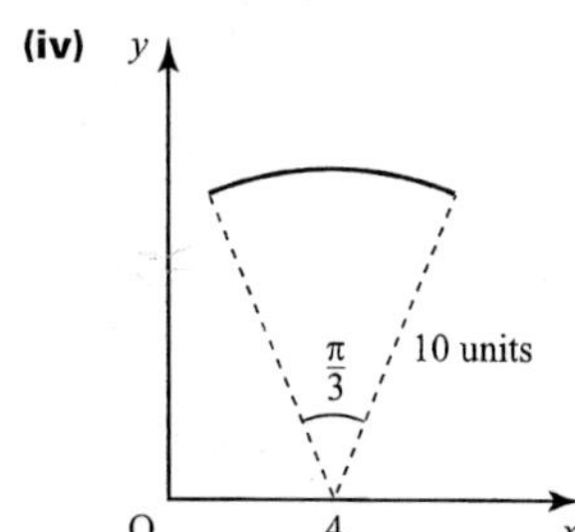

6 The following shapes are made out of uniform card.

For each shape find the co-ordinates of the centre of mass relative to O.

(i)

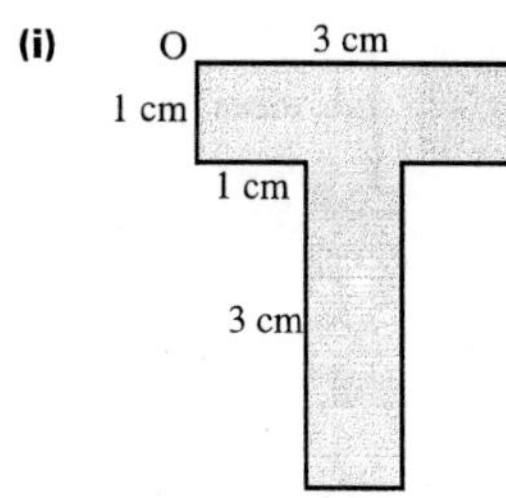

(ii)

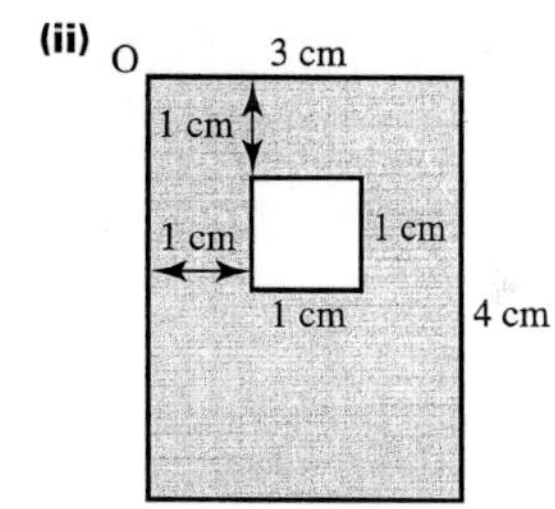

7 A pendant is made from a uniform circular disc of mass $4m$ and radius 2 cm with a decorative edging of mass m as shown. The centre of mass of the decoration is 1 cm below the centre, O, of the disc. The pendant is symmetrical about the diameter AB.

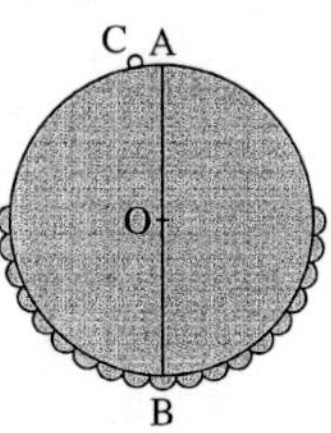

(i) Find the position of the centre of mass of the pendant.

The pendant should be hung from A but the light ring for hanging it is attached at C where angle AOC is 10°.

(ii) Find the angle between AB and the vertical when the pendant is hung from C.

8 A uniform rectangular lamina, ABCD, where AB is of length a and BC of length $2a$, has a mass $10m$. Further point masses m, $2m$, $3m$ and $4m$ are fixed to the points A, B, C and D, respectively.

(i) Find the centre of mass of the system relative to x and y axes along AB and AD respectively.

(ii) If the lamina is suspended from the point A find the angle that the diagonal AC makes with the vertical.

(iii) To what must the mass at point D be altered if this diagonal is to hang vertically?

[MEI]

9 The diagram gives the dimensions of the design of a uniform metal plate.

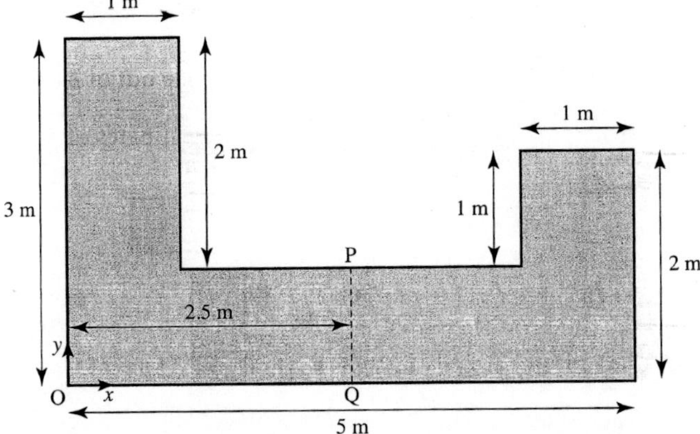

Using a co-ordinate system with O as origin, the x and y axes as shown and 1 metre as 1 unit,

(i) show that the centre of mass has y co-ordinate 1 and find its x co-ordinate.

The design requires the plate to have its centre of mass half-way across (i.e. on the line PQ in the diagram), and in order to achieve this a circular hole centred on $\left(\frac{1}{2}, \frac{1}{2}\right)$ is considered.

(ii) Find the appropriate radius for such a hole and explain why this idea is not feasible.

It is then decided to cut two circular holes each of radius r, both centred on the line $x = \frac{1}{2}$. The first hole is centred at $\left(\frac{1}{2}, \frac{1}{2}\right)$ and the centre of mass of the plate is to be at P.

(iii) Find the value of r and the co-ordinates of the centre of the second hole.

[MEI]

10 A uniform triangular lamina ABC is right-angled at B and has sides AB = 0.6 m and BC = 0.8 m. The mass of the lamina is 4 kg. One end of a light inextensible rope is attached to the lamina at C. The other end of the rope is attached to a fixed point D on a vertical wall. The lamina is in equilibrium with A in contact with the wall at a point vertically below D. The lamina is in a vertical plane perpendicular to the wall, and AB is horizontal. The rope is taut and at right angles to AC (see diagram).

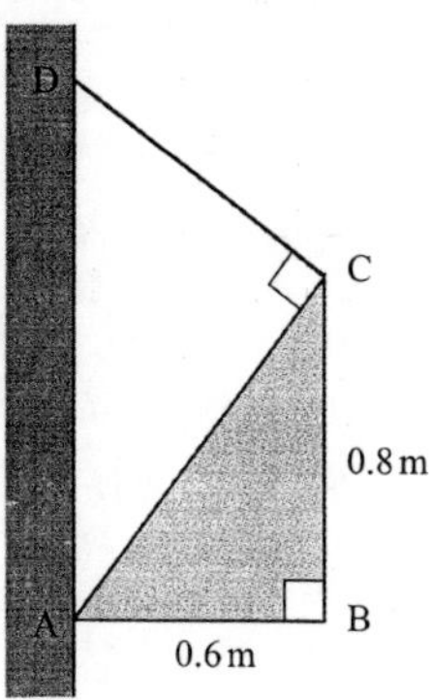

Find

(i) the tension in the rope,

(ii) the horizontal and vertical components of the force exerted at A on the lamina by the wall.

[Cambridge AS and A Level Mathematics 9709, Paper 5 Q4 June 2007]

11 A uniform rigid wire AB is in the form of a circular arc of radius 1.5 m with centre O. The angle AOB is a right angle. The wire is in equilibrium, freely suspended from the end A. The chord AB makes an angle of $\theta°$ with the vertical (see diagram).

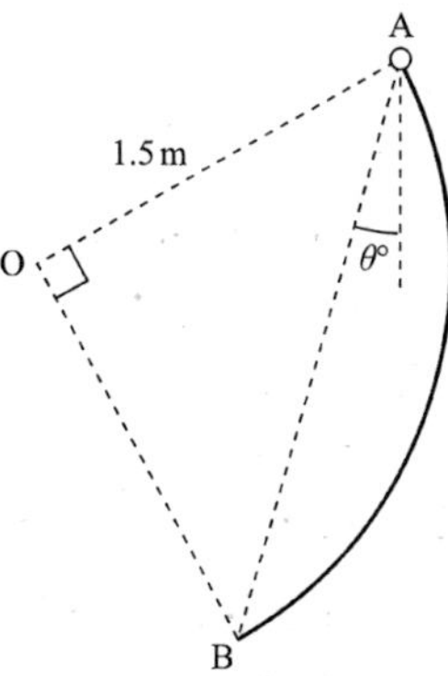

(i) Show that the distance of the centre of mass of the arc from O is 1.35 m, correct to 3 significant figures.

(ii) Find the value of θ.

[Cambridge AS and A Level Mathematics 9709, Paper 5 Q2 June 2008]

12 A uniform lamina ABCD is in the form of a trapezium in which AB and DC are parallel and have lengths 2 m and 3 m respectively. BD is perpendicular to the parallel sides and has length 1 m (see diagram).

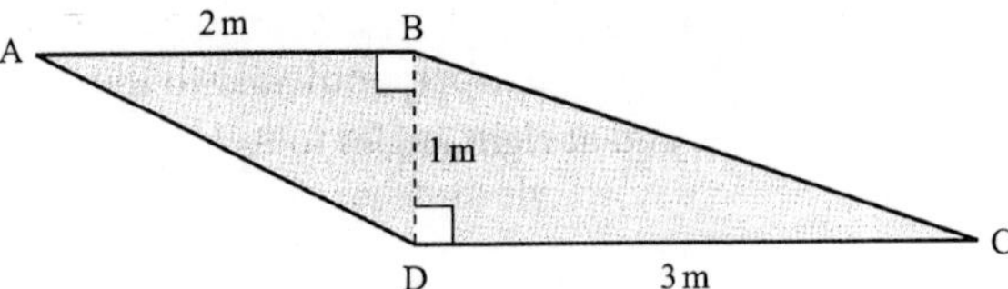

(i) Find the distance of the centre of mass of the lamina from BD.

The lamina has weight W N and is in equilibrium, suspended by a vertical string attached to the lamina at B. The lamina rests on a vertical support at C. The lamina is in a vertical plane with AB and DC horizontal.

(ii) Find, in terms of W, the tension in the string and the magnitude of the force exerted on the lamina at C.

[Cambridge AS and A Level Mathematics 9709, Paper 5 Q3 November 2005]

13 P is the vertex of a uniform solid cone of mass 5 kg, and O is the centre of its base. Strings are attached to the cone at P and at O. The cone hangs in equilibrium with PO horizontal and the strings taut. The strings attached at P and O make angles of $\theta°$ and 20°, respectively, with the vertical (see diagram, which shows a cross-section).

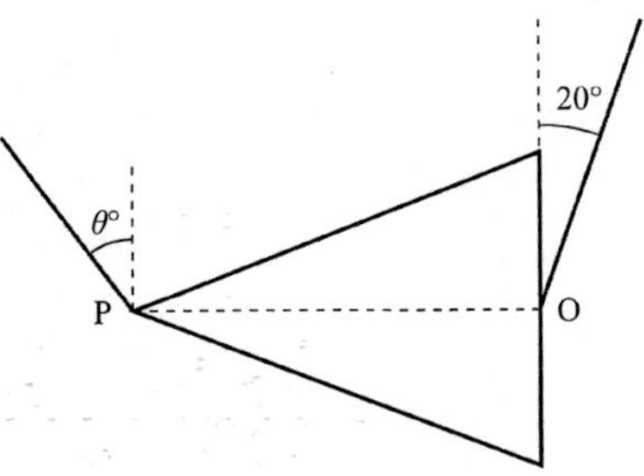

(i) By taking moments about P for the cone, find the tension in the string attached at O.

(ii) Find the value of θ and the tension in the string attached at P.

[Cambridge AS and A Level Mathematics 9709, Paper 52 Q6 November 2009]

14 A uniform lamina of weight 15 N has dimensions as shown in the diagram.

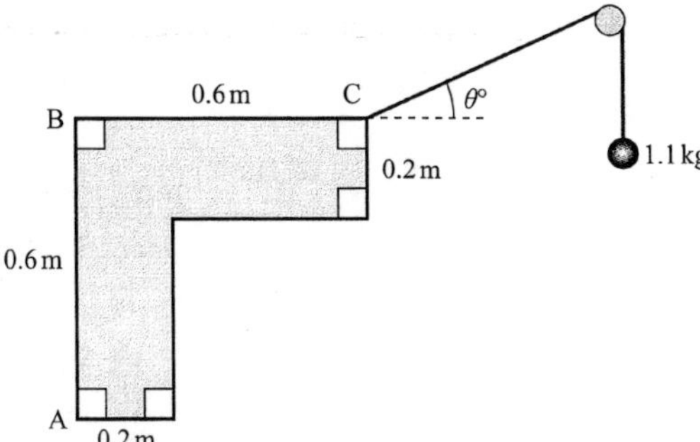

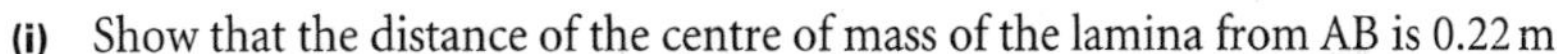

(i) Show that the distance of the centre of mass of the lamina from AB is 0.22 m.

The lamina is freely hinged at B to a fixed point. One end of a light inextensible string is attached to the lamina at C. The string passes over a fixed smooth pulley and a particle of mass 1.1 kg is attached to the other end of the string. The lamina is in equilibrium with BC horizontal. The string is taut and makes an angle of $\theta°$ with the horizontal at C, and the particle hangs freely below the pulley (see diagram).

(ii) Find the value of θ.

[Cambridge AS and A Level Mathematics 9709, Paper 5 Q5 June 2006]

Sliding and toppling

The photograph shows a double decker bus on a test ramp. The angle of the ramp to the horizontal is slowly increased.

[Photo courtesy of Millbrook Proving Ground Ltd]

? What happens to the bus? Would a loaded bus behave differently from the empty bus in the photograph?

EXPERIMENT

The diagrams show a force being applied in different positions to a cereal packet.

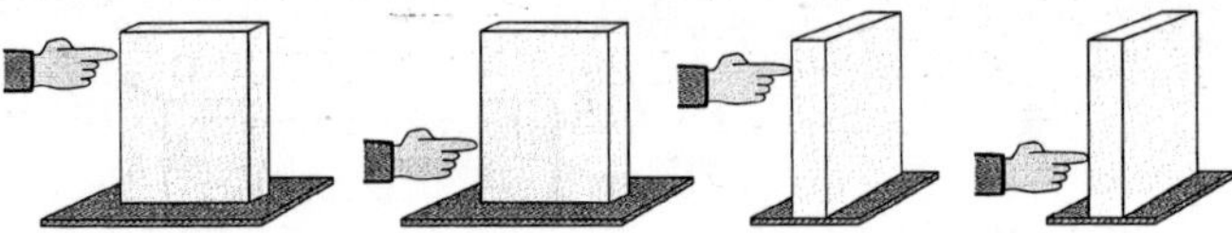

Figure 12.17

In which case do you think the packet is most likely to fall over? In which case is it most likely to slide? Investigate your answers practically, using boxes of different shapes.

Figure 12.18 shows the cereal packet placed on a slope. Is the box more likely to topple or slide as the angle of the slope to the horizontal increases?

Figure 12.18

To what extent is this situation comparable to that of the bus on the test ramp?

Two critical cases

When an object stands on a surface, the only forces acting are its weight W and the *resultant* of all the contact forces between the surfaces which must act through a point on both surfaces. This resultant contact force is often resolved into two components: the friction, F, parallel to any possible sliding and the normal reaction, R, perpendicular to F as in figures 12.19–12.21.

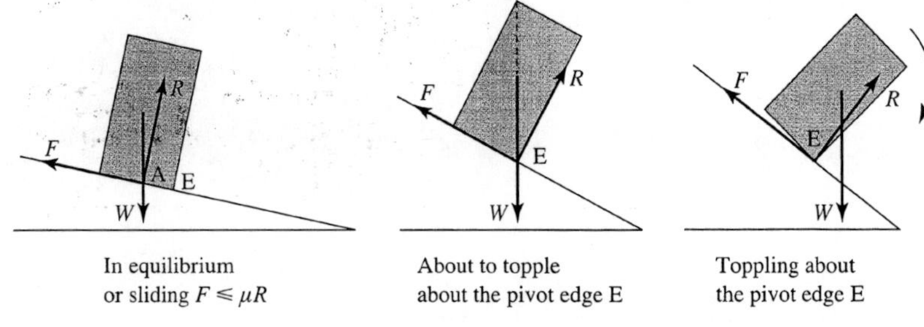

In equilibrium or sliding $F \leqslant \mu R$

Figure 12.19

About to topple about the pivot edge E

Figure 12.20

Toppling about the pivot edge E

Figure 12.21

Equilibrium can be broken in two ways:

(i) *The object is on the point of sliding*; then $F = \mu R$ according to our model.

(ii) *The object is on the point of toppling.* The pivot is at the lowest point of contact which is the point E in figure 12.20. In this critical case:

- the centre of mass is directly above E so the weight acts vertically downwards through E;
- the resultant reaction of the plane on the object acts through E, vertically upwards. This is the resultant of F and R.

? Why does the object topple in figure 12.21?

When three non-parallel forces are in equilibrium, their lines of action must be concurrent (they must all pass through one point). Otherwise there is a resultant moment about the point where two of them meet as in figure 12.21.

EXAMPLE 12.8

An increasing force P N is applied to a block, as shown in figure 12.22, until the block moves. The coefficient of friction between the block and the plane is 0.4. Does it slide or topple?

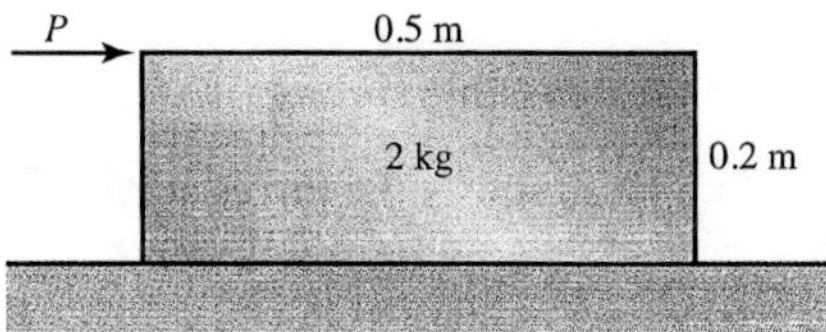

Figure 12.22

SOLUTION

The forces acting are shown in figure 12.23. The normal reaction may be thought of as a single force acting somewhere within the area of contact. When toppling occurs (or is about to occur) the line of action is through the edge about which it topples.

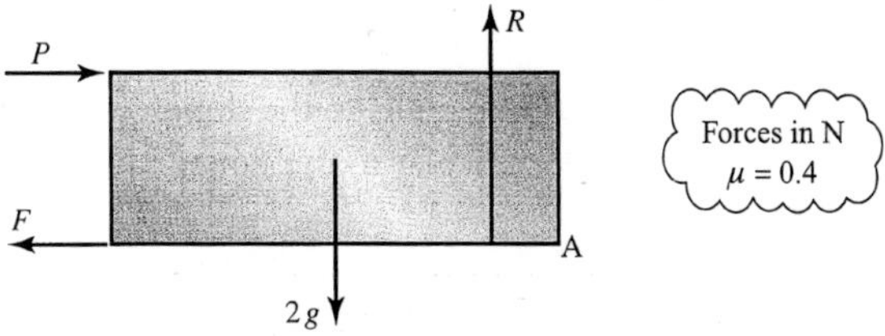

Figure 12.23

Until the block moves, it is in equilibrium.

Horizontally:	$P = F$	①
Vertically:	$R = 2g$	②
If *sliding* is about to occur	$F = \mu R$	
From ①	$P = \mu R = 0.4 \times 2g$	
	$= 8$	

If the block is about to *topple*, then A is the pivot point and the reaction of the plane on the block acts at A. Taking moments about A gives

$$(\curvearrowleft) \quad 2g \times 0.25 - P \times 0.2 = 0$$

$$P = 25$$

So to slide P needs to exceed 8 N but to topple it needs to exceed 25 N: the block will slide before it topples.

EXAMPLE 12.9

A rectangular block of mass 3 kg is placed on a slope as shown. The angle α is gradually increased. What happens to the block, given that the coefficient of friction between the block and slope is 0.6?

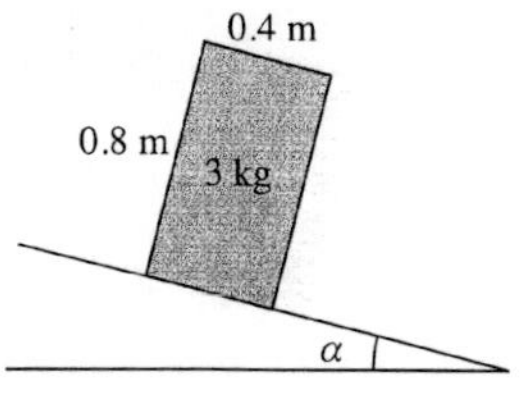

Figure 12.24

SOLUTION

Check for possible sliding

Figure 12.25 shows the forces acting when the block is in equilibrium.

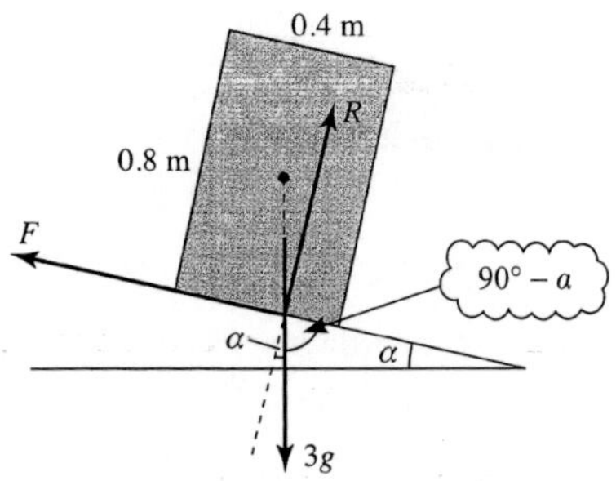

Figure 12.25

Resolve parallel to the slope:	$F = 3g \sin \alpha$
Perpendicular to the slope:	$R = 3g \cos \alpha$

When the block is on the point of sliding $F = \mu R$ so

$$3g \sin \alpha = \mu \times 3g \cos \alpha$$

$$\Rightarrow \quad \tan \alpha = \mu = 0.6$$

$$\Rightarrow \quad \alpha = 31°$$

The block is on the point of sliding when $\alpha = 31°$.

Check for possible toppling

When the block is on the point of toppling about the edge E the centre of mass is vertically above E, as shown in figure 12.26.

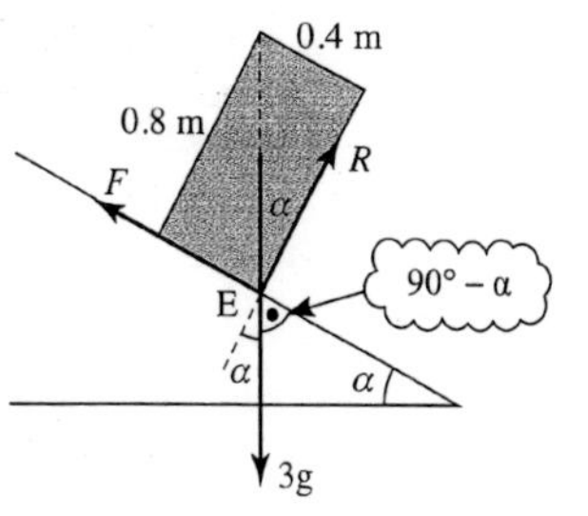

Figure 12.26

Then the angle α is given by:

$$\tan \alpha = \frac{0.4}{0.8}$$

$$\alpha = 26.6°$$

The block topples when $\alpha = 26.6°$.

The angle for sliding (31°) is greater than the angle for toppling (26.6°), so the block topples without sliding when $\alpha = 26.6°$.

Is it possible for sliding and toppling to occur for the same angle?

EXERCISE 12C

1 A force of magnitude PN acts as shown on a block resting on a horizontal plane. The coefficient of friction between the block and the plane is 0.7.

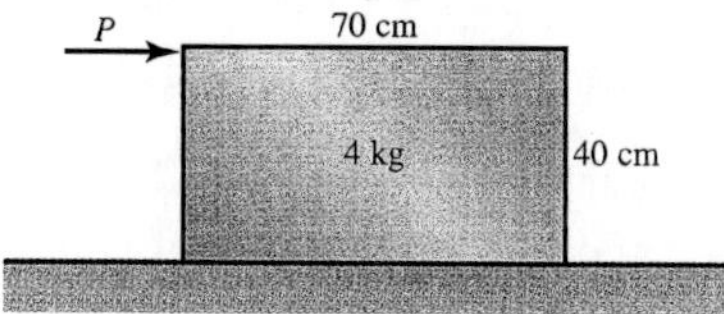

The magnitude of the force P is gradually increased from zero.

(i) Find the magnitude of P if the block is on the point of sliding, assuming it does not topple.

(ii) Find the magnitude of P if the block is on the point of toppling, assuming it does not slide.

(iii) Does the block slide or topple?

2 A solid uniform cuboid is placed on a horizontal surface. A force P is applied as shown in the diagram.

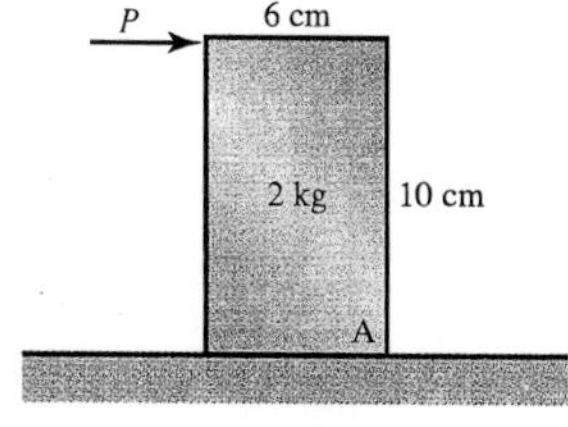

(i) If the block is on the point of sliding express P in terms of μ, the coefficient of friction between the block and the plane.

(ii) Find the magnitude of P if the cuboid is on the point of toppling.

(iii) For what values of μ will the block slide before it topples?

(iv) For what values of μ will the block topple before it slides?

3 A horizontal force of increasing magnitude is applied to the middle of the face of a 50 cm uniform cube, at right angles to the face. The coefficient of friction between the cube and the surface is 0.4 and the cube is on a level surface. What happens to the cube?

4 A solid uniform cube of side 4 cm and weight 60 N is situated on a rough horizontal plane. The coefficient of friction between the cube and the plane is 0.4. A force PN acts in the middle of one of the edges of the top of the cube, as shown in the diagram.

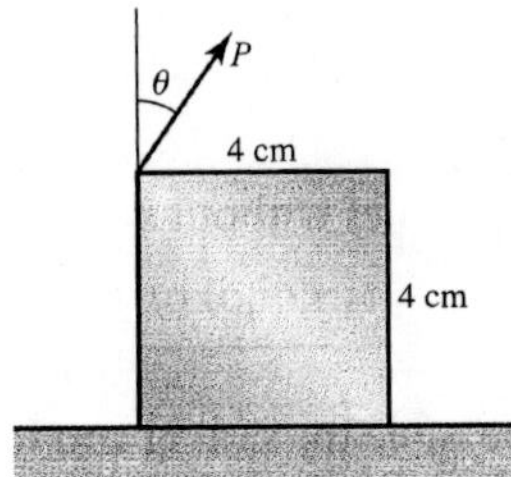

In the cases when the value of θ is **(a)** 60° **(b)** 80°, find

(i) the force P needed to make the cube slide, assuming it does not topple

(ii) the force P needed to make the cube topple, assuming it does not slide

(iii) whether it first slides or topples as the force P is increased.

For what value of θ do toppling and sliding occur for the same value of P, and what is that value of P?

5 A uniform rectangular block of height 30 cm and width 10 cm is placed on a rough plane inclined at an angle α to the horizontal. The block lies on the plane with its length horizontal. The coefficient of friction between the block and the plane is 0.25.

(i) Assuming that it does not topple, for what value of α does the block just slide?

(ii) Assuming that it does not slide, for what value of α does the block just topple?

(iii) The angle α is increased slowly from an initial value of 0°. Which happens first, sliding or toppling?

6 A solid uniform cuboid, 10 cm × 20 cm × 50 cm, is to stand on an inclined plane, which makes an angle α with the horizontal. One edge of the cuboid is to be parallel to the line of the slope. The coefficient of friction between the cuboid and the plane is μ.

(i) Which face of the cuboid should be placed on the slope to make it

(a) least likely and **(b)** most likely to topple?

(ii) How does the cuboid's orientation influence the likelihood of it sliding?

(iii) Find the range of values of μ in the situations where

(a) it will slide first whatever its orientation

(b) it will topple first whatever its orientation.

7 A cube of side 4 cm and mass 100 g is acted on by a force as shown in the diagram.

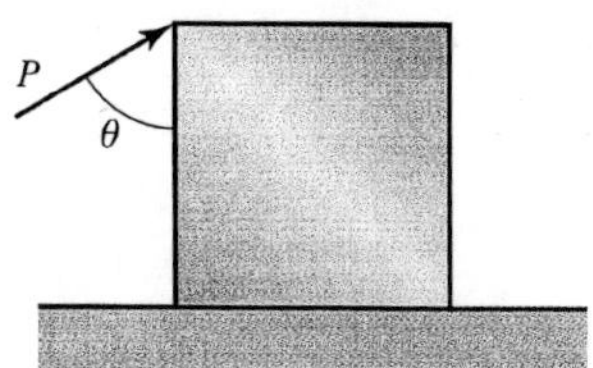

The coefficient of friction between the cube and the plane is 0.3. What happens to the cube if

(i) $\theta = 45°$ and $P = 0.3$ N?

(ii) $\theta = 15°$ and $P = 0.45$ N?

8 A packing case is in the form of a cube of side 1 m. Its weight W newtons may be taken as acting at the centre of the cube. A man is trying to push the case up uniformly sloping ground inclined at an angle α to the horizontal, with a force P newtons applied to the middle of the top edge of the case, as shown in the diagram, in a direction parallel to the slope and at right angles to the edge of the case. The coefficient of friction between the case and the ground is μ.

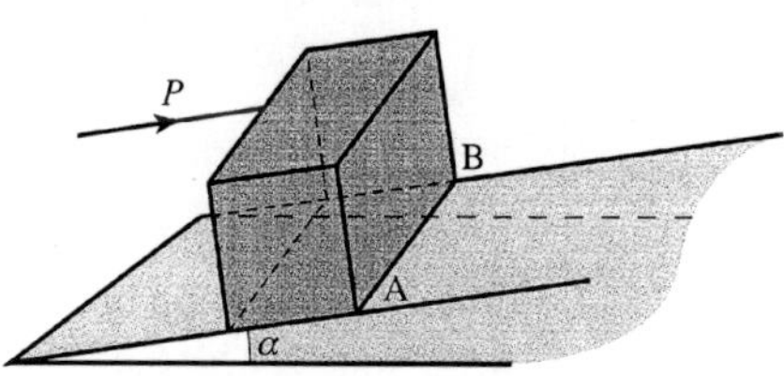

(i) Find the normal reaction of the ground on the case in terms of some or all of W, P, μ, g and α.

Take the value of W to be 200 and that of α to be 30°. Assuming that the case does not turn about the edge AB,

(ii) show that the case will slip if $P > 100(1 + \sqrt{3}\mu)$.

It is possible that the case turns about the line AB before it slips. Assume that this happens and that the case is on the point of turning.

(iii) Find the moment of the weight about the line AB and hence, or otherwise, find the values of P for which the case will turn.

The man applies the least force P necessary to move the case.

(iv) For what values of μ will the case slip and not turn?

[MEI]

9 A filing cabinet has the dimensions shown in the diagram. The body of the cabinet has mass 20 kg and its construction is such that its centre of mass is at a height of 60 cm, and is 25 cm from the back of the cabinet. The mass of a drawer and its contents may be taken to be 10 kg and its centre of mass to be 10 cm above its base and 10 cm from its front face.

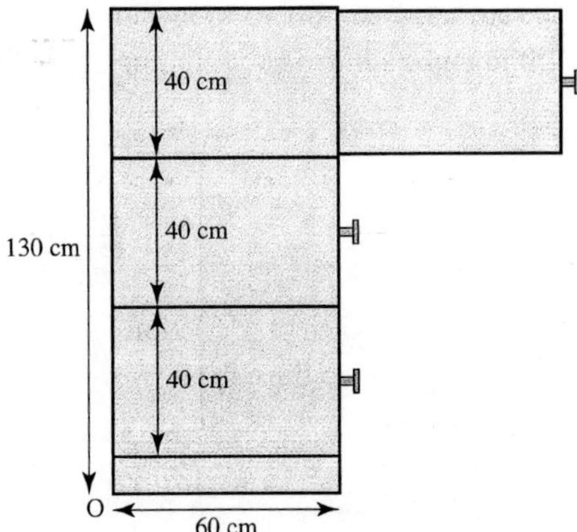

(i) Find the position of the centre of mass when all the drawers are closed.

(ii) Find the position of the centre of mass when the top two drawers are fully open.

(iii) Show that when all three drawers are fully opened the filing cabinet will tip over.

(iv) Two drawers are fully open. How far can the third one be opened without the cabinet tipping over?

10 A bird table is made from a uniform square base of side 0.3 m with mass 5 kg, a uniform square top of side 0.5 m and mass 2 kg, and a uniform thin rod of length 1.6 m and mass 1 kg connecting the centre of the top and base. The top and base have negligible thickness.

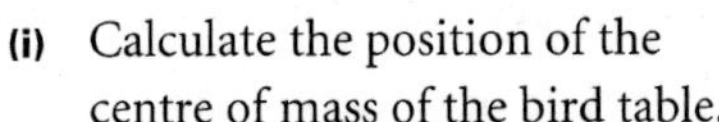

(i) Calculate the position of the centre of mass of the bird table.

(ii) At what angle can the bird table be turned about an edge of the base before it will topple?

It is decided to make the base heavier so that the bird table can be tipped at 40° to the horizontal before it topples. The base still has negligible thickness.

(iii) Show that the centre of mass must now be about 0.18 m above the base.

(iv) What is the new mass of the base?

[MEI]

11 Uniform wooden bricks have length 20 cm and height 5 cm. They are glued together as shown in the diagram with each brick 5 cm to the right of the one below it. The origin is taken to be at O.

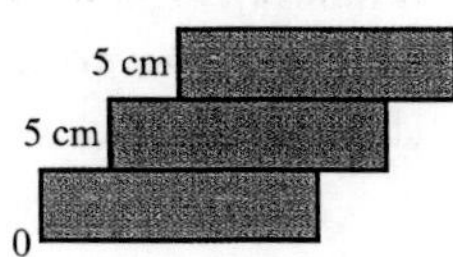

(i) Find the co-ordinates of the centre of mass for

(a) 1 **(b)** 2 **(c)** 3 **(d)** 4 **(e)** 5 bricks.

(ii) How many bricks is it possible to assemble in this way without them tipping over?

(iii) If the displacement is changed from 5 cm to 2 cm find the co-ordinates of the centre of mass for n bricks. How many bricks can now be assembled?

(iv) If the displacement is $\frac{1}{2}$ cm, what is the maximum height possible for the centre of mass of such an assembly of bricks without them tipping over?

12 A uniform solid cone has height 30 cm and base radius r cm. The cone is placed with its axis vertical on a rough horizontal plane. The plane is slowly tilted and the cone remains in equilibrium until the angle of inclination of the plane reaches 35°, when the cone topples. The diagram shows a cross-section of the cone.

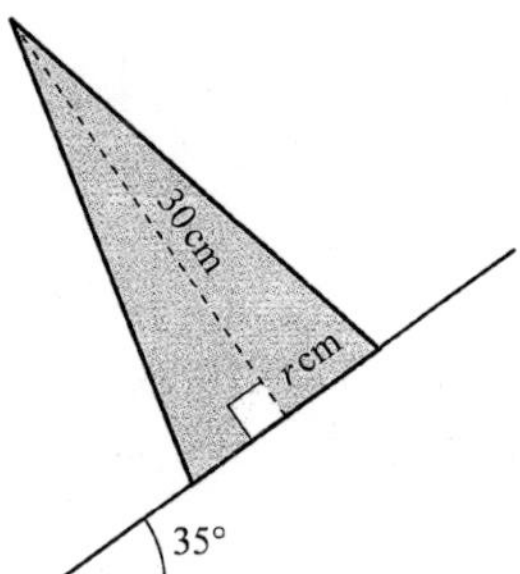

(i) Find the value of r.

(ii) Show that the coefficient of friction between the cone and the plane is greater than 0.7.

[Cambridge AS and A Level Mathematics 9709, Paper 51 Q2 June 2010]

13 A uniform prism has a cross-section in the form of a triangle ABC which is right-angled at A. The sides AB and AC have lengths 4 cm and 3 cm respectively. The prism is held with the edge containing C in contact with a horizontal surface and with AC making an angle of 60° with the horizontal (see diagram). The prism is now released. Determine whether it falls on the face containing AC or the face containing BC.

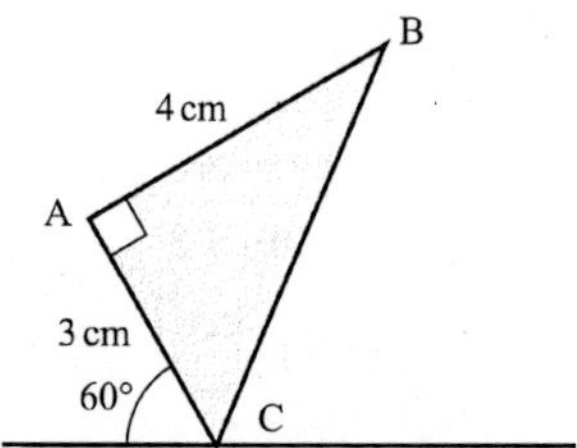

[Cambridge AS and A Level Mathematics 9709, Paper 52 Q1 November 2009]

14 Figure **(A)** shows the cross-section of a uniform solid. The cross-section has the shape and dimensions shown. The centre of mass C of the solid lies in the plane of this cross-section. The distance of C from DE is y cm.

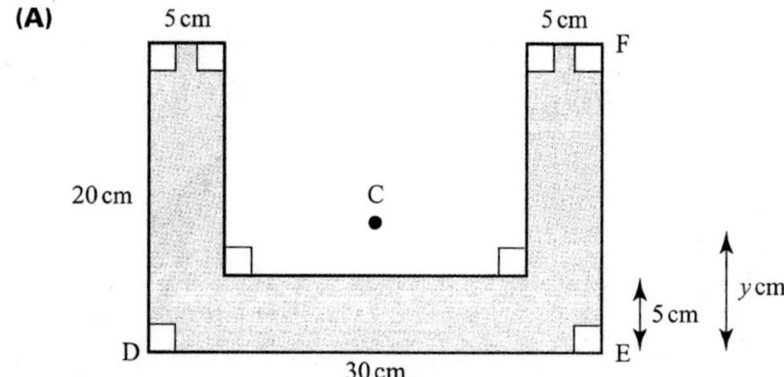

(i) Find the value of y.

The solid is placed on a rough plane. The coefficient of friction between the solid and the plane is μ. The plane is tilted so that EF lies along a line of greatest slope.

(ii) The solid is placed so that F is higher up the plane than E (see figure **(B)**). When the angle of inclination is sufficiently great the solid starts to topple (without sliding). Show that $\mu > \frac{1}{2}$.

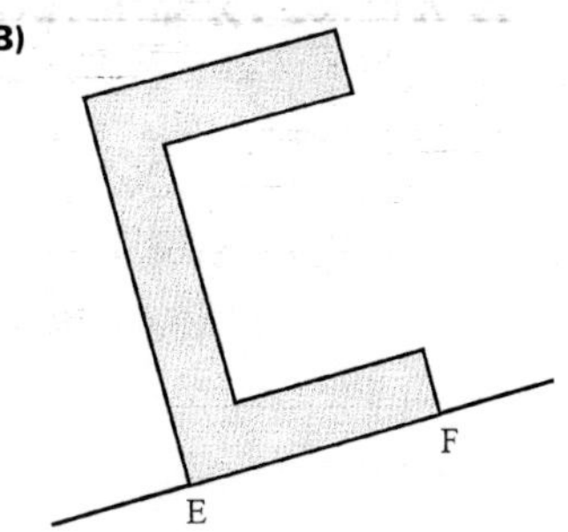

(iii) The solid is now placed so that E is higher up the plane than F (see figure **(C)**). When the angle of inclination is sufficiently great the solid starts to slide (without toppling). Show that $\mu < \frac{5}{6}$.

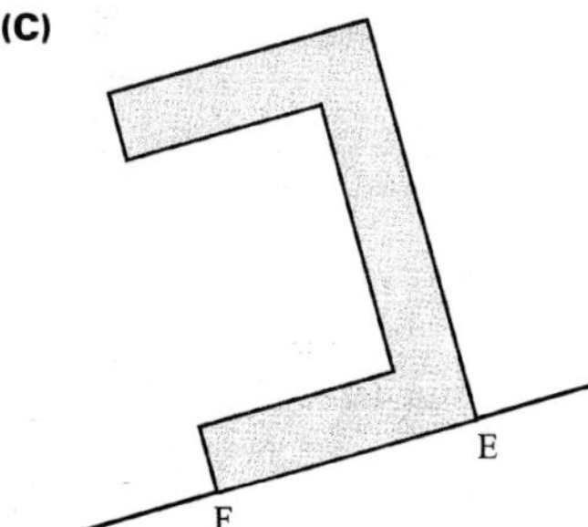

[Cambridge AS and A Level Mathematics 9709, Paper 5 Q7 November 2007]

15 A uniform solid cylinder has height 24 cm and radius r cm. A uniform solid cone has base radius r cm and height h cm. The cylinder and the cone are both placed with their axes vertical on a rough horizontal plane (see diagram, which shows cross-sections of the solids). The plane is slowly tilted and both solids remain in equilibrium until the angle of inclination of the plane reaches $\alpha°$, when both solids topple simultaneously.

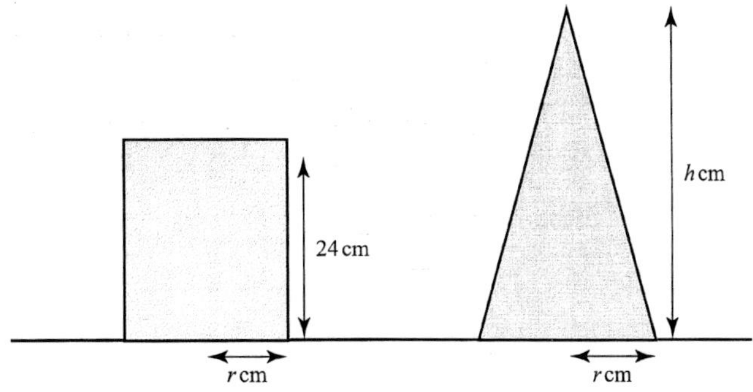

(i) Find the value of h.

(ii) Given that $r = 10$, find the value of α.

[Cambridge AS and A Level Mathematics 9709, Paper 5 Q2 November 2008]

Baby buggy

Borrow a baby buggy and investigate its stability.

How stable is it when you hang some shopping on its handle?

How could the design of the buggy be altered to improve its stability?

Think about the handling of the buggy in other situations. Would your changes cause any problems?

Sliding and toppling

Make a pile of rough bricks on a board, then raise one edge of the board so that it slopes. Investigate what happens as the angle of the slope is increased.

Drink can

A drink can is cylindrical. When the can is full the centre of mass is clearly half-way up. The same is true when it is completely empty. In between these two extremes, the centre of mass is below the middle.

Find the minimum height of the centre of mass.

Bridge

A bridge is made by placing identical bricks on top of each other as shown in the diagram. No glue or cement is used. How far can the bridge be extended without toppling over? You may use as many bricks as you like but only one is allowed at each level.

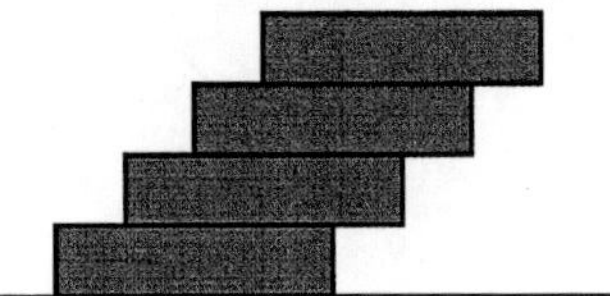

Figure 12.27

Finding the centre of mass

Collect a number of flat (but not necessarily uniform) objects, and investigate, for each of them, which is the most accurate method of determining its centre of mass.

(i) Calculation.

(ii) Balancing it on a pin.

(iii) Hanging it from two (or more) corners.

(iv) Balancing it on the edge of a table in a number of different orientations.

KEY POINTS

1 The centre of mass of a body has the property that the moment, about any point, of the whole mass of the body taken at the centre of mass is equal to the sum of the moments of the various particles comprising the body.

$$M\bar{\mathbf{r}} = \sum m_i \mathbf{r}_i \text{ where } M = \sum m_i$$

2 In one dimension

$$M\bar{x} = \sum m_i x_i$$

3 In two dimensions

$$M\begin{pmatrix}\bar{x}\\ \bar{y}\end{pmatrix} = \sum m_i \begin{pmatrix}x_i\\ y_i\end{pmatrix}$$

4 In three dimensions

$$M\begin{pmatrix}\bar{x}\\ \bar{y}\\ \bar{z}\end{pmatrix} = \sum m_i \begin{pmatrix}x_i\\ y_i\\ z_i\end{pmatrix}$$

13 Uniform motion in a circle

Whirlpools and storms his circling arm invest
With all the might of gravitation blest.

Alexander Pope

These pictures show some objects which move in circular paths. What other examples can you think of?

What makes objects move in circles?
Why does the moon circle the earth?
What happens to the 'hammer' when the athlete lets it go?
Does the pilot of the plane need to be strapped into his seat at the top of a loop in order not to fall out?

The answers to these questions lie in the nature of circular motion. Even if an object is moving at constant speed in a circle, its velocity keeps changing because its direction of motion keeps changing. Consequently the object is accelerating and so, according to Newton's first law, there must be a force acting on it. The force required to keep an object moving in a circle can be provided in many ways.

Without the earth's gravitational force, the moon would move off at constant speed in a straight line into space. The wire attached to the athlete's hammer provides a tension force which keeps the ball moving in a circle. When the athlete lets go, the ball flies off at a tangent because the tension has disappeared.

Although it would be sensible for the pilot to be strapped in, no upward force is necessary to stop him falling out of the plane because his weight contributes to the force required for motion in a circle.

In this chapter, these effects are explained.

Notation

To describe circular motion (or indeed any other topic) mathematically you need a suitable notation. It will be helpful in this chapter to use the notation (attributed to Newton) for differentiation with respect to time in which, for example, $\frac{ds}{dt}$ is written as $\dot{s}$, and $\frac{d^2\theta}{dt^2}$ as $\ddot{\theta}$.

Figure 13.1 shows a particle P moving round the circumference of a circle of radius r, centre O. At time t, the position vector $\overrightarrow{OP}$ of the particle makes an angle θ (in radians) with the fixed direction $\overrightarrow{OA}$. The arc length AP is denoted by s.

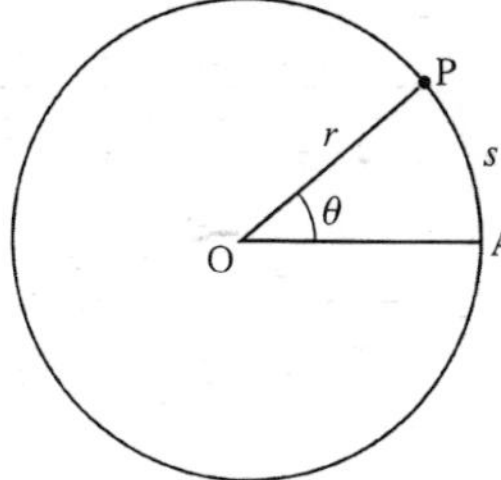

Figure 13.1

Angular speed

Using this notation,

$$s = r\theta$$

Differentiating this with respect to time using the product rule gives:

$$\frac{ds}{dt} = r\frac{d\theta}{dt} + \theta\frac{dr}{dt}.$$

Since r is constant for a circle, $\frac{dr}{dt} = 0$, so the rate at which the arc length increases is

$$\frac{ds}{dt} = r\frac{d\theta}{dt} \quad \text{or} \quad \dot{s} = r\dot{\theta}. \qquad ①$$

In this equation $\dot{s}$ is the speed at which P is moving round the circle (often denoted by v), and $\dot{\theta}$ is the rate at which the angle θ is increasing, i.e. the rate at which the position vector $\overrightarrow{OP}$ is rotating.

The quantity $\frac{d\theta}{dt}$, or $\dot{\theta}$, can be called the *angular velocity* or the *angular speed* of P. In more advanced work, angular velocity is treated as a vector, whose direction is taken to be that of the axis of rotation. In this book, $\frac{d\theta}{dt}$ is often referred to as angular speed, but is given a sign: positive when θ is increasing (usually anticlockwise) and negative when θ is decreasing (usually clockwise).

Angular speed is often denoted by ω, the Greek letter omega. So the equation $\dot{s} = r\dot{\theta}$ may be written as

$$v = r\omega.$$

Notice that for this equation to hold, θ must be measured in radians, so the angular speed is measured in *radians per second* or rad s^{-1}.

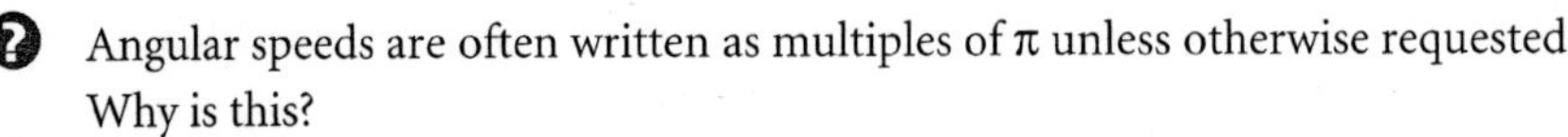

Angular speeds are often written as multiples of π unless otherwise requested. Why is this?

Figure 13.2 shows a disc rotating about its centre, O, with angular speed ω. The line OP represents any radius.

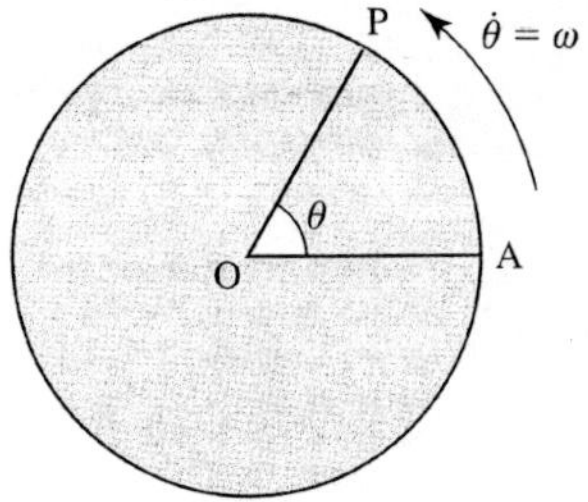

Figure 13.2

Every point on the disc describes a circular path, and all points have the same angular speed. However the *actual* speed of any point depends on its distance from the centre: increasing r in the equation $v = r\omega$ increases v. You will appreciate this if you have ever been at the end of a rotating line of people in a dance or watched a body of marching soldiers wheeling round a corner.

Angular speeds are sometimes measured in revolutions per second or revolutions per minute (rpm) where one revolution is equal to 2π radians. For example, turntables for vinyl records used to rotate at 45 or $33\frac{1}{3}$ rpm while a computer hard disc might spin at 7200 rpm or more. At cruising speeds, crankshafts in car engines typically rotate at 3000 to 4000 rpm.

EXAMPLE 13.1

A police car drives at $64\,\text{km}\,\text{h}^{-1}$ around a circular bend of radius 16 m. A second car moves so that it has the same angular speed as the police car but in a circle of radius 12 m. Is the second car breaking the $50\,\text{km}\,\text{h}^{-1}$ speed limit?

SOLUTION

Converting kilometres per hour to metres per second gives

$$64\,\text{km}\,\text{h}^{-1} = \frac{64 \times 1000}{3600}\,\text{m}\,\text{s}^{-1}$$
$$= \frac{160}{9}\,\text{m}\,\text{s}^{-1}$$

Using $v = r\omega$,

$$\omega = \frac{160}{9 \times 16}\,\text{rad s}^{-1}$$
$$= \frac{10}{9}\,\text{rad s}^{-1}$$

The speed of the second car is

$$v = 12\omega$$
$$= \frac{10}{9} \times 12\,\text{m}\,\text{s}^{-1}$$
$$= \frac{120 \times 3600}{9 \times 1000}\,\text{km}\,\text{h}^{-1}$$
$$= 48\,\text{km}\,\text{h}^{-1}$$

The second car is just below the speed limit.

Notes

1 Notice that working in fractions gives an exact answer.

2 A quicker way to do this question would be to notice that, because the cars have the same angular speed, the actual speeds of the cars are proportional to the radii of the circles in which they are moving. Using this method it is possible to stay in $\text{km}\,\text{h}^{-1}$. The ratio of the two radii is $\frac{12}{16}$ so the speed of the second car is $\frac{12}{16} \times 64\,\text{km}\,\text{h}^{-1} = 48\,\text{km}\,\text{h}^{-1}$.

EXERCISE 13A

1 Find the angular speed, in radians per second correct to one decimal place, of records rotating at

(i) 78 rpm

(ii) 45 rpm

(iii) $33\frac{1}{3}$ rpm.

2 A flywheel is rotating at $300\,\text{rad}\,\text{s}^{-1}$. Express this angular speed in rpm, correct to the nearest whole number.

3 The London Eye observation wheel has a diameter of 135 m and completes one revolution in 30 minutes.

(i) Calculate its angular speed in

(a) rpm **(b)** radians per second.

(ii) Calculate the speed of the point on the circumference where passengers board the moving wheel.

4 A lawnmower engine is started by pulling a rope that has been wound round a cylinder of radius 4 cm. Find the angular speed of the cylinder at a moment when the rope is being pulled with a speed of $1.3\,\mathrm{m\,s^{-1}}$. Give your answer in radians per second, correct to one decimal place.

5 The wheels of a car have radius 20 cm. What is the angular speed, in radians per second correct to one decimal place, of a wheel when the car is travelling at

(i) $10\,\mathrm{m\,s^{-1}}$ **(ii)** $30\,\mathrm{m\,s^{-1}}$?

6 The angular speed of an audio CD changes continuously so that a laser can read the data at a constant speed of $12\,\mathrm{m\,s^{-1}}$. Find the angular speed (in rpm) when the distance of the laser from the centre is

(i) 30 mm **(ii)** 55 mm.

7 What is the average angular speed of the earth in radians per second as it

(i) orbits the sun?

(ii) rotates about its own axis?

The radius of the earth is 6400 km.

(iii) At what speed is someone on the equator travelling relative to the centre of the earth?

(iv) At what speed are you travelling relative to the centre of the earth?

8 A tractor has front wheels of diameter 70 cm and back wheels of diameter 1.6 m. What is the ratio of their angular speeds when the tractor is being driven along a straight road?

9 **(i)** Find the kinetic energy of a 50 kg person riding a big wheel with radius 5 m when the ride is rotating at 3 rpm. You should assume that the person can be modelled as a particle.

(ii) Explain why this modelling assumption is necessary.

10 The minute hand of a clock is 1.2 m long and the hour hand is 0.8 m long.

(i) Find the speeds of the tips of the hands.

(ii) Find the ratio of the speeds of the tips of the hands and explain why this is not the same as the ratio of the angular speeds of the hands.

11 The diagram represents a 'Chairoplane' ride at a fair. It completes one revolution every 2.5 seconds.

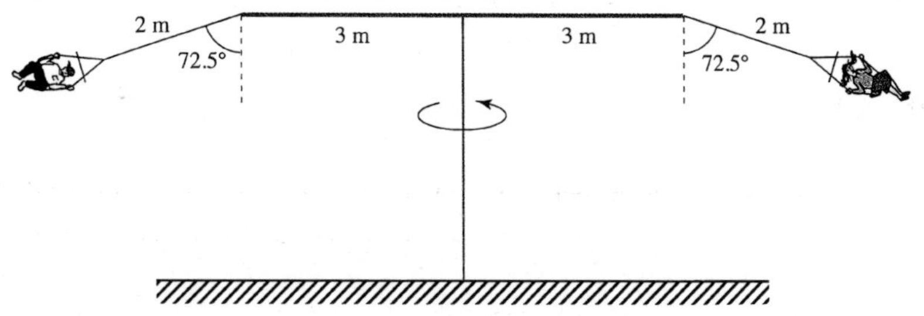

(i) Find the radius of the circular path which a rider follows.

(ii) Find the speed of a rider.

12 The diagram shows a roundabout in a playground, seen from above. It is rotating clockwise. A child on the roundabout at X, aims a ball at a friend sitting opposite at Y.

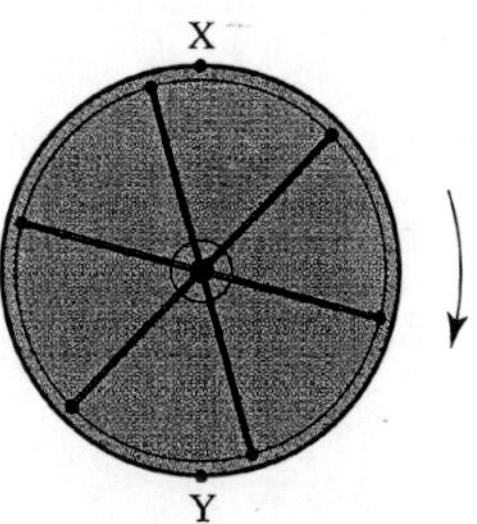

(i) Once the ball is thrown, can the friend catch it?

(ii) Draw a plan of the path of the ball after it has been thrown.

Velocity and acceleration

Velocity and acceleration are both vector quantities. They can be expressed either in magnitude–direction form, or in components. When describing circular motion or other orbits it is most convenient to take components in directions along the radius (*radial* direction) and at right angles to it (*transverse* direction).

For a particle moving round a circle of radius r, the velocity has:

radial component: 0
transverse component: $r\dot{\theta}$ or $r\omega$.

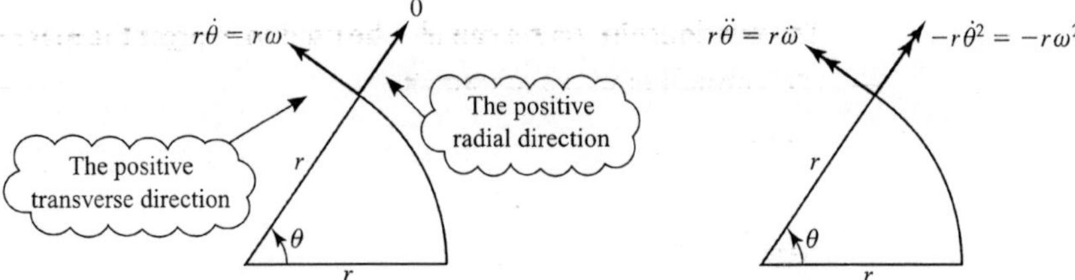

Figure 13.3 *Velocity*

Figure 13.4 *Acceleration*

The acceleration of a particle moving round a circle of radius r has:

radial component:	$-r\dot{\theta}^2$	or	$-r\omega^2$
transverse component:	$r\ddot{\theta}$	or	$r\dot{\omega}$.

The transverse component is just what you would expect: the radius multiplied by the angular acceleration, $\ddot{\theta}$. If the particle has constant angular speed, its angular acceleration is zero and so the transverse component of its acceleration is also zero.

In contrast, the radial component of the acceleration, $-r\omega^2$, is almost certainly not a result you would have expected intuitively. It tells you that a particle travelling in a circle is always accelerating towards the centre of the circle, but without ever getting any closer to the centre. If this seems a strange idea, you may find it helpful to remember that circular motion is not a natural state; left to itself a particle will travel in a straight line. To keep a particle in the unnatural state of circular motion it must be given an acceleration at right angles to its motion, i.e. towards the centre of the circle.

Circular motion with constant speed

In this chapter, the circular motion is assumed to be uniform and so have no transverse component of acceleration.

Problems involving circular motion often refer to the actual speed of the object, rather than its angular speed. It is easy to convert the one into the other using the relationship $v = r\omega$.

The relationship $v = r\omega$ can also be used to express the magnitude of the acceleration in terms of v and r:

$$\omega = \frac{v}{r}$$

$$a = r\omega^2 = r\left(\frac{v}{r}\right)^2$$

$$\Rightarrow \quad a = \frac{v^2}{r} \text{ towards the centre.}$$

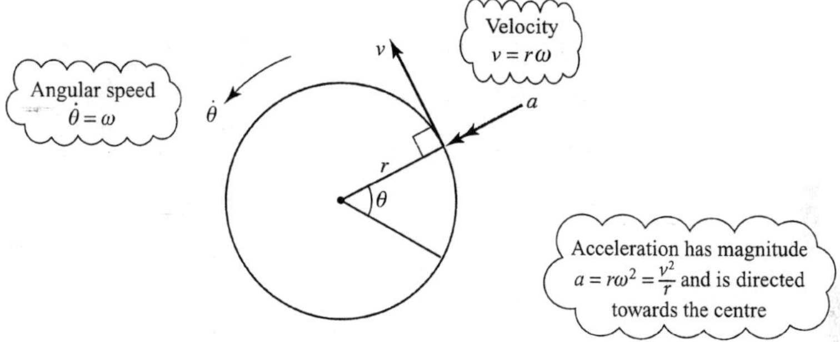

Figure 13.5

EXAMPLE 13.2

A fly is standing on a small turntable at a distance of 8 cm from the centre. If the turntable is rotating at 45 rpm, find

(i) the angular speed of the fly in radians per second

(ii) the speed of the fly in metres per second

(iii) the acceleration of the fly.

SOLUTION

(i) One revolution is 2π rad so

$$\begin{aligned} 45 \text{ rpm} &= 45 \times 2\pi \text{ rad min}^{-1} \\ &= \frac{45 \times 2\pi}{60} \text{ rad s}^{-1} \\ &= \frac{3\pi}{2} \text{ rad s}^{-1}. \end{aligned}$$

(ii) If the speed of the fly is v m s^{-1}, v can be found using

$$\begin{aligned} v &= r\omega \\ &= 0.08 \times \frac{3\pi}{2} \\ &= 0.377... \end{aligned}$$

So the speed of the fly is 0.38 m s^{-1} (to 2 d.p.).

(iii) The acceleration of the fly is given by

$$r\omega^2 = 0.08 \times \left(\frac{3\pi}{2}\right)^2$$

$$= 1.78$$

The acceleration of the fly is $1.78\,\mathrm{m\,s^{-2}}$ directed towards the centre of the turntable.

? A wheel of radius r m is rolling in a straight line with forward speed $u\,\mathrm{m\,s^{-1}}$. What are

(i) the speed of the point which is instantaneously in contact with the ground?

(ii) the angular speed of the wheel?

(iii) the velocities of the highest point and the point on the edge of the wheel which is level with and behind the axle?

The forces required for circular motion

Newton's first law of motion states that a body will continue in a state of rest or uniform motion in a straight line unless acted upon by an external force. Any object moving in a circle, such as the police car and the fly in Examples 13.1 and 13.2 must therefore be acted upon by a resultant force in order to produce the required acceleration towards the centre.

A force towards the centre is called a *centripetal* (centre-seeking) force. A resultant centripetal force is necessary for a particle to move in a circular path.

Examples of circular motion

You are now in a position to use Newton's second law to determine theoretical answers to some of the questions which were posed at the beginning of this chapter. These will, as usual, be obtained using models of the true motion which will be based on simplifying assumptions, for example zero air resistance. Large objects are assumed to be particles concentrated at their centres of mass.

EXAMPLE 13.3

A coin is placed on a rotating turntable. Its centre is 5 cm from the centre of rotation and the coefficient of friction, μ, between the coin and the turntable is 0.5.

(i) If the speed of rotation of the turntable is gradually increased, at what angular speed will the coin begin to slide?

(ii) What happens next?

SOLUTION

(i) Because the speed of the turntable is increased only gradually, it can be assumed that the coin will not slip tangentially.

Figure 13.6 shows the forces acting on the coin, and its acceleration.

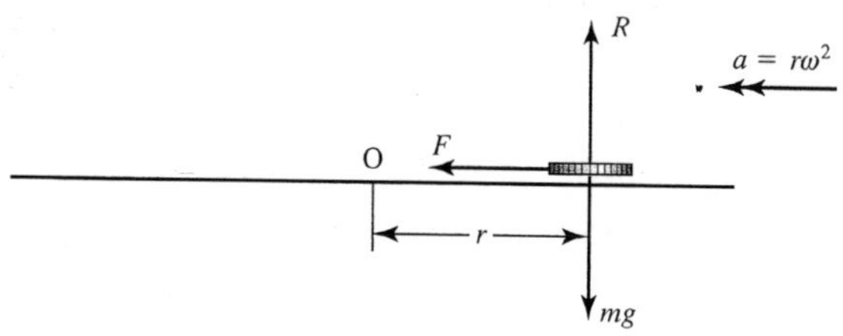

Figure 13.6

The acceleration is towards the centre, O, of the circular path so there must be a frictional force F in that direction.

There is no vertical component of acceleration, so the resultant force acting on the coin has no vertical component.

Therefore $\quad R - mg = 0$

$$R = mg \qquad ①$$

By Newton's second law towards the centre of the circle:

$$\text{Force } F = ma = mr\omega^2 \qquad ②$$

The coin will not slide so long as $F \leqslant \mu R$.

Substituting from ② and ① this gives

$$mr\omega^2 \leqslant \mu mg$$

$$\Rightarrow \quad r\omega^2 \leqslant \mu g$$

Notice that the mass, m, has been eliminated at this stage, so that the answer does not depend upon it

Taking g in $\mathrm{m\,s^{-2}}$ as 10 and substituting $r = 0.05$ and $\mu = 0.5$

$$\omega^2 \leqslant 100$$

$$\omega \leqslant 10$$

The coin will move in a circle provided that the angular speed is less than $10\,\mathrm{rad\,s^{-1}}$, and this speed is independent of the mass of the coin.

(ii) When the angular speed increases beyond this, the coin slips to a new position. If the angular speed continues to increase the coin will slip right off the turntable. When it reaches the edge it will fly off in the direction of the tangent.

The conical pendulum

A conical pendulum consists of a small bob tied to one end of a string. The other end of the string is fixed and the bob is made to rotate in a horizontal circle below the fixed point so that the string describes a cone as in figure 13.7.

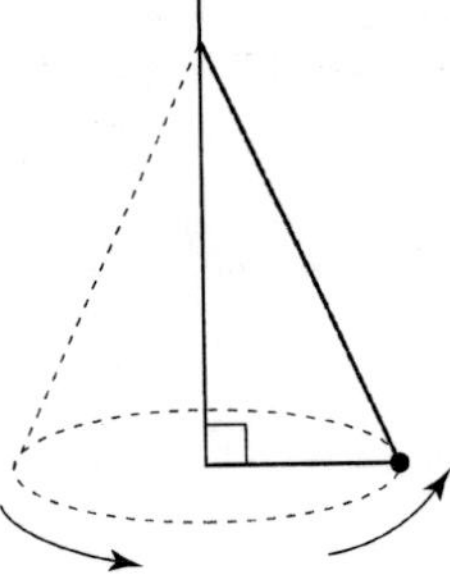

Figure 13.7

EXPERIMENT

1 Draw a diagram showing the magnitude and direction of the acceleration of a bob and the forces acting on it.

2 In the case that the radius of the circle remains constant, try to predict the effect on the angular speed when the length of the string is increased or when the mass of the bob is increased. What might happen when the angular speed increases?

3 Draw two circles of equal diameter on horizontal surfaces so that two people can make the bobs of conical pendulums rotate in circles of the same radius.

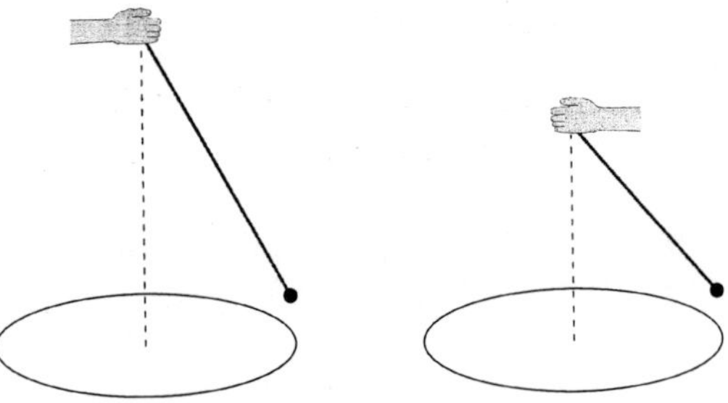

Figure 13.8

(i) Compare pendulums of different lengths with bobs of equal mass.
(ii) Compare pendulums of the same length but with bobs of different masses.

Does the angular speed depend on the length of the pendulum or the mass of the bob?

4 What happens when somebody makes the speed of the bob increase?

5 Can a bob be made to rotate with the string horizontal?

Theoretical model for the conical pendulum

A conical pendulum may be modelled as a particle of mass m attached to a light, inextensible string of length l. The mass is rotating in a horizontal circle with angular speed ω and the string makes an angle α with the downward vertical. The radius of the circle is r and the tension in the string is T, all in consistent units (e.g. S.I. units). The situation is shown in figure 13.9.

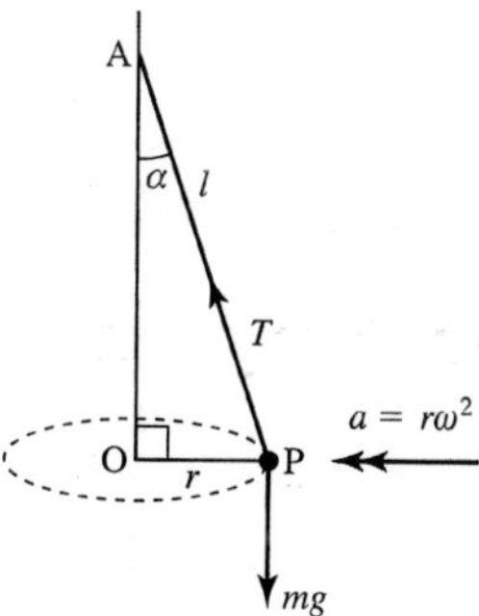

Figure 13.9

The magnitude of the acceleration is $r\omega^2$. The acceleration acts in a horizontal direction towards the centre of the circle. This means that there must be a resultant force acting towards the centre of the circle.

There are two forces acting on this particle, its weight mg and the tension T in the string.

As the acceleration of the particle has no vertical component, the resultant force has no vertical component, so

$$T\cos\alpha - mg = 0 \qquad ①$$

Using Newton's second law towards the centre, O, of the circle

$$T\sin\alpha = ma = mr\omega^2 \qquad ②$$

In triangle AOP

$$r = l\sin\alpha$$

Substituting for r in ② gives

$$T\sin\alpha = m(l\sin\alpha)\omega^2$$

$$\Rightarrow \quad T = ml\omega^2$$

Substituting this in ① gives

$$ml\omega^2\cos\alpha - mg = 0$$

$$\Rightarrow \qquad l\cos\alpha = \frac{g}{\omega^2} \qquad ③$$

This equation provides sufficient information to give theoretical answers to the questions in the experiment.

- When r is kept constant and the length of the string is increased, the length $AO = l\cos\alpha$ increases. Equation ③ indicates that the value of $\frac{g}{\omega^2}$ increases and so the angular speed ω decreases. Conversely, the angular speed increases when the string is shortened.
- The mass of the particle does not appear in equation ③, so it has no effect on the angular speed, ω.
- When the length of the pendulum is unchanged, but the angular speed is increased, $\cos\alpha$ decreases, leading to an increase in the angle α and hence in r.
- If $\alpha \geqslant 90°$, $\cos\alpha \leqslant 0$, so $\frac{g}{\omega^2} \leqslant 0$, which is impossible. You can see from figure 13.9 that the tension in the string must have a vertical component to balance the weight of the particle.

EXAMPLE 13.4

The diagram on the right represents one of several arms of a fairground ride, shown on the left. The arms rotate about an axis and riders sit in chairs linked to the arms by chains.

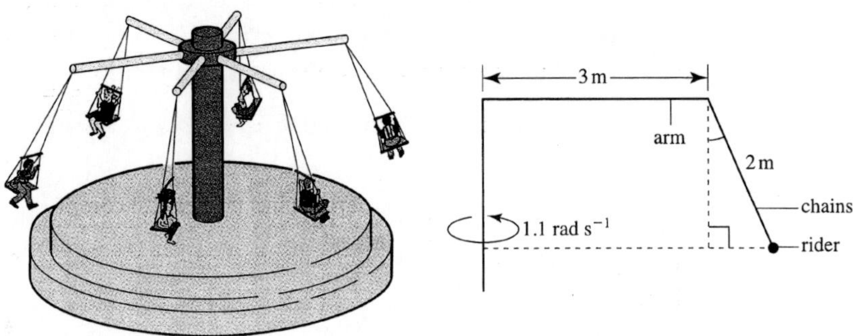

Figure 13.10

The chains are 2 m long and the arms are 3 m long. Find the angle that the chains make with the vertical when the rider rotates at 1.1 rad s^{-1}.

SOLUTION

Let T N be the resultant tension in the chains holding a chair, and m kg the mass of chair and rider.

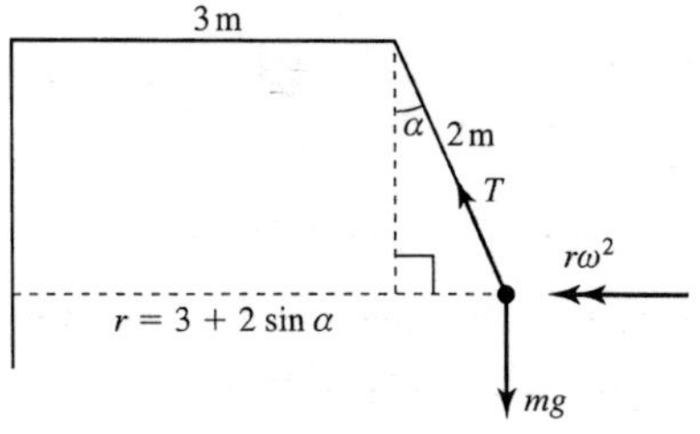

Figure 13.11

If the chains make an angle α with the vertical, the motion is in a horizontal circle with radius given by

$$r = 3 + 2\sin\alpha.$$

The magnitude of the acceleration is given by

$$r\omega^2 = (3 + 2\sin\alpha) \times 1.1^2.$$

It is in a horizontal direction towards the centre of the circle. Using Newton's second law in this direction gives

$$\text{Force} = mr\omega^2$$

$$\Rightarrow T\sin\alpha = m(3 + 2\sin\alpha) \times 1.1^2 \qquad ①$$

$$= 1.21m(3 + 2\sin\alpha)$$

Vertically: $\quad T\cos\alpha - mg = 0$

$$\Rightarrow \quad T = \frac{mg}{\cos\alpha}$$

Substituting for T in equation ①:

$$\frac{mg}{\cos\alpha}\sin\alpha = 1.21m(3 + 2\sin\alpha)$$

$$\Rightarrow \quad 10\tan\alpha = 3.63 + 2.42\sin\alpha$$

Since m cancels out at this stage, the angle does not depend on the mass of the rider

This equation cannot be solved directly, but a numerical method will give you the solution 25° correct to the nearest degree. You might like to solve the equation yourself or check that this solution does in fact satisfy the equation.

Note

Since the answer does not depend on the mass of the rider and chair, when riders of different masses, or even no riders, are on the equipment all the chains should make the same angle with the vertical.

Banked tracks

ACTIVITY 13.1

Keep away from other people and breakable objects when carrying out this activity.

Place a coin on a piece of stiff A4 card and hold it horizontally at arm's length with the coin near your hand.

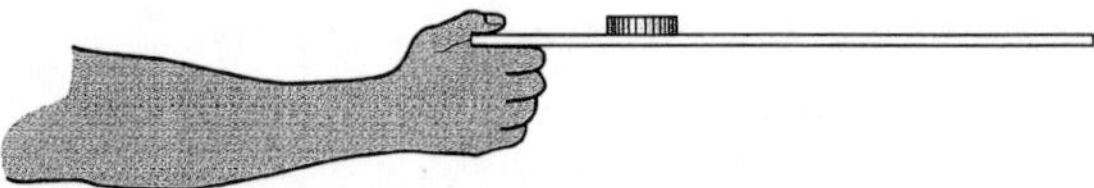

Figure 13.12

Turn round slowly so that your hand moves in a horizontal circle. Now gradually speed up. The outcome will probably not surprise you.

What happens, though, if you tilt the card?

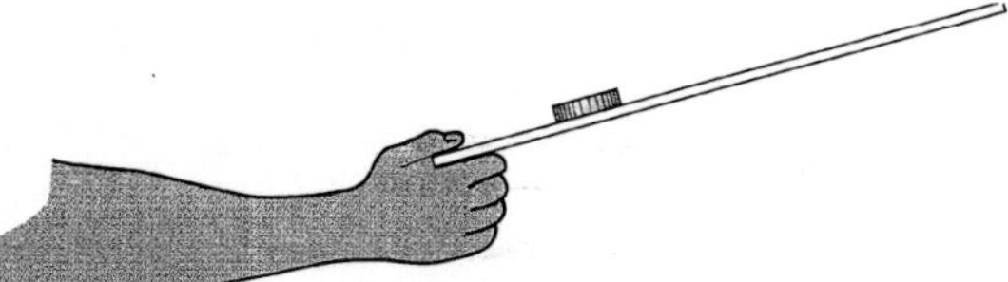

Figure 13.13

You may have noticed that when they curve round bends, most roads are banked so that the edge at the outside of the bend is slightly higher than that at the inside. For the same reason the outer rail of a railway track is slightly higher than the inner rail when it goes round a bend. On bobsleigh tracks the bends are almost bowl shaped, with a much greater gradient on the outside.

Figure 13.14 shows a car rounding a bend on a road which is banked so that the cross-section makes an angle α with the horizontal.

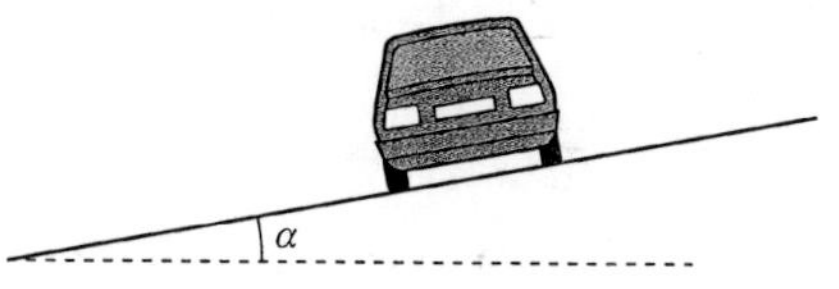

Figure 13.14

In modelling such situations, it is usual to treat the bend as part of a horizontal circle whose radius is large compared to the width of the car. In this case, the radius of the circle is taken to be r metres, and the speed of the car constant at v metres per second. The car is modelled as a particle which has an acceleration of $\frac{v^2}{r}$ m s^{-2} in a horizontal direction towards the centre of the circle. The forces and acceleration are shown in figure 13.15.

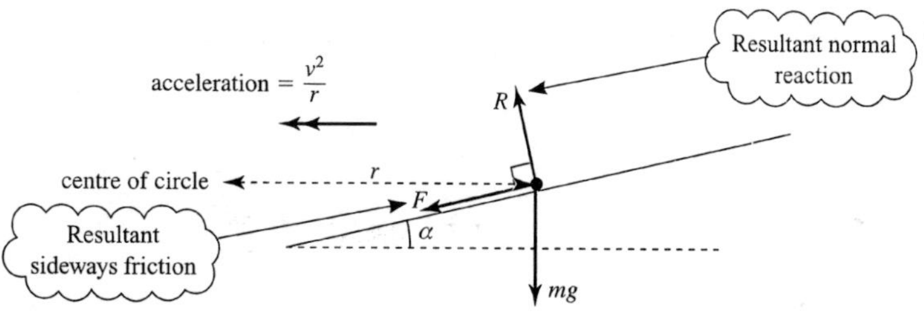

Figure 13.15

The direction of the frictional force F will be up or down the slope depending on whether the car has a tendency to slip sideways towards the inside or outside of the bend.

 Under what conditions do you think each of these will occur?

EXAMPLE 13.5

A car is rounding a bend of radius 100 m which is banked at an angle of 10° to the horizontal. At what speed must the car travel to ensure it has no tendency to slip sideways?

SOLUTION

When there is no tendency to slip there is no frictional force, so in the plane perpendicular to the direction of motion of the car, the forces and acceleration are as shown in figure 13.16. The only horizontal force is provided by the horizontal component of the normal reaction of the road on the car.

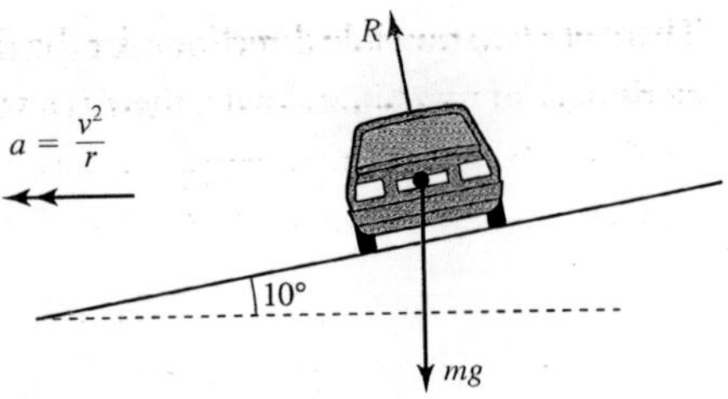

Figure 13.16

Vertically, there is no acceleration so there is no resultant force.

$$R\cos 10° - mg = 0$$

$$\Rightarrow \quad R = \frac{mg}{\cos 10°} \qquad ①$$

By Newton's second law in the horizontal direction towards the centre of the circle,

$$R\sin 10° = ma = \frac{mv^2}{r}$$

$$= \frac{mv^2}{100}$$

Substituting for R from: ①:

$$\left(\frac{mg}{\cos 10°}\right)\sin 10° = \frac{mv^2}{100}$$

$$\Rightarrow \quad v^2 = 100\,g\tan 10°$$

$$\Rightarrow \quad v = 13.3 \text{ (to 3 s.f.)}$$

The mass, m, cancels out at this stage, so the answer does not depend on it

The speed of the car must be about 13.3 m s^{-1}.

There are two important points to notice in this example.

- The speed is the same whatever the mass of the car.
- The example looks at the situation when the car does not tend to slide, and finds the speed at which this is the case. At this speed the car does not depend on friction to keep it from sliding, and indeed it could travel safely round the bend at this speed even in very icy conditions. However, at other speeds there is a tendency to slide, and friction actually helps the car to follow its intended path.

Safe speeds on a bend

What would happen in the previous example if the car travelled either more slowly than 13.3 m s^{-1} or more quickly?

The answer is that there would be a frictional force acting so as to prevent the car from sliding across the road.

There are two possible directions for the frictional force. When the vehicle is stationary or travelling slowly, there is a tendency to slide down the slope and the friction acts up the slope to prevent this. When it is travelling quickly round the bend, the car is more likely to slide up the slope, so the friction acts down the slope.

Fortunately, under most road conditions, the coefficient of friction between tyres and the road is large, typically about 0.8. This means that there is a range of speeds that are safe for negotiating any particular bend.

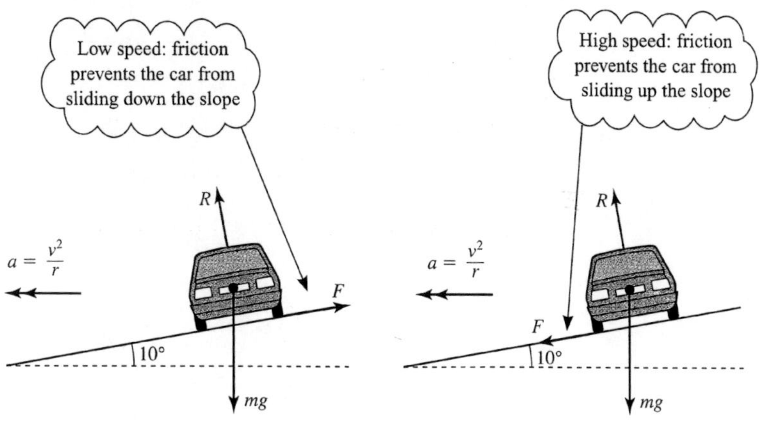

Figure 13.17

---

1 Using a particle model for the car, show that it will not slide up or down the slope provided

$$\sqrt{rg\frac{(\sin\alpha - \mu\cos\alpha)}{(\cos\alpha + \mu\sin\alpha)}} < v < \sqrt{rg\frac{(\sin\alpha + \mu\cos\alpha)}{(\cos\alpha - \mu\sin\alpha)}}$$

If $r = 100$ and $\alpha = 10°$ (so that $\tan\alpha = 0.176$) the minimum and maximum safe speeds (in km h^{-1}) for different values of μ are given in the following table.

μ	0	0.1	0.2	0.3	0.4	0.5	0.6	0.7	0.8	0.9	1.0	1.1	1.2
Minimum safe speed	48	31	0	0	0	0	0	0	0	0	0	0	0
Maximum safe speed	48	60	71	81	90	98	106	114	121	129	136	143	150

2 Would you regard this bend as safe? How, by changing the values of r and α, could you make it safer?

EXAMPLE 13.6

A bend on a railway track has a radius of 500 m and is to be banked so that a train can negotiate it at $96\,\text{km}\,\text{h}^{-1}$ without the need for a lateral force between its wheels and the rail. The distance between the rails is 1.43 m.

How much higher should the outside rail be than the inside one?

SOLUTION

There is very little friction between the track and the wheels of a train. Any sideways force required is provided by the 'lateral thrust' between the wheels and the rail. The ideal speed for the bend is such that the lateral thrust is zero.

Figure 13.18 shows the forces acting on the train and its acceleration when the track is banked at an angle α to the horizontal.

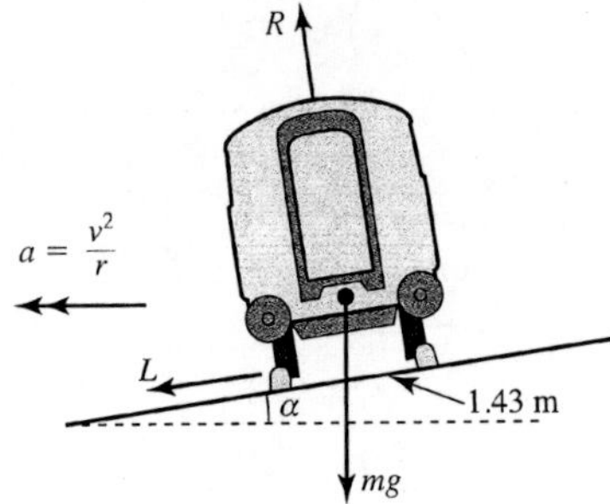

Figure 13.18

When there is no lateral thrust, $L = 0$.

Horizontally: $$R\sin\alpha = \frac{mv^2}{r} \quad ①$$

Vertically: $$R\cos\alpha = mg \quad ②$$

Dividing ① by ② gives $$\tan\alpha = \frac{v^2}{rg}$$

Using the fact that $96\,\text{km}\,\text{h}^{-1} = 26\frac{2}{3}\,\text{m}\,\text{s}^{-1}$ this becomes

$$\tan\alpha = \frac{32}{225}$$

$$\Rightarrow \quad \alpha = 8.1° \text{ (to 2 s.f.)}$$

The outside rail should be raised by $1.43\sin\alpha$ metres, i.e. by about 20 cm.

EXERCISE 13B

1 The diagram shows two cars, A and B, travelling at constant speeds in different lanes (radii 24 m and 20 m) round a circular traffic island. Car A has speed $18\,\mathrm{m\,s^{-1}}$ and car B has speed $15\,\mathrm{m\,s^{-1}}$.

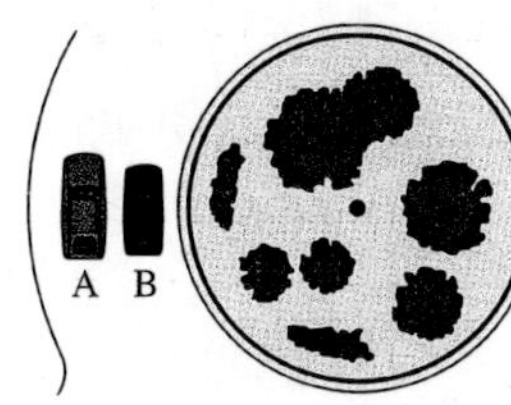

Answer the following questions, giving reasons for your answers.

(i) Which car has the greater angular speed?

(ii) Is one car overtaking the other?

(iii) Find the magnitude of the acceleration of each car.

(iv) In which direction is the resultant force on each car acting?

2 Two coins are placed on a horizontal turntable. Coin A has mass 15 g and is placed 5 cm from the centre; coin B has mass 10 g and is placed 7.5 cm from the centre. The coefficient of friction between each coin and the turntable is 0.4.

(i) Describe what happens to the coins when the turntable turns at

(a) $6\,\mathrm{rad\,s^{-1}}$ **(b)** $8\,\mathrm{rad\,s^{-1}}$ **(c)** $10\,\mathrm{rad\,s^{-1}}$.

(ii) What would happen if the coins were interchanged?

3 A car is travelling at a steady speed of $15\,\mathrm{m\,s^{-1}}$ round a roundabout of radius 20 m.

(i) Criticise this false argument:

The car is travelling at a steady speed and so its speed is neither increasing nor decreasing and therefore the car has no acceleration.

(ii) Calculate the magnitude of the acceleration of the car.

(iii) The car has mass 800 kg. Calculate the sideways force on each wheel assuming it to be the same for all four wheels.

(iv) Is the assumption in part **(iii)** realistic?

4 A fairground ride has seats at 3 m and at 4.5 m from the centre of rotation. Each rider travels in a horizontal circle. Say whether each of the following statements is true, giving your reasons.

(i) Riders in the two positions have the same angular speed at any time.

(ii) Riders in the two positions have the same speed at any time.

(iii) Riders in the two positions have the same magnitude of acceleration at any time.

5 A skater of mass 60 kg follows a circular path of radius 4 m, moving at $2\,\text{m}\,\text{s}^{-1}$.

(i) Calculate:

(a) the angular speed of the skater

(b) the magnitude of the acceleration of the skater

(c) the resultant force acting on the skater.

(ii) What modelling assumptions have you made?

6 Two spin driers, both of which rotate about a vertical axis, have different specifications as given in the table below.

Model	Rate of rotation	Drum diameter
A	600 rpm	60 cm
B	800 rpm	40 cm

State, with reasons, which model you would expect to be the more effective.

7 A satellite of mass M_s is in a circular orbit around the earth, with a radius of r metres. The force of attraction between the earth and the satellite is given by

$$F = \frac{GM_e M_s}{r^2}$$

where $G = 6.67 \times 10^{-11}$ in S.I. units. The mass of the earth M_e is 5.97×10^{24} kg.

(i) Find, in terms of r, expressions for

(a) the speed of the satellite, $v\,\text{m}\,\text{s}^{-1}$

(b) the time, T s, it takes to complete one revolution.

(ii) Hence show that, for all satellites, T^2 is proportional to r^3.

A geostationary satellite orbits the earth so that it is always above the same place on the equator.

(iii) How far is it from the centre of the earth?

(The law found in part **(ii)** was discovered experimentally by Johannes Kepler (1571–1630) to hold true for the planets as they orbit the sun, and is commonly known as Kepler's third law.)

8 In this question you should assume that the orbit of the earth around the sun is circular, with radius 1.44×10^{11} m, and that the sun is fixed.

(i) Find the magnitude of the acceleration of the earth as it orbits the sun.

The force of attraction between the earth and the sun is given by

$$F = \frac{GM_eM_s}{r^2}$$

where M_e is the mass of the earth, M_s is the mass of the sun, r the radius of the earth's orbit and G the universal constant of gravitation (6.67×10^{-11} S.I. units).

(ii) Calculate the mass of the sun.

(iii) Comment on the significance of the fact that you cannot calculate the mass of the earth from the radius of its orbit.

9 Sarah ties a model plane of mass 180 g to the end of a piece of string 80 cm long and then swings it round so that the plane travels in a horizontal circle. The plane is not designed to fly and there is no lift force acting on its wings.

(i) Explain why it is not possible for the string to be horizontal.

Sarah gives the plane an angular speed of 120 rpm.

(ii) What is the angular speed in radians per second?

(iii) Copy the diagram below and mark in the tension in the string, the weight of the plane and the direction of the acceleration.

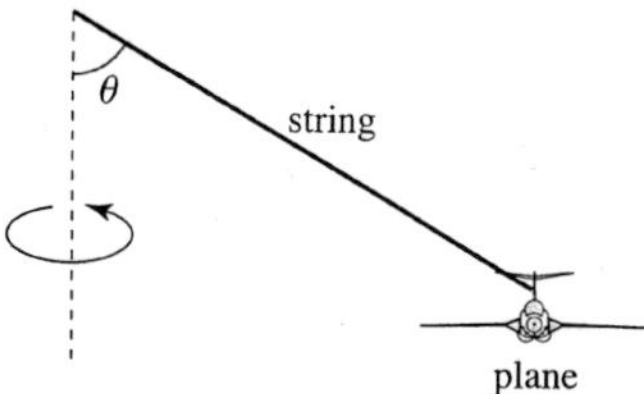

(iv) Write down the horizontal radial equation of motion for the plane and the vertical equilibrium equation in terms of the angle θ.

(v) Show that under these conditions θ has a value between 85° and 86°.

(vi) Find the tension in the string.

10 A rotary lawn mower uses a piece of light nylon string with a small metal sphere on the end to cut the grass. The string is 20 cm in length and the mass of the sphere is 30 g.

(i) Find the tension in the string when the sphere is rotating at 2000 rpm, assuming the string is horizontal.

(ii) Explain why it is reasonable to assume that the string is horizontal.

(iii) Find the speed of the sphere when the tension in the string is 80 N.

11 The coefficient of friction between the tyres of a car and the road is 0.8. The mass of the car and its passengers is 800 kg. Model the car as a particle.

(i) Find the maximum frictional force the road can exert on the car and describe what might be happening when this maximum force is acting

(a) at right angles to the line of motion

(b) along the line of motion.

(ii) What is the maximum speed that the car can travel without skidding on level ground round a circular bend of radius 120 m?

The diagram shows the car, now travelling around a bend of radius 120 m on a road banked at an angle α to the horizontal. The car's speed is such that there is no sideways force (up or down the slope) exerted on its tyres by the road.

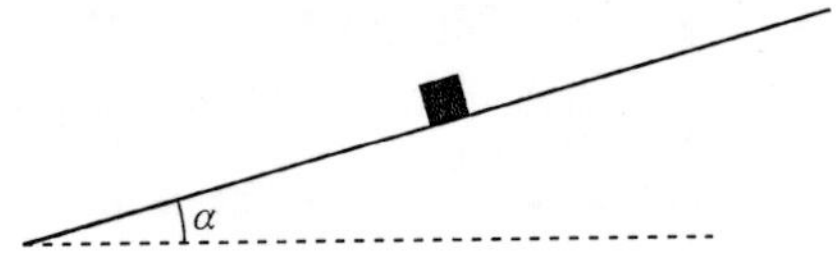

(iii) Draw a diagram showing the weight of the car, the normal reaction of the road on it and the direction of its acceleration.

(iv) Resolve the forces in the horizontal radial and vertical directions and write down the horizontal equation of motion and the vertical equilibrium equation.

(v) Show that $\tan\alpha = \dfrac{v^2}{120g}$ where v is the speed of the car in metres per second.

(vi) On this particular bend, vehicles are expected to travel at $15\,\mathrm{m\,s^{-1}}$. At what angle, α, should the road be banked?

12 Experiments carried out by the police accident investigation department suggest that a typical value for a coefficient of friction between the tyres of a car and a road surface is 0.8.

(i) Using this information, find the maximum safe speed on a level circular motorway slip road of radius 50 m.

(ii) How much faster could cars travel if the road were banked at an angle of 5° to the horizontal?

13 An astronaut's training includes periods in a centrifuge. This may be modelled as a cage on the end of a rotating arm of length 5 m.

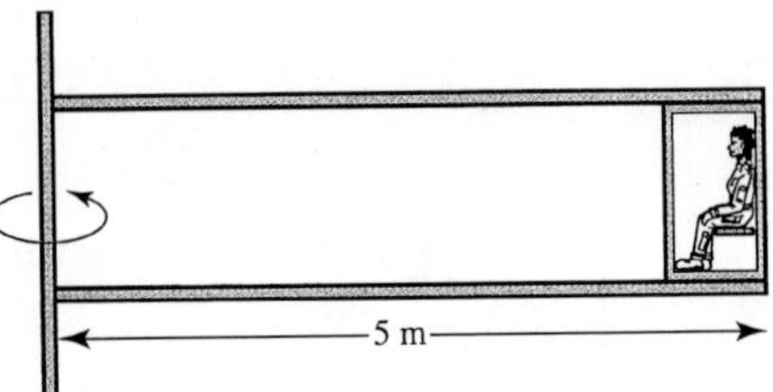

At a certain time, the arm is rotating at 30 rpm.

(i) Find the angular velocity of the astronaut in radians per second and her speed in metres per second.

(ii) Show that under these circumstances the astronaut is subject to an acceleration of magnitude about $5g$.

At a later stage in the training, the astronaut blacks out when her acceleration is $9g$.

(iii) Find her angular velocity (in rpm) when she blacks out.

The training is criticised on the grounds that, in flight, astronauts are not subject to rotation and the angular speed is too great. An alternative design is considered in which the astronaut is situated in a carriage driven round a circular railway track. The device must be able to simulate accelerations of up to $10g$ and the carriage can be driven at up to $100\,\mathrm{m\,s^{-1}}$.

(iv) What should be the radius of the circular railway track?

14 A light, inelastic string of length $2a$ is attached to fixed points A and B where A is vertically above B and the distance AB $< 2a$. A small, smooth ring, P, of mass m slides on the string and is moving in a horizontal circle at a constant angular speed ω. The string sections AP and PB are straight and there is the same tension, T, in each section. The distance AP is x and AP and PB make angles α and β respectively with the vertical, as shown in the diagram.

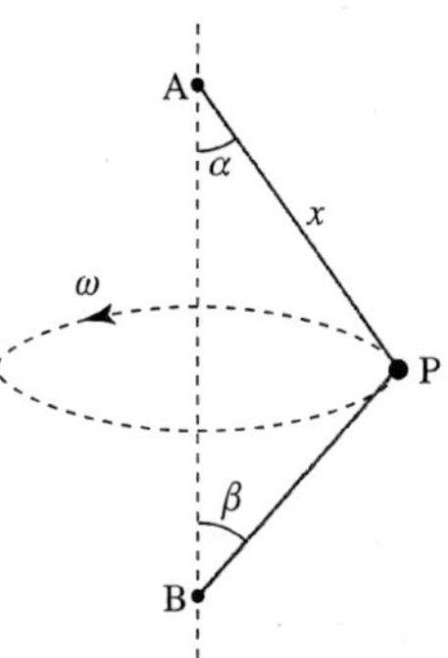

(i) Show that $x\sin\alpha = (2a - x)\sin\beta$.

(ii) By considering the vertical components of forces on the ring, explain why $x > a$.

(iii) By considering the radial motion of the ring, show that $T(\sin\alpha + \sin\beta) = mx\omega^2\sin\alpha$.

(iv) Using your answer to part **(i)**, show that the tension in the string is

$$\frac{mx\omega^2(2a-x)}{2a}.$$

[MEI]

15 A particle of mass 0.15 kg is attached to one end of a light inextensible string of length 2 m. The other end of the string is attached to a fixed point. The particle moves with constant speed in a horizontal circle. The magnitude of the acceleration of the particle is 7 m s^{-2}. The string makes an angle of $\theta°$ with the downward vertical, as shown in the diagram.

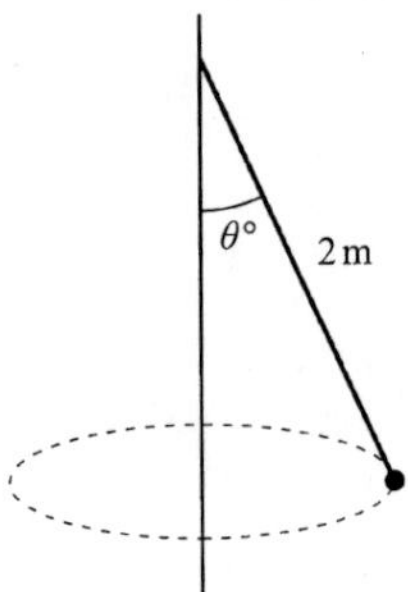

Find

(i) the value of θ to the nearest whole number,

(ii) the tension in the string,

(iii) the speed of the particle.

[Cambridge AS and A Level Mathematics 9709, Paper 5 Q2 June 2005]

16 A hollow container consists of a smooth circular cylinder of radius 0.5 m, and a smooth hollow cone of semi-vertical angle 65° and radius 0.5 m. The container is fixed with its axis vertical and with the cone below the cylinder. A steel ball of weight 1 N moves with constant speed 2.5 m s^{-1} in a horizontal circle inside the container. The ball is in contact with both the cylinder and the cone (see figure **(A)**). Figure **(B)** shows the forces acting on the ball, i.e. its weight and the forces of magnitudes R N and S N exerted by the container at the points of contact. Given that the radius of the ball is negligible compared with the radius of the cylinder, find R and S.

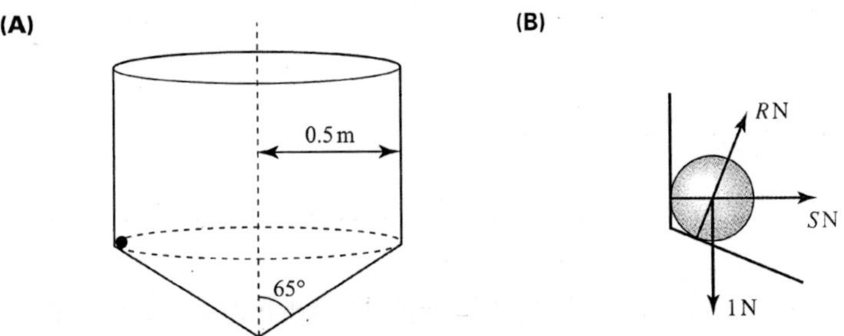

[Cambridge AS and A Level Mathematics 9709, Paper 5 Q3 June 2007]

17 One end of a light inextensible string is attached to a point C. The other end is attached to a point D, which is 1.1 m vertically below C. A small smooth ring R, of mass 0.2 kg, is threaded on the string and moves with constant speed $v\,\text{m s}^{-1}$ in a horizontal circle, with centre at O and radius 1.2 m, where O is 0.5 m vertically below D (see diagram).

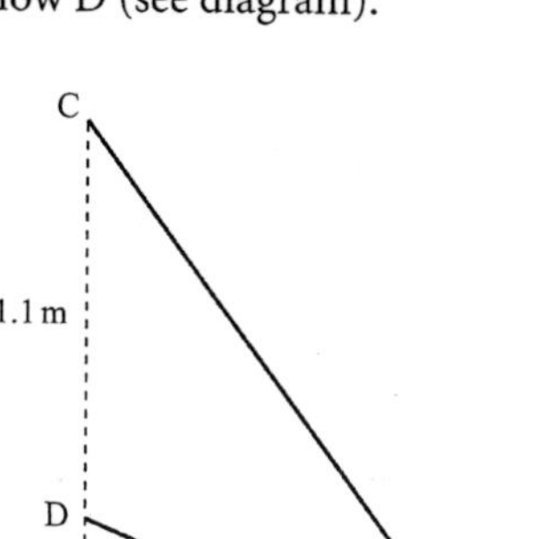

(i) Show that the tension in the string is 1.69 N, correct to 3 significant figures.

(ii) Find the value of v.

[Cambridge AS and A Level Mathematics 9709, Paper 5 Q3 June 2008]

18 A particle of mass 0.12 kg is moving on the smooth inside surface of a fixed hollow sphere of radius 0.5 m. The particle moves in a horizontal circle whose centre is 0.3 m below the centre of the sphere (see diagram).

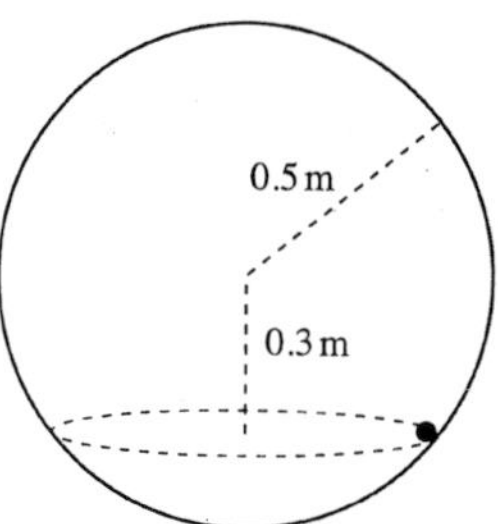

(i) Show that the force exerted by the sphere on the particle has magnitude 2 N.

(ii) Find the speed of the particle.

(iii) Find the time taken for the particle to complete one revolution.

[Cambridge AS and A Level Mathematics 9709, Paper 5 Q4 June 2009]

19 A horizontal circular disc of radius 4 m is free to rotate about a vertical axis through its centre O. One end of a light inextensible rope of length 5 m is attached to a point A of the circumference of the disc, and an object P of mass 24 kg is attached to the other end of the rope. When the disc rotates with constant angular speed ω rad s^{-1}, the rope makes an angle of θ radians with the vertical and the tension in the rope is T N (see diagram). You may assume that the rope is always in the same vertical plane as the radius OA of the disc.

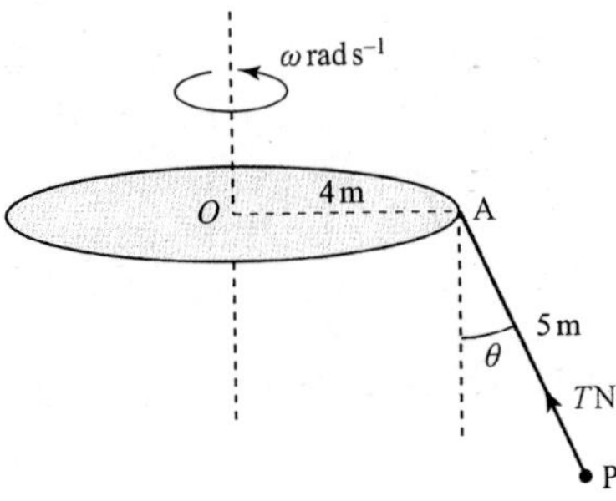

(i) Given that $\cos\theta = \frac{24}{25}$, find the value of ω.

(ii) Given instead that the speed of P is twice the speed of the point A, find

(a) the value of T,

(b) the speed of P.

[Cambridge AS and A Level Mathematics 9709, Paper 5 Q6 November 2005]

20 A particle of mass 0.24 kg is attached to one end of a light inextensible string of length 2 m. The other end of the string is attached to a fixed point. The particle moves with constant speed in a horizontal circle. The string makes an angle θ with the vertical (see diagram), and the tension in the string is T N. The acceleration of the particle has magnitude 7.5 m s^{-2}.

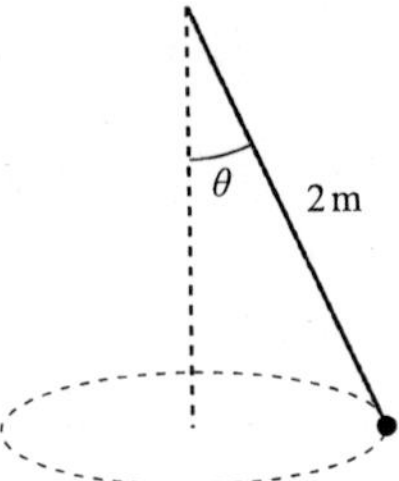

(i) Show that $\tan\theta = 0.75$ and find the value of T.

(ii) Find the speed of the particle.

[Cambridge AS and A Level Mathematics 9709, Paper 51 Q3 June 2010]

Hammer

Investigate the action of throwing the hammer. Estimate the maximum tension in the wire for a top class athlete. (Data: the hammer is a ball of mass 3 kg attached to a light wire of length 1.8 m. A throw of 80 m is world class.)

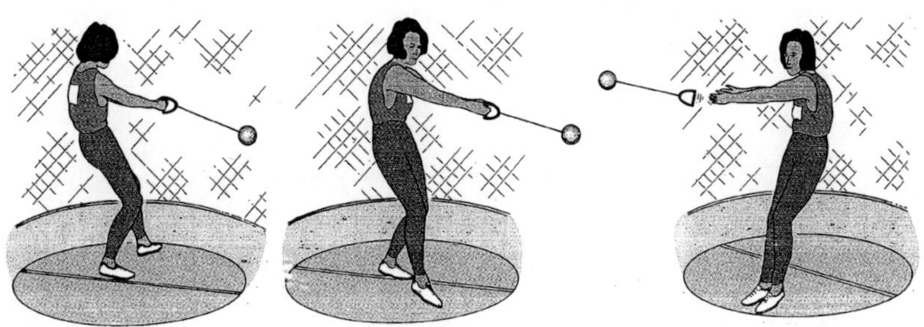

Figure 13.19

Mountain biking

Why do those taking part in mountain bike rallies go home with mud on their backs?

KEY POINTS

1 Position, velocity and acceleration of a particle moving on a circle of radius r.

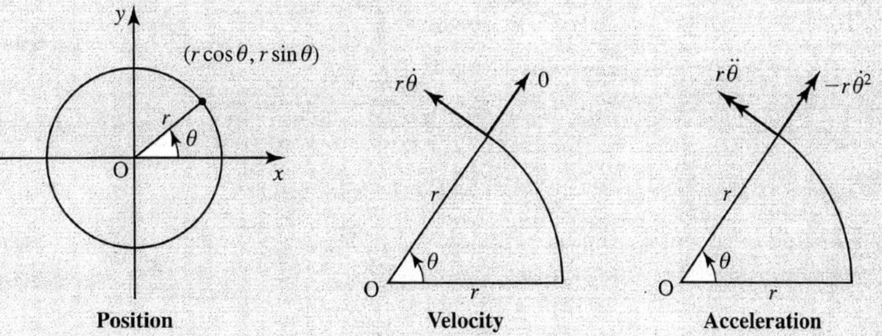

- position $(r\cos\theta, r\sin\theta)$
- velocity transverse component: $v = r\dot{\theta} = r\omega$
  radial component: 0

  where $\dot{\theta}$ or ω is the angular velocity of the particle.
- acceleration transverse component: $r\ddot{\theta} = r\dot{\omega}$
  radial component: $-r\dot{\theta}^2 = -r\omega^2 = -\dfrac{v^2}{r}$

  where $\ddot{\theta}$ or $\dot{\omega}$ is the angular acceleration of the particle.

2 By Newton's second law the forces acting on a particle of mass m in circular motion are equal to

- transverse component: $mr\dot{\omega} = mr\ddot{\theta}$
- radial component: $-\dfrac{mv^2}{r} = -mr\omega^2$
- or radial component: $+\dfrac{mv^2}{r} = +mr\omega^2$ towards the centre.

3 Circular motion breaks down when the available force towards the centre is $< mr\omega^2$ or $\dfrac{mv^2}{r}$.

14 Hooke's law

The only way of finding the limits of the possible is by going beyond them into the impossible.

Arthur C. Clarke

The picture shows someone taking part in the sport of bungee jumping. This is an extreme sport which originated in the South Sea islands where creepers were used rather than ropes. In the more modern version, people jump off a high bridge or crane to which they are attached by elastic ropes round their ankles.

? If somebody bungee jumping from a bridge wants the excitement of just reaching the surface of the water below, how would you calculate the length of rope required?

The answer to this question clearly depends on the height of the bridge, the mass of the person jumping and the elasticity of the rope. All ropes are elastic to some extent, but it would be extremely dangerous to use an ordinary rope for this sport because the impulse necessary to stop somebody falling would involve a very large tension acting in the rope for a short time and this would provide too great a shock to the body. A bungee is a strong elastic rope, similar to those used to secure loads on cycles, cars or lorries, with the essential property for this sport that it allows the impulse to act over a much longer time so that the rope exerts a smaller force on the jumper.

Generally in mechanics, the word *string* is used to represent such things as ropes which can be in tension but not in compression. In this chapter you will be studying some of the properties of elastic strings and springs and will return to the problem of the bungee jumper as a final investigation.

Strings and springs

So far in situations involving strings it has been assumed that they do not stretch when they are under tension. Such strings are called *inextensible.* For some materials this is a good assumption, but for others the length of the string increases significantly under tension. Strings and springs which stretch are said to be *elastic. Open coiled springs* are springs which can also be compressed. In this book springs are assumed to be open coiled.

The length of a string or spring when there is no force applied to it is called its *natural length* (figure 14.1**(a)**). If it is stretched, the increase in length is called its *extension.* If a spring is compressed it is said to have a *negative extension* or *compression.*

When stretched, a spring exerts an inward force, or *tension*, on whatever is attached to its ends (figure 14.1**(b)**).When compressed it exerts an outward force, or *thrust*, on its ends (figure 14.1**(c)**). An elastic string exerts a tension when stretched, but when slack exerts no force.

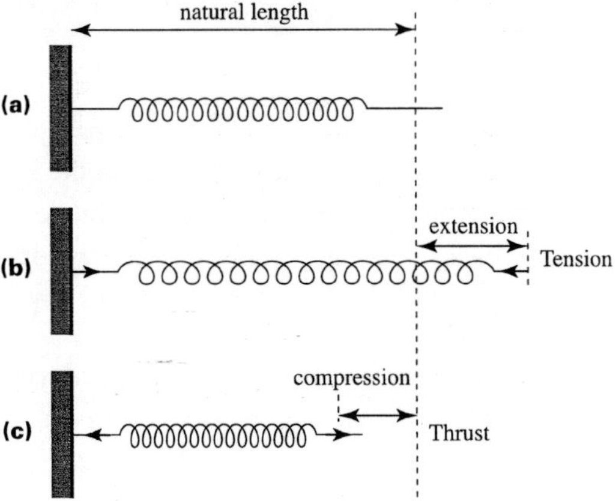

Figure 14.1

EXPERIMENT

You will need some elastic strings, some open coiled springs, some weights and a support stand. Set up the apparatus as shown.

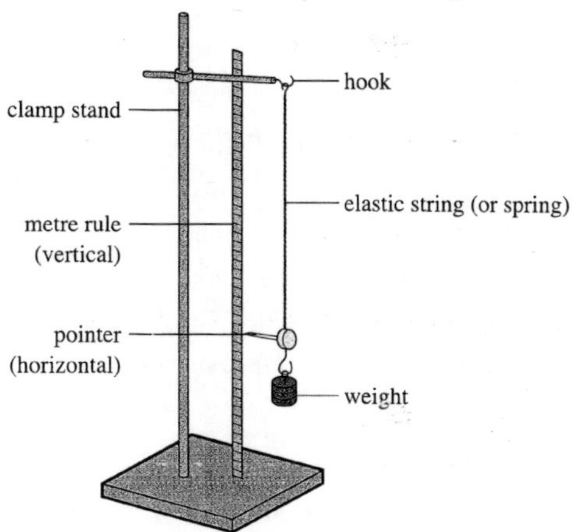

Figure 14.2

Before doing any experiments, predict the answers to the following questions:

1 How are the extension of the string and the weight hanging on it related?

2 If a string of the same material but twice the natural length has the same weight attached, how does the extension change?

3 Does the string return to its original length when unloaded

(i) if the weight of the object is small?
(ii) if the weight of the object is large?

Now use the apparatus to plot a graph, for each string, of tension, i.e. the weight of the object (vertical axis), against the extension (horizontal axis) to help you to answer these questions.

Design and carry out an experiment which will investigate the relationship between the thrust in an open coiled spring and the decrease in its length.

From your experiments you should have made the following observations:

- Each string or spring returned to its original length once the object was removed, up to a certain limit.
- The graph of tension or thrust against extension for each string or spring was a straight line for all or part of the data. Strings or springs which exhibit this linear behaviour are said to be *perfectly elastic*.
- The gradient of the linear part of the graph was roughly halved when the string was doubled in length.

- If you kept increasing the weight, the string or spring might have stopped stretching or might have stretched without returning to its original length. In this case the graph would no longer be a straight line: the material had passed its *elastic limit.*
- During your experiment using an open coiled spring you might have found it necessary to prevent the spring from buckling. You might also have found that there came a point when the coils were completely closed and a further decrease in length was impossible.

Hooke's law

In 1678 Robert Hooke formulated a *Rule or law or nature in every springing body* which, for small extensions relative to the length of the string or spring, can be stated as follows:

- The tension in an elastic spring or string is proportional to the extension. If a spring is compressed the thrust is proportional to the decrease in length of the spring.

When a string or spring is described as elastic, it means that it is reasonable to apply the modelling assumption that it obeys Hooke's law. A further assumption, that it is light (i.e. has zero mass), is usual and is made in this book.

There are three ways in which Hooke's law is commonly expressed for a string. Which one you use depends on the extent to which you are interested in the string itself rather than just its overall properties. Denoting the natural length of the string by l_0 and its area of cross-section by A, the different forms are as follows.

- $T = \frac{EA}{l_0}x$ In this form E is called the *Young modulus* and is a property of the material out of which the string is formed. This form is commonly used in physics and engineering, subjects in which properties of materials are studied. It is rarely used in mathematics. The S.I. unit for the Young modulus is $\mathrm{N\,m^{-2}}$.
- $T = \frac{\lambda}{l_0}x$ The constant λ is called the *modulus of elasticity* of the string and will be the same for any string of a given cross-section made out of the same material. Many situations require knowledge of the natural length of a string and this form may well be the most appropriate in such cases. The S.I. unit for the modulus of elasticity is N.
- $T = kx$ In this simplest form, k is called the *stiffness* of the string. It is a property of the string as a whole. You may choose to use this form if neither the natural length nor the cross-sectional area of the string is relevant to the situation. The S.I. unit for stiffness is $\mathrm{N\,m^{-1}}$.

Notice that $k = \frac{\lambda}{l_0} = \frac{EA}{l_0}$

In this book only the form using the modulus of elasticity is used, and this can be applied to springs as well as strings.

EXAMPLE 14.1

A light elastic string of natural length 0.7 m and modulus of elasticity 50 N has one end fixed and a particle of mass 1.4 kg attached to the other. The system hangs vertically in equilibrium. Find the extension of the string.

SOLUTION

The forces acting on the particle are the tension, T N, upwards and the weight, $1.4g$ N, downwards.

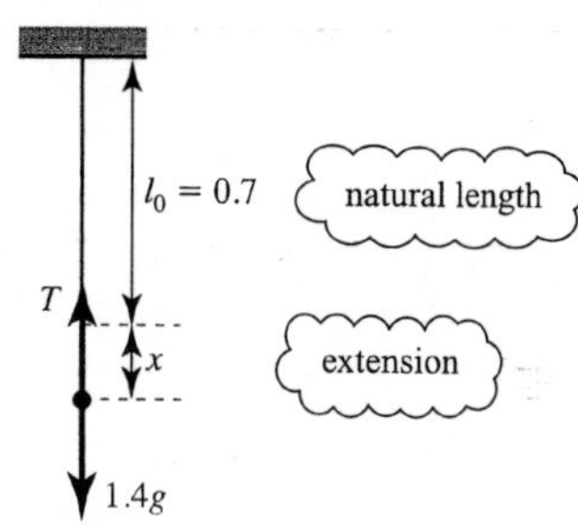

Figure 14.3

Since the particle is in equilibrium

$$T = 1.4g$$

Using Hooke's law:
$$T = \frac{\lambda}{l_0}x$$
$$\Rightarrow \quad 1.4g = \frac{50}{0.7}x$$
$$\Rightarrow \quad x = \frac{0.7 \times 1.4g}{50}$$
$$= 0.196$$

The extension in the string is 0.196 m.

EXERCISE 14A

1 A light elastic spring of natural length 1.5 m is attached to the ceiling. A block of mass 2 kg hangs in equilibrium, attached to the other end of the spring and the spring is extended by 30 cm.

(i) Draw a diagram showing the forces acting on the block.

(ii) Find the modulus of elasticity of the spring.

2 **(i)** An elastic string has natural length 20 cm. The string is fixed at one end. When a force of 20 N is applied to the other end the string doubles in length. Find the modulus of elasticity.

(ii) Another elastic string also has natural length 20 cm. When a force of 20 N is applied to each end the string doubles in length. Find the modulus of elasticity.

(iii) Explain the connection between the answers to parts **(i)** and **(ii)**.

3 A light spring has modulus of elasticity 0.4 N and natural length 50 cm. One end is attached to a ceiling, the other to a particle of weight 0.03 N which hangs in equilibrium below the ceiling.

(i) Find the tension in the spring.

(ii) Find the extension of the spring.

The particle is removed and replaced with one of weight w N. When this hangs in equilibrium the spring has length 60 cm.

(iii) What is the value of w?

4 An object of mass 0.5 kg is attached to an elastic string with natural length 1.2 m and causes an extension of 8 cm when the system hangs vertically in equilibrium.

(i) What is the tension in the spring?

(ii) What is the modulus of elasticity of the spring?

(iii) What is the mass of an object which causes an extension of 10 cm?

5 The diagram shows a spring of natural length 60 cm which is being compressed under the weight of a block of mass m kg. Smooth supports constrain the block to move only in the vertical direction.

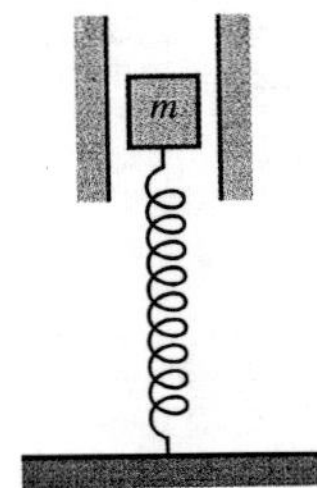

The modulus of elasticity of the spring is 180 N. The system is in equilbrium and the length of the spring is 50 cm. Find

(i) the thrust in the spring

(ii) the value of m

More blocks are piled on.

(iii) Describe the situation when there are seven blocks in total, all identical to the first one.

6 A small sphere, A, of mass m kg moves in a circle with centre B on a smooth horizontal table. A is joined to a smoothly rotating vertical axle at B by an elastic string of natural length a m and modulus of elasticity λ N and has constant angular speed ω rad s^{-1}. Find an expression for the radius of the circle in terms of m, a, λ and ω.

Using Hooke's law with more than one spring or string

Hooke's law allows you to investigate situations involving two or more springs or strings in various configurations.

EXAMPLE 14.2

A particle of mass 0.4 kg is attached to the mid-point of a light elastic string of natural length 1 m and modulus of elasticity λ N. The string is then stretched between a point A at the top of a doorway and a point B which is on the floor 2 m vertically below A.

(i) Find, in terms of λ, the extensions of the two parts of the string.

(ii) Calculate their values in the case where $\lambda = 10$.

(iii) Find the minimum value of λ which will ensure that the lower half of the string is not slack.

SOLUTION

For a question like this it is helpful to draw two diagrams, one showing the relevant natural lengths and extensions, and the other showing the forces acting on the particle.

Since the force of gravity acts downwards on the particle, its equilibrium position will be below the mid-point of AB. This is also shown in the diagram.

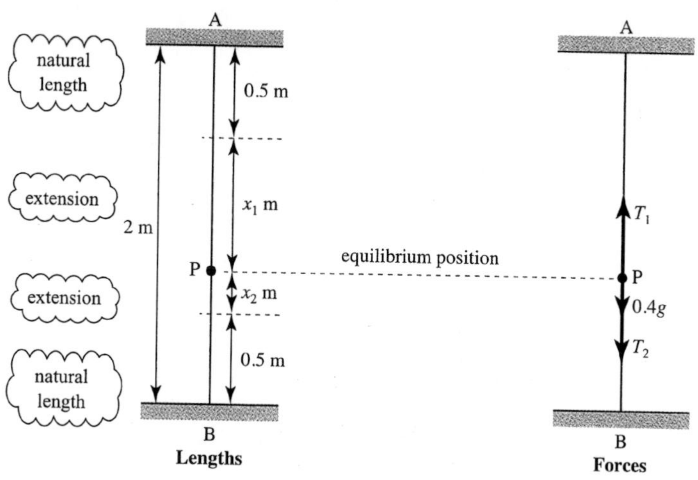

Figure 14.4

(i) The particle is in equilibrium, so the resultant vertical force acting on it is zero.

Therefore $\quad T_1 = T_2 + 0.4g \qquad ①$

Hooke's law can be applied to each part of the string.

For AP: $\quad T_1 = \dfrac{\lambda}{0.5} x_1 \qquad ②$

For BP: $T_2 = \dfrac{\lambda}{0.5} x_2$ ③

Substituting these expressions in equation ① gives

Alternatively, you can use x for x_1 and $(1 - x)$ for x_2.

$$\frac{\lambda}{0.5} x_1 = \frac{\lambda}{0.5} x_2 + 0.4g$$

$$\Rightarrow \quad \lambda x_1 - \lambda x_2 = 0.5 \times 0.4g$$

$$\Rightarrow \quad x_1 - x_2 = 0.2\frac{g}{\lambda} \quad ④$$

But from the first diagram it can be seen that

$$x_1 + x_2 = 1 \quad ⑤$$

Adding ④ and ⑤ gives:

$$2x_1 = 1 + 0.2\frac{g}{\lambda}$$

$$\Rightarrow \quad x_1 = 0.5 + 0.1\frac{g}{\lambda}$$

Similarly, subtracting ④ from ⑤ gives:

$$x_2 = 0.5 - 0.1\frac{g}{\lambda} \quad ⑥$$

(ii) Since $\lambda = 10$ the extensions are 0.6 m and 0.4 m.

(iii) The lower part of the string will not become slack providing $x_2 > 0$. It follows from equation ⑥ that:

$$0.5 - 0.1\frac{g}{\lambda} > 0$$

$$\Rightarrow \quad 0.5 > 0.1\frac{g}{\lambda}$$

$$\Rightarrow \quad \lambda > 0.2g$$

The minimum value of λ for which the lower part of the string is not slack is 2 N, and in this case BP has zero tension.

Historical note

If you search for Robert Hooke (1635–1703) on the internet, you will find that he was a man of many parts. He was one of a talented group of polymaths (which included his rival Newton) who have had an enormous impact on scientific thought and practice. Among other things, he designed and built Robert Boyle's air pump, discovered the red spot on Jupiter and invented the balanced spring mechanism for watches. His work on microscopy led to his becoming the father of microbiology and he was the first to use the term 'cell' with respect to living things. Hooke worked closely with his friend Sir Christopher Wren in the rebuilding of the City of London after the great fire, and was responsible for the realisation of many of his designs including the Royal Greenwich Observatory. Both Hooke and Wren were astronomers and architects and they designed the Monument to the fire with a trapdoor at the top and a laboratory in the basement so that it could be used as an enormous 62 m telescope. Hooke, the great practical man, also used the column for experiments on air pressure and pendulums.

EXERCISE 14B

1 The diagram shows a uniform plank of weight 120 N symmetrically suspended in equilibrium by two identical elastic strings, each of natural length 0.8 m and modulus of elasticity 1200 N.

Find

(i) the tension in each string

(ii) the extension of each string.

The two strings are replaced by a single string, also of natural length 0.8 m, attached to the middle of the plank. The plank is in the same position.

(iii) Find the modulus of elasticity of this string and comment on its relationship to that of the original strings.

2 The manufacturer of a sports car specifies the coil spring for the front suspension as a spring of 10 coils with a natural length 0.3 m and a compression 0.1 m when under a load of 4000 N.

(i) Calculate the modulus of elasticity of the spring.

(ii) If the spring were cut into two equal parts, what would be the modulus of elasticity of each part?

The weight of a car is 8000 N and half of this weight is taken by two such 10-coil front springs so that each bears a load of 2000 N.

(iii) Find the compression of each spring.

(iv) Two people each of weight 800 N get into the front of the car. How much further are the springs compressed? (Assume that their weight is carried equally by the front springs.)

3 The coach of an impoverished rugby club decides to construct a scrummaging machine as illustrated in the diagrams below. It is to consist of a vertical board, supported in horizontal runners at the top and bottom of each end. The board is held away from the wall by springs, as shown, and the players push the board with their shoulders, against the thrust of the springs.

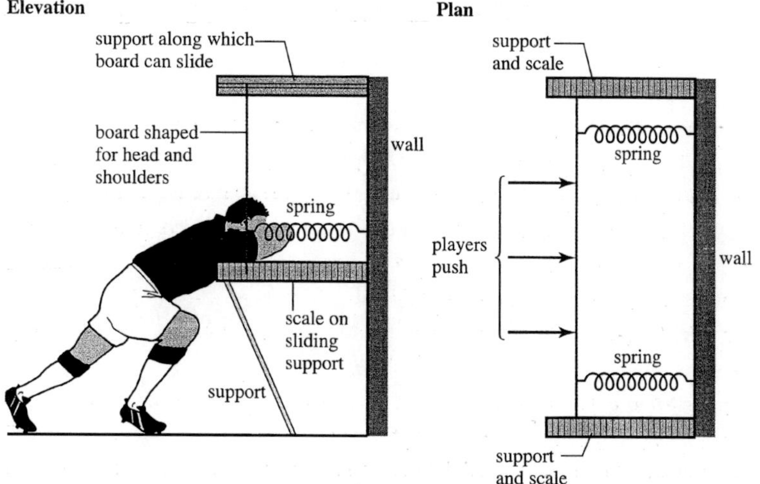

The coach has one spring of length 1.4 m and modulus of elasticity 7000 N, which he cuts into two pieces of equal length.

(i) Find the modulus of elasticity of the original spring.

(ii) Find the modulus of elasticity of each of the half-length springs.

(iii) On one occasion the coach observes that the players compress the springs by 20 cm. What total force do they produce in the forward direction?

4 The diagram shows the rear view of a load of weight 300 N in the back of a pick-up truck of width 2 m.

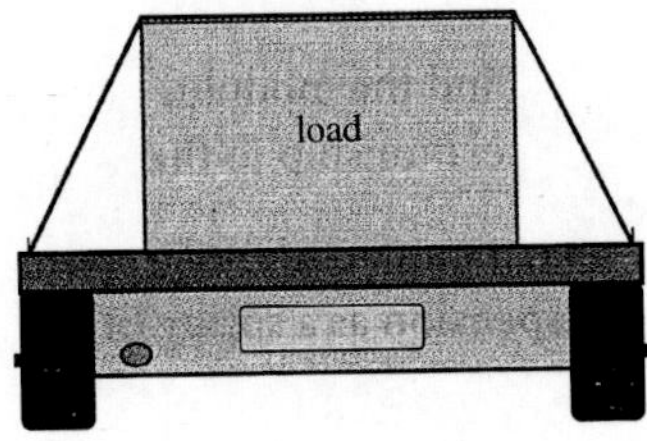

The load is 1.2 m wide, 0.8 m high and is situated centrally on the truck. The coefficient of friction between the load and the truck is 0.4. The load is held down by an elastic rope of natural length 2 m and modulus of elasticity 400 N which may be assumed to pass smoothly over the corners and across the top of the load. The rope is secured at the edges of the truck platform. Find

(i) the tension in the rope

(ii) the normal reaction of the truck on the load

(iii) the percentage by which the maximum possible frictional force is increased by using the rope

(iv) the shortest stopping distance for which the load does not slide, given that the truck is travelling at $30\,\text{m}\,\text{s}^{-1}$ initially. (Assume constant deceleration.)

5 The diagram shows two light springs, AP and BP, connected at P. The ends A and B are secured firmly and the system is in equilibrium.

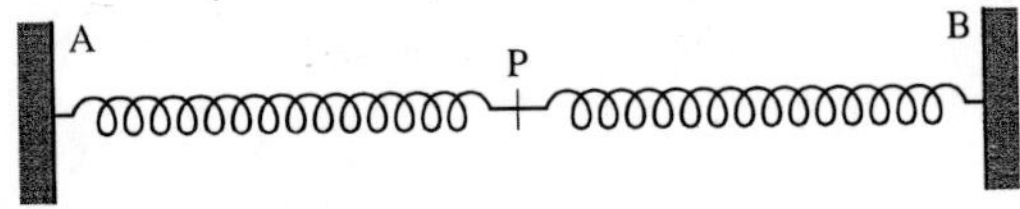

The spring AP has natural length 1 m and modulus of elasiticity 16 N.
The spring BP has natural length 1.2 m and modulus of elasticity 30 N.
The distance AB is 2.5 m and the extension of the spring AP is x m.

(i) Write down an expression, in terms of x, for the extension of the spring BP.

(ii) Find expressions, in terms of x, for the tensions in both springs.

(iii) Find the value of x.

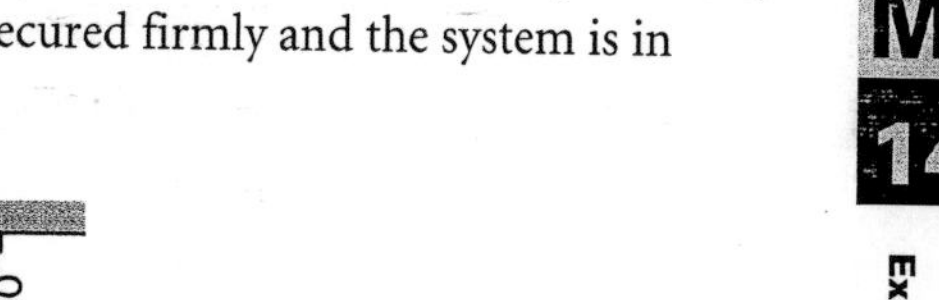

6 The diagram shows two light springs, CQ and DQ, connected to a particle, Q, of weight 20 N. The ends C and D are secured firmly and the system is in equilibrium, lying in a vertical line.

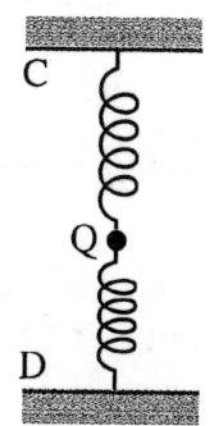

The spring CQ has natural length 0.8 m and modulus of elasticity 16 N. DQ has natural length 1.2 m and modulus of elasticity 36 N. The distance CD is 3 m and QD is h m.

(i) Write down expressions, in terms of h, for the extensions of the two springs.

(ii) Find expressions, in terms of h, for the tensions in the two springs.

(iii) Use these results to find the value of h.

(iv) Find the forces the system exerts at C and at D.

7 The diagram shows a block of wood of mass m lying on a plane inclined at an angle α to the horizontal. The block is attached to a fixed peg by means of a light elastic string of natural length l_0 and modulus of elasticity λ; the string lies parallel to the line of greatest slope. The block is in equilibrium.

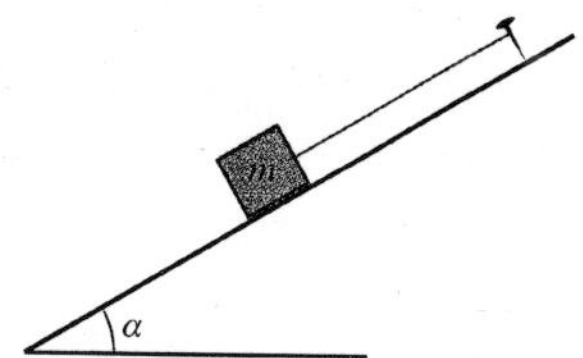

Find the extension of the string in the following cases.

(i) The plane is smooth.

(ii) The coefficient of friction between the plane and the block is μ $(\mu \neq 0)$ and the block is about to slide **(a)** up the plane **(b)** down the plane.

8 A strong elastic band of natural length 1 m and modulus of elasticity 12 N is stretched round two pegs P and Q which are in a horizontal line a distance 1 m apart. A bag of mass 1.5 kg is hooked on to the band at H and hangs in equilibrium so that PH and QH make angles of θ with the horizontal. Make the modelling assumptions that the elastic band is light and runs smoothly over the pegs.

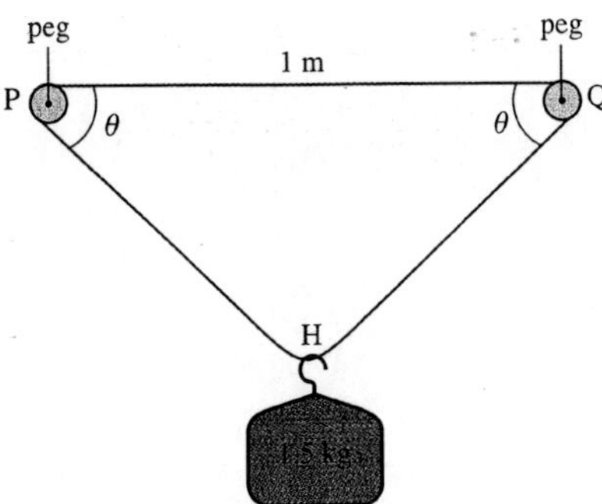

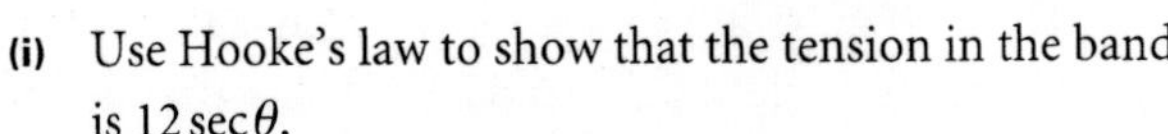

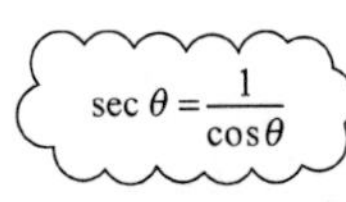

(i) Use Hooke's law to show that the tension in the band is $12\sec\theta$.

(ii) Find the depth of the hook below the horizontal line PQ.

(iii) Is the modelling in this question realistic?

9 A particle A and a block B are attached to opposite ends of a light elastic string of natural length 2 m and modulus of elasticity 6 N. The block is at rest on a rough horizontal table. The string passes over a small smooth pulley P at the edge of the table, with the part BP of the string horizontal and of length 1.2 m. The frictional force acting on B is 1.5 N and the system is in equilibrium (see diagram). Find the distance PA.

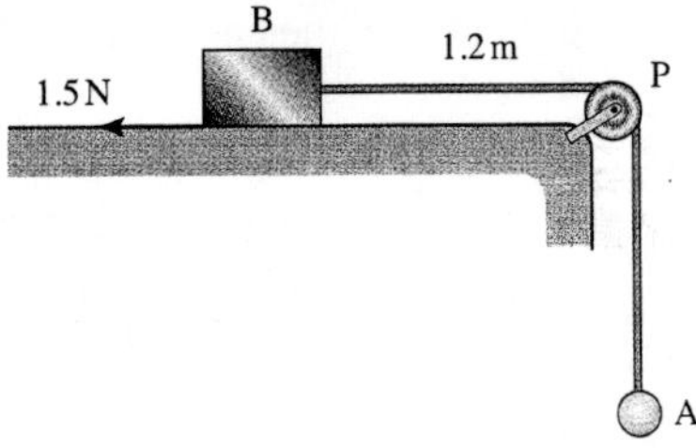

[Cambridge AS and A Level Mathematics 9709, Paper 5 Q1 June 2008]

10 A light elastic string has natural length 0.6 m and modulus of elasticity λ N. The ends of the string are attached to fixed points A and B, which are at the same horizontal level and 0.63 m apart. A particle P of mass 0.064 kg is attached to the mid-point of the string and hangs in equilibrium at a point 0.08 m below AB (see diagram).

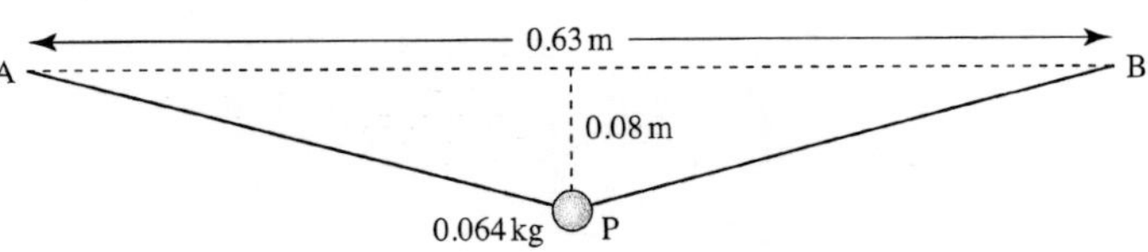

Find

(i) the tension in the string,

(ii) the value of λ.

[Cambridge AS and A Level Mathematics 9709, Paper 5 Q1 June 2006]

Work and energy

In order to stretch an elastic spring a force must do work on it. In the case of the muscle exerciser in figure 14.5, this force is provided by the muscles working against the tension. When the exerciser is pulled at constant speed, at any given time the force F applied at each end is equal to the tension in the spring; consequently it changes as the spring stretches.

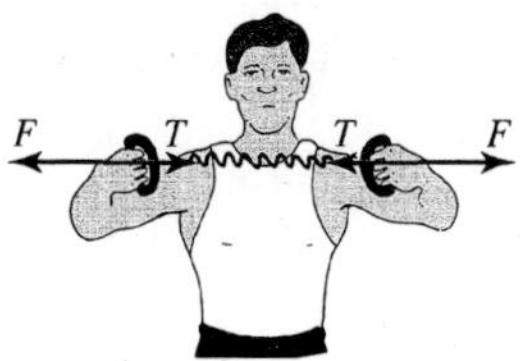

Figure 14.5

Suppose that one end of the spring is stationary and the extension is x as in figure 14.6. By Hooke's law the tension is given by

$$T = \frac{\lambda}{2_{\iota}}x, \qquad \text{and so} \qquad F = \frac{\lambda}{2_{\iota}}x$$

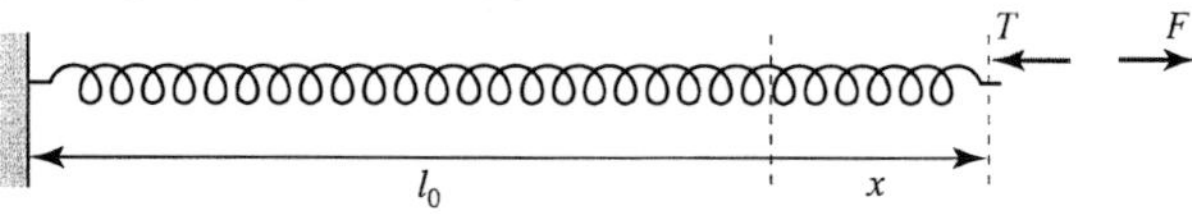

Figure 14.6

The work done by a *constant* force F in moving a distance d in its own direction is given by Fd. To find the work done by a *variable* force the process has to be considered in small stages.

Now imagine that the force extends the string a small distance δx. The work done is given by $\quad F\delta x = \frac{\lambda}{l_0}x\delta x.$

The total work done in stretching the spring many small distances is

$$\sum F\delta x = \sum k\frac{\lambda}{l_0}x\delta x$$

In the limit as $\delta x \to 0$, the work done is:

$$\int F\,\mathrm{d}x = \int \frac{\lambda}{l_0} x\,\mathrm{d}x$$

$$= \frac{\lambda}{2l_0} x^2 + c$$

λ and l_0 are constants for a given spring.

When the extension $x = 0$, the work done is zero, so $c = 0$.

The total work done in stretching the spring an extension x *from its natural length* l_0 is therefore given by $\frac{\lambda}{2l_0} x^2$.

The result is the same for the work done in compressing a spring.

Elastic potential energy

The tensions and thrusts in perfectly elastic springs and strings are conservative forces, since any work done against them can be recovered in the form of kinetic energy. A catapult and a jack-in-a-box use this property.

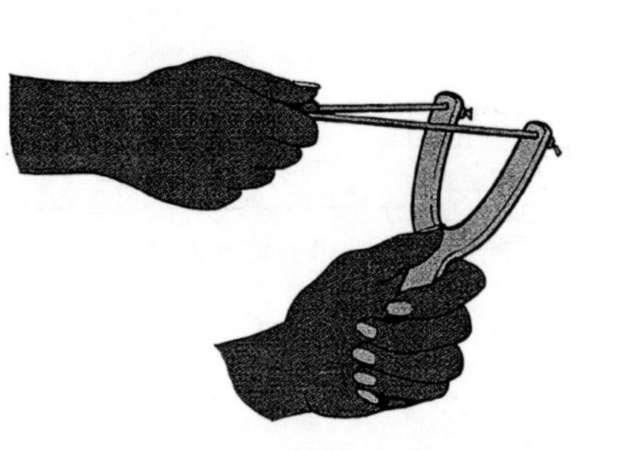

Figure 14.7

The work done in stretching or compressing a string or spring can therefore be regarded as potential energy. It is known as *elastic potential energy.*

The elastic potential energy stored in a spring which is stretched or compressed by an amount x is

$$\frac{\lambda}{2l_0} x^2.$$

EXAMPLE 14.3

An elastic rope of natural length 0.6 m is extended to a length of 0.8 m. The modulus of elasticity of the rope is 25 N. Find

(i) the elastic potential energy stored in the rope

(ii) the further energy required to stretch it to a length of 1.65 m over a roof-rack.

SOLUTION

(i) The extension of the elastic is (0.8 – 0.6) m = 0.2 m.

The energy stored in the rope is

$$\frac{\lambda}{2l_0}x^2 = \frac{25}{2 \times 0.6}(0.2)^2$$
$$= 0.83\text{ J (to 2 d.p.)}.$$

(ii) The extension of the elastic rope is now 1.65 – 0.6 = 1.05 m

The elastic energy stored in the rope is

$$\frac{25}{2 \times 0.6}(1.05)^2 = 22.97\text{ J}$$

The extra energy required to stretch the rope is 22.14 J (correct to 2 d.p.).

⚠ In the example above, the string is stretched so that its extension changes from x_1 to x_2 (in this case, from 0.2 m to 1.05 m). The work required to do this is

$$\frac{\lambda}{2l_0}x_2^2 - \frac{\lambda}{2l_0}x_1^2 = \frac{\lambda}{2l_0}(x_2^2 - x_1^2)$$

You can see by using algebra that this expression is *not* the same as $\frac{\lambda}{2l_0}(x_2 - x_1)^2$, so it is *not* possible to use the extra extension $(x_2 - x_1)$ directly in the energy expression to calculate the extra energy stored in the string.

EXAMPLE 14.4

A catapult has prongs which are 16 cm apart and the elastic string is 20 cm long. A marble of mass 70 g is placed in the centre of the elastic string and pulled back so that the string is just taut. The marble is then pulled back a further 9 cm and the force required to keep it in this position is 60 N. Find

(i) the stretched length of the string

(ii) the tension in the string and its modulus of elasticity

(iii) the elastic potential energy stored in the string and the speed of the marble when the string regains its natural length, assuming they remain in contact.

SOLUTION

To solve this problem it is necessary to assume that there is no elasticity in the frame of the catapult, and that the motion takes place in a horizontal plane. In addition, any air resistance is ignored.

In figure 14.8, A and B are the ends of the elastic string and M_1 and M_2 are the two positions of the marble (before and after the string is stretched). D is the mid-point of AB.

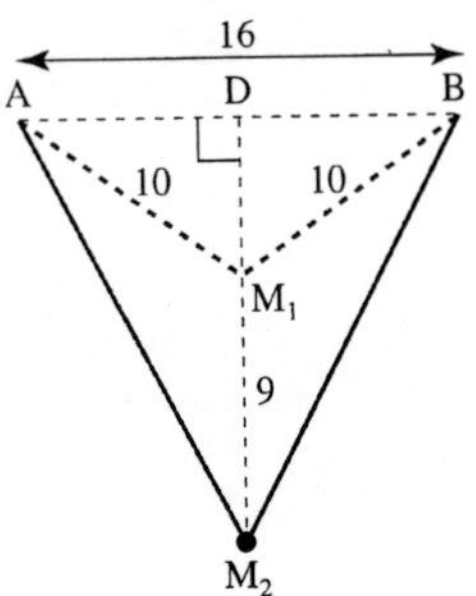

Figure 14.8

(i) Using Pythagoras' theorem in triangle DBM_1 gives

$$DM_1 = \sqrt{10^2 - 8^2} = 6\,\text{cm}.$$

So

$$DM_2 = 9 + 6 = 15\,\text{cm}.$$

Using Pythagoras' theorem in triangle DBM_2 gives

$$BM_2 = \sqrt{15^2 + 8^2} = 17\,\text{cm}.$$

The stretched length of the string is $2 \times 17\,\text{cm} = 0.34\,\text{m}$.

(ii) Take the tension in the string to be TN.

Resolving parallel to M_2D:

$$2T\cos\alpha = 60$$

Now $$\cos\alpha = \frac{DM_2}{BM_2} = \frac{0.15}{0.17}$$

so $$T = 34$$

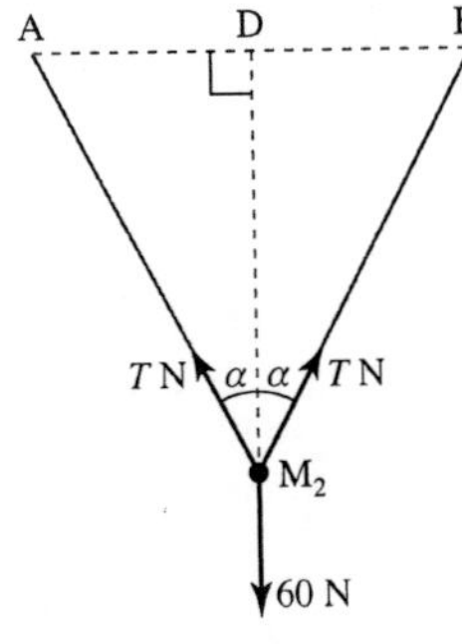

Figure 14.9

The extension of the string is $(0.34 - 0.2)\,\text{m} = 0.14\,\text{m}$.

By Hooke's law the modulus of elasticity λ is given by $\frac{\lambda}{l_0}x = T$

$$\lambda = \frac{34}{0.14} \times 0.2 = 48.57...$$

The modulus of elasticity of the string is 48.6 N (to 3 s.f.).

(iii) The elastic potential energy stored in the string is

$$\frac{\lambda}{2l_0}x^2 = \frac{48.57...}{2 \times 0.2} \times (0.14)^2 = 2.38\,\text{J}$$

By the principle of conservation of energy, this is equal to the kinetic energy given to the marble. The mass of the marble is 0.07 kg, so

$$\tfrac{1}{2} \times 0.07\,v^2 = 2.38$$
$$\Rightarrow \quad v = 8.25...$$

The speed of the marble is $8.3\,\text{m}\,\text{s}^{-1}$.

EXERCISE 14C

1 An open coiled spring has natural length 0.3 m and modulus of elasticity 6 N. Find the elastic potential energy in the spring when

(i) it is extended by 0.1 m
(ii) it is compressed by 0.01 m
(iii) its length is 0.5 m
(iv) its length is 0.3 m.

2 A spring has natural length 0.4 m and modulus of elasticity 20 N. Find the elastic energy stored in the spring when

(i) it is extended by 0.4 m
(ii) it is compressed by 0.1 m
(iii) its length is 0.2 m
(iv) its length is 0.45 m.

3 A pinball machine fires small balls of mass 50 g by means of a spring of natural length 20 cm and a light plunger. The spring and the ball move in a horizontal plane. The spring has modulus of elasticity 120 N and is compressed by 5 cm to fire a ball.

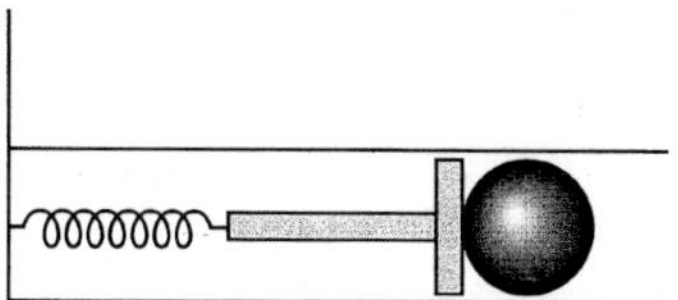

(i) Find the energy stored in the spring immediately before the ball is fired.
(ii) Find the speed of the ball when it is fired.

4 A catapult is made from elastic string with modulus of elasticity 5 N. The string is attached to two prongs which are 15 cm apart, and is just taut. A pebble of mass 40 g is placed in the centre of the string and is pulled back 4 cm and then released in a horizontal direction.

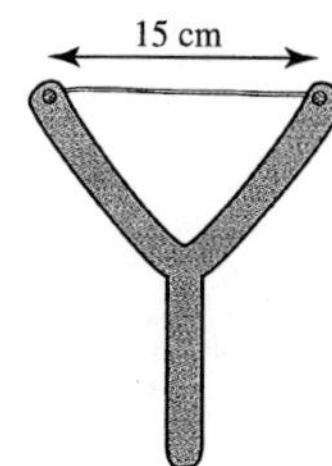

(i) Calculate the work done in stretching the string.
(ii) Calculate the speed of the pebble on leaving the catapult.

5 A simple mathematical model of a railway buffer consists of a horizontal open coiled spring attached to a fixed point. The modulus of elasticity of the spring is 2×10^5 N and its natural length is 2 m.

The buffer is designed to stop a railway truck before the spring is compressed to half its natural length, otherwise the truck will be damaged.

(i) Find the elastic energy stored in the spring when it is half its natural length.

(ii) Find the maximum speed at which a truck of mass 2 tonnes can approach the buffer safely. Neglect any other reasons for loss of energy of the truck.

A truck of mass 2 tonnes approaches the buffer at $5\,\text{m s}^{-1}$.

(iii) Calculate the minimum length of the spring during the subsequent period of contact.

(iv) Find the thrust in the spring and the acceleration of the truck when the spring is at its minimum length.

(v) What happens next?

6 Two identical springs are attached to a sphere of mass 0.5 kg that rests on a smooth horizontal surface as shown. The other ends of the springs are attached to fixed points A and B.

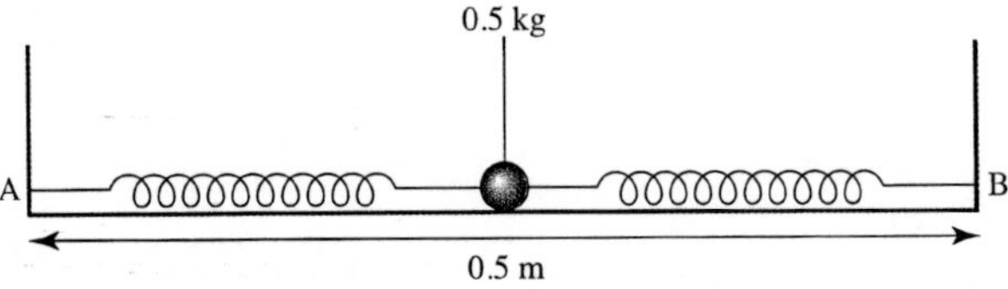

The springs each have modulus of elasticity 7.5 N and natural length 25 cm. The sphere is at rest at the mid-point when it is projected with speed $2\,\text{m s}^{-1}$ along the line of the springs towards B. Calculate the length of each spring when the sphere first comes to rest.

7 Two light springs are joined and stretched between two fixed points A and C which are 2 m apart as shown in the diagram. The spring AB has natural length 0.5 m and modulus of elasticity 10 N. The spring BC has natural length 0.6 m and modulus of elasticity 6 N. The system is in equilibrium.

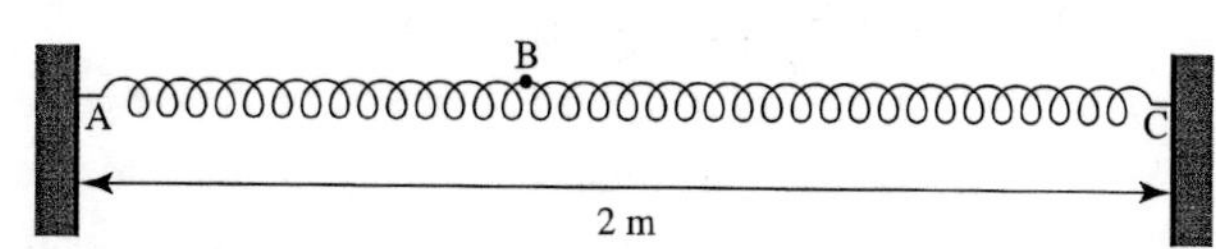

(i) Explain why the tensions in the two springs are the same.

(ii) Find the distance AB and the tension in each spring.

(iii) How much work must be done to stretch the springs from their natural length to connect them as described above?

A small object of mass 0.012 kg is attached at B and is supported on a smooth horizontal table. A, B and C lie in a straight horizontal line and the mass is released from rest at the mid-point of AC.

(iv) What is the speed of the mass when it passes through the equilibrium position of the system?

[MEI]

8 A particle P of mass 0.4 kg is attached to one end of a light elastic string of natural length 1.5 m and modulus of elasticity 6 N. The other end of the string is attached to a fixed point O on a rough horizontal table. P is released from rest at a point on the table 3.5 m from O. The speed of P at the instant the string becomes slack is 6 m s^{-1}. Find

(i) the work done against friction during the period from the release of P until the string becomes slack,

(ii) the coefficient of friction between P and the table.

[Cambridge AS and A Level Mathematics 9709, Paper 5 Q4 June 2005]

9 A and B are fixed points on a smooth horizontal table. The distance AB is 2.5 m. An elastic string of natural length 0.6 m and modulus of elasticity 24 N has one end attached to the table at A, and the other end attached to a particle P of mass 0.95 kg. Another elastic string of natural length 0.9 m and modulus of elasticity 18 N has one end attached to the table at B, and the other end attached to P. The particle P is held at rest at the mid-point of AB (see diagram).

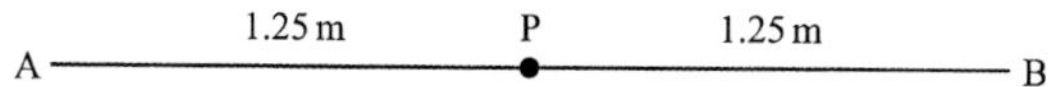

(i) Find the tensions in the strings.

The particle is released from rest.

(ii) Find the acceleration of P immediately after its release.

(iii) P reaches its maximum speed at the point C. Find the distance AC.

[Cambridge AS and A Level Mathematics 9709, Paper 5 Q6 June 2007]

10 A particle P of mass 1.6 kg is attached to one end of each of two light elastic strings. The other ends of the strings are attached to fixed points A and B which are 2 m apart on a smooth horizontal table. The string attached to A has natural length 0.25 m and modulus of elasticity 4 N, and the string attached to B has natural length 0.25 m and modulus of elasticity 8 N. The particle is held at the mid-point M of AB (see diagram).

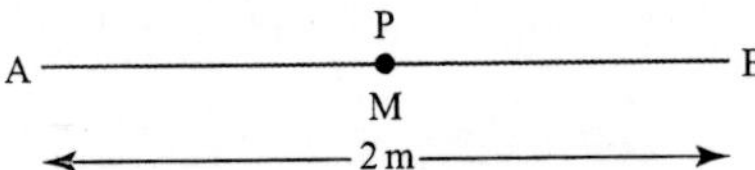

(i) Find the tensions in the strings.

(ii) Show that the total elastic potential energy in the two strings is 13.5 J.

P is released from rest and in the subsequent motion both strings remain taut. The displacement of P from M is denoted by x m. Find

(iii) the initial acceleration of P,

(iv) the non-zero value of x at which the speed of P is zero.

[Cambridge AS and A Level Mathematics 9709, Paper 5 Q6 June 2009]

Vertical motion

This chapter began with a bungee jumper undergoing vertical motion at the end of an elastic rope. The next example involves a particle in vertical motion at the end of a spring. This, along with the questions in the following exercise, covers the essential work involved in modelling the bungee jump, which you are then invited to investigate.

EXAMPLE 14.5

A particle of mass 0.2 kg is attached to the end A of a perfectly elastic spring OA which has natural length 0.5 m and modulus of elasticity 20 N. The spring is suspended from O and the particle is pulled down and released from rest when the length of the spring is 0.7 m. In the subsequent motion the extension of the spring is denoted by x m.

(i) Write down expressions for the increase in the particle's gravitational potential energy and the decrease in the energy stored in the spring when the extension is x m.

(ii) Hence find an expression for the speed of the particle in terms of x.

(iii) Calculate the length of the spring when the particle is at its highest point.

SOLUTION

(i)

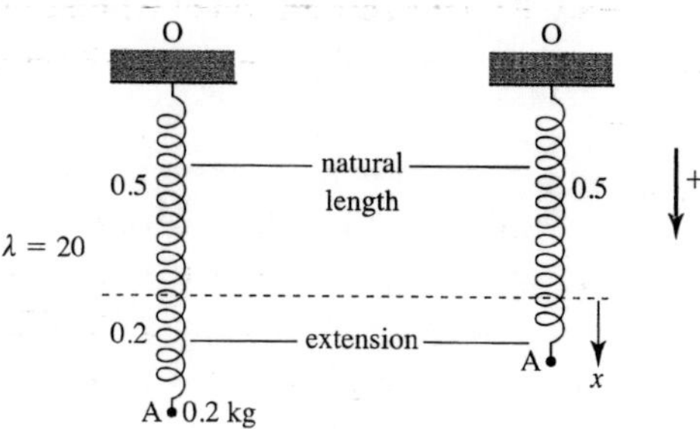

Figure 14.10

The particle has risen a distance $(0.2 - x)$ m

$$\text{Increase in gravitational P.E.} = mgh = 2 \times (0.2 - x)$$

$$\text{Stored energy} = \frac{\lambda}{2l_0}x^2 = \frac{20}{2 \times 0.5}x^2 = 20x^2$$

$$\text{Initial stored energy} = 20 \times 0.2^2$$

$$\text{Decrease in stored energy} = 20(0.2^2 - x^2)$$

(ii) The initial K.E. is zero.

$$\text{Increase in K.E.} = \tfrac{1}{2} \times 0.2 \times \dot{x}^2 - 0 = 0.1\dot{x}^2$$

$\dot{x}$ is the same as v, the speed in the positive direction

Using the law of conservation of mechanical energy,

$$\text{Increase in K.E.} + \text{P.E.} = \text{Decrease in stored energy}$$

$$0.1\dot{x}^2 + 2(0.2 - x) = 20(0.2^2 - x^2)$$

and so $$v = \dot{x} = \sqrt{200(0.2^2 - x^2) - 20(0.2 - x)}$$

(iii) At the highest point, $v = \dot{x} = 0$,

$$200(0.2^2 - x^2) - 20(0.2 - x) = 0$$
$$(0.2 - x)[200(0.2 + x) - 20] = 0$$
$$\Rightarrow \quad x = 0.2 \text{ or } 40 + 200x = 20$$
$$\Rightarrow \quad x = 0.2 \text{ or } x = -0.1$$

but $x = 0.2$ at the lowest position

so $x = -0.1$ at the highest point

This negative value of x indicates a compression rather than an extension, so at its highest point the spring has length $(0.5 - 0.1)$ m $= 0.4$ m.

Note

For this question it is important that you are dealing with a spring, which still obeys Hooke's law when it contracts, rather than a string which becomes slack.

EXERCISE 14D

1 A particle of mass 0.2 kg is attached to one end of a light elastic spring of modulus of elasticity 10 N and natural length 1 m. The system hangs vertically and the particle is released from rest when the spring is at its natural length. The particle comes to rest when it has fallen a distance h m.

(i) Write down an expression in terms of h for the energy stored in the spring when the particle comes to rest at its lowest point.

(ii) Write down an expression in terms of h for the gravitational potential energy lost by the particle when it comes to rest at its lowest point.

(iii) Find the value of h.

2 A particle of mass m is attached to one end of a light vertical spring of natural length l_0 and modulus of elasticity $2mg$. The particle is released from rest when the spring is at its natural length. Find, in terms of l_0, the maximum length of the spring in the subsequent motion.

3 A block of mass m is placed on a smooth plane inclined at 30° to the horizontal. The block is attached to the top of the plane by a spring of natural length l_0 and modulus λ. The system is released from rest with the spring at its natural length. Find an expression for the maximum length of the spring in the subsequent motion.

4 A particle of mass 0.1 kg is attached to one end of a spring of natural length 0.3 m and modulus of elasticity 20 N. The other end is attached to a fixed point and the system hangs vertically. The particle is released from rest when the length of the spring is 0.2 m. In the subsequent motion the extension of the spring is denoted by x m.

(i) Show that $\quad 0.05\dot{x}^2 + \dfrac{10}{0.3}(x^2 - 0.1^2) - (x + 0.1) = 0$

(ii) Find the maximum value of x.

5 A small apple of mass 0.1 kg is attached to one end of an elastic string of natural length 25 cm and modulus of elasticity 5 N. David is asleep under a tree and Sam fixes the free end of the string to the branch of the tree just above David's head. Sam releases the apple level with the branch and it just touches David's head in the subsequent motion. How high above his head is the branch?

6 A block of mass 0.5 kg lies on a light scale pan which is supported on a vertical spring of natural length 0.4 m and modulus of elasticity 40 N. Initially the spring is at its natural length and the block is moving downwards with a speed of 2 m s^{-1}. Gravitational potential energy is measured relative to the initial position.

(i) Find the initial mechanical energy of the system.

(ii) Show that the speed v m s^{-1} of the block when the compression of the spring is m is given by $v = 2\sqrt{1 + 5x - 50x^2}$.

(iii) Find the minimum length of the spring during the oscillations.

7 A scale pan of mass 0.5 kg is suspended from a fixed point by a spring of modulus of elasticity 50 N and natural length 10 cm.

(i) Calculate the length of the spring when the scale pan is in equilibrium.

(ii) A bag of sugar of mass 1 kg is gently placed on the pan and the system is released from rest. Find the maximum length of the spring in the subsequent motion.

8 A bungee jump is carried out by a person of mass m kg using an elastic rope which can be taken to obey Hooke's law. It is known that the jump operator does not exceed the total length limit of four times the original length of the rope in any jump. Prove that the tension in the rope is at most $\frac{8}{3}mg$ N.

9 A conical pendulum consists of a bob of mass m attached to an inextensible string of length l. The bob describes a circle of radius r with angular speed ω, and the string makes an angle θ with the vertical as shown.

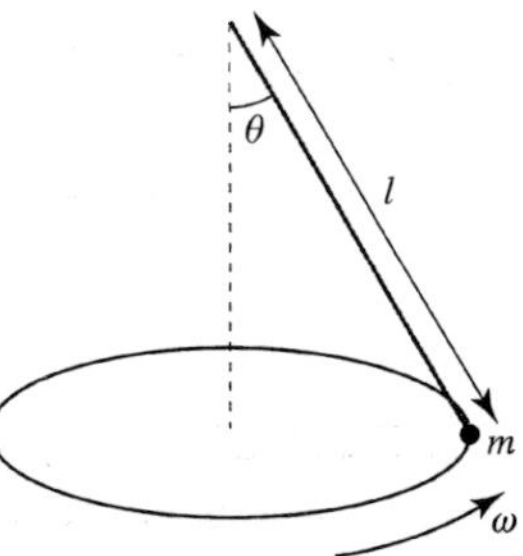

(i) Find an expression for θ in terms of ω, l and g.

The string is replaced with an elastic string of modulus of elasticity λ and natural length l_0.

(ii) Find an expression for the new value of θ in terms of ω, m, g, l_0 and λ.

10 Use $g = 9.8\,\text{m}\,\text{s}^{-2}$ in this question.

A light, elastic string has natural length 0.5 m and modulus of elasticity 49 N. The end A is attached to a point on a ceiling. A small object of mass 3 kg is attached to the end B of the string and hangs in equilibrium.

(i) Calculate the length AB.

A second string, identical to the first one, is now attached to the object at B and to a point C on the floor, 2.5 m vertically below the point A. The system is equilibrium with B a distance x m below A, as shown in the diagram below.

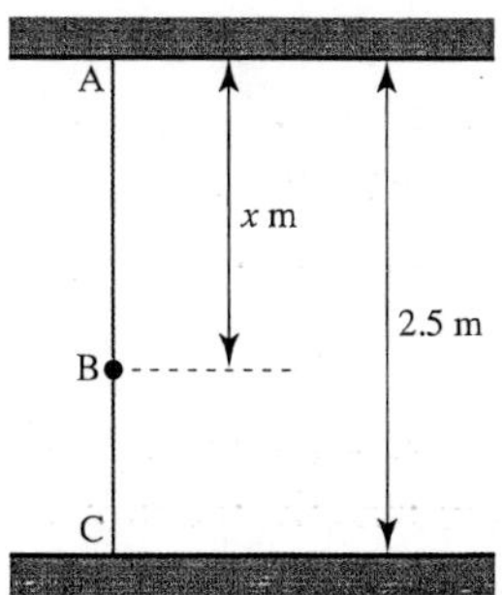

(ii) Find the tension in each of the strings in terms of x and hence show that $x = 1.4$.

(iii) Calculate the total elastic potential energy in the strings when the object hangs in equilibrium.

The object is now pulled down 0.1 m from its equilibrium position and released from rest.

(iv) Calculate the speed of the object when it passes through the equilibrium position. Any resistances to motion may be neglected.

[MEI]

11 A light elastic string has natural length 4 m and modulus of elasticity 2 N. One end of the string is attached to a fixed point O of a smooth plane which is inclined at 30° to the horizontal. The other end of the string is attached to a particle P of mass 0.1 kg. P is held at rest at O and then released. The speed of P is $v\,\text{m s}^{-1}$ when the extension of the string is x m.

(i) Show that $v^2 = 45 - 5(x - 1)^2$.

Hence find

(i) the distance of P from O when P is at its lowest point,

(iii) the maximum speed of P.

[Cambridge AS and A Level Mathematics 9709, Paper 5 Q6 November 2008]

12 A particle P of mass 0.35 kg is attached to the mid-point of a light elastic string of natural length 4 m. The ends of the string are attached to fixed points A and B which are 4.8 m apart at the same horizontal level. P hangs in equilibrium at a point 0.7 m vertically below the mid-point M of AB (see diagram).

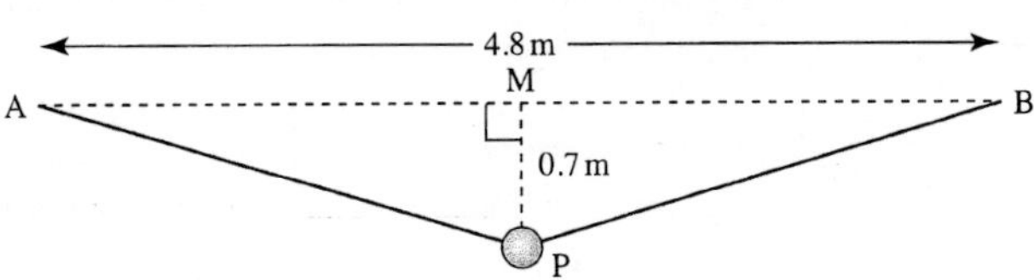

(i) Find the tension in the string and hence show that the modulus of elasticity of the string is 25 N.

P is now held at rest at a point 1.8 m vertically below M, and is then released.

(ii) Find the speed with which P passes through M.

[Cambridge AS and A Level Mathematics 9709, Paper 51 Q6 November 2010]

The bungee jump

Figure 14.11

(i) *Experiment*

Use a weight to represent the jumper and a piece of elastic for the bungee. Measure l_0 and find the value of λ by suspending the weight in equilibrium. Try to predict the lowest point reached by the weight when it is dropped. Can you estimate a suitable length of elastic for any given weight to fall a standard height?

(ii) *Modelling*

Typical parameters for a mobile crane bungee jump are:

Height of jump station: 55 m
Bottom safety space: 5 m
Static line length: 5 m (non-elastic straps etc.)
Unstretched elastic rope length: 12 m
Modulus of elasticity: 1000 N

? Calculate the maximum deceleration of the heaviest person who can jump safely. Would a lighter person experience a greater or lesser deceleration?

In practice, bungee jumpers usually use a braided rope. The braiding not only keeps the elastic core stretched, it also prevents the rope from stretching too much. As the rope begins to approach its maximum length, the modulus of elasticity gradually increases until 'lock out' occurs at maximum extension. This rope then no longer obeys Hooke's law. How would the jump feel different using a braided rope?

KEY POINTS

1 **Hooke's law**

The tension T in an elastic string or spring and its extension x are related by:

$$T = \frac{\lambda}{l_0}x$$

where λ is the modulus of elasticity and l_0 is the natural length of the string or spring.

2 When a spring is compressed, x is negative and the tension becomes a thrust.

3 **Elastic potential energy**

The elastic potential energy stored in a stretched spring or string, or in a compressed spring, is given by E.P.E. where:

$$\text{E.P.E.} = \frac{\lambda}{2l_0}x^2.$$

This is the work done in stretching a spring or string or compressing a spring starting at its natural length.

4 The tension or thrust in an elastic string or spring is a conservative force and so the elastic potential energy is recoverable.

5 When no frictional or other dissipative forces are involved, elastic potential energy can be used with kinetic energy and gravitational potential energy to form equations using the principle of conservation of energy.

15 Linear motion under a variable force

Is it possible to fire a projectile up to the moon?

The Earth to the Moon *by Jules Verne (1865)*

In his book, Jules Verne says that this is possible ... 'provided it possesses an initial velocity of 12 000 yards per second. In proportion as we recede from the Earth the action of gravitation diminishes in the inverse ratio of the square of the distance; that is to say at three times a given distance the action is nine times less. Consequently the weight of a shot will decrease and will become reduced to zero at the instant that the attraction of the moon exactly counterpoises that of the Earth; at $\frac{47}{52}$ of its journey. There the projectile will have no weight whatever; and if it passes that point it will fall into the moon by the sole effect of lunar attraction.'

? If an *unpowered* projectile could be launched from the earth with a high enough speed in the right direction, it would reach the moon.

What forces act on the projectile during its journey?

How near to the moon will it get if its initial speed is not *quite* enough?

In Jules Verne's story, three men and two dogs were sent to the moon inside a projectile fired from an enormous gun. Although this is completely impracticable, the basic mathematical ideas in the passage above are correct. As a projectile moves further from the earth and nearer to the moon, the gravitational attraction of the earth decreases and that of the moon increases. In many of the dynamics problems you have met so far it has been assumed that forces are *constant*, whereas on Jules Verne's space missile the total force *varies* continuously as the motion proceeds.

You may have already met problems involving variable force. When an object is suspended on a spring and bounces up and down, the varying tension in the spring leads to *simple harmonic motion*. You will also be aware that air resistance depends on velocity.

Gravitation, spring tension and air resistance all give rise to variable force problems, the subject of this chapter.

Newton's second law as a differential equation

Calculus techniques are used extensively in mechanics and you will already have used differentiation and integration in earlier work. In this chapter you will see how essential calculus methods are in the solution of a variety of problems.

To solve variable force problems, you can use Newton's second law to give an equation for the *instantaneous* value of the acceleration. When the mass of a body is constant, this can be written in the form of a *differential equation.*

$$F = m\frac{\mathrm{d}v}{\mathrm{d}t}$$

It can also be written as

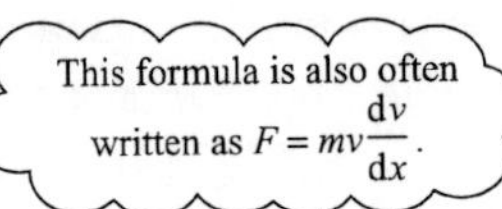

$$F = mv\frac{\mathrm{d}v}{\mathrm{d}s}$$

This follows from the chain rule for differentiation.

$$\frac{\mathrm{d}v}{\mathrm{d}t} = \frac{\mathrm{d}v}{\mathrm{d}s} \times \frac{\mathrm{d}s}{\mathrm{d}t}$$
$$= v\frac{\mathrm{d}v}{\mathrm{d}s}$$

Note

Here and throughout this chapter the mass, *m*, is assumed to be constant. Jules Verne's spacecraft was a projectile fired from a gun. It was not a *rocket* whose mass varies, due to ejection of fuel.

Deriving the constant acceleration formulae

To see the difference in use between the $\frac{\mathrm{d}v}{\mathrm{d}t}$ and $v\frac{\mathrm{d}v}{\mathrm{d}s}$ forms of acceleration, it is worth looking at the case where the force, and therefore the acceleration, $\frac{F}{m}$ is *constant* (say a). Starting from the $\frac{\mathrm{d}v}{\mathrm{d}t}$ form,

$$\frac{\mathrm{d}v}{\mathrm{d}t} = a$$

Integrating gives

$$v = u + at$$

where u is the constant of integration ($v = u$ when $t = 0$).

Since $v = \frac{\mathrm{d}s}{\mathrm{d}t}$, integrating again gives

$$s = ut + \tfrac{1}{2}at^2 + s_0$$

assuming the displacement is s_0 when $t = 0$.

These are the familiar formulae for motion under constant acceleration.

Starting from the $v\dfrac{dv}{ds}$ form,

$$v\frac{dv}{ds} = a$$

Separating the variables and integrating gives

$$\int v\,dv = \int a\,ds$$

$$\Rightarrow \qquad \tfrac{1}{2}v^2 = as + k$$

where k is the constant of integration.

Assuming $v = u$ when $s = 0$, $k = \tfrac{1}{2}u^2$, so the formula becomes

$$v^2 = u^2 + 2as$$

This is another of the standard constant acceleration formulae. Notice that *time is not involved* when you start from the $v\dfrac{dv}{ds}$ form of acceleration.

Solving *F* = *ma* for variable force

When the force is continuously *variable*, you write Newton's second law in the form of a differential equation and then solve it using one of the forms of acceleration, $v\dfrac{dv}{ds}$ or $\dfrac{dv}{dt}$. The choice depends on the particular problem. Some guidelines are given below and you should check these with the examples which follow.

Normally, the resulting differential equation can be solved by separating the variables.

The force is a function of time

When the force is a function, $F(t)$, of time you use $a = \dfrac{dv}{dt}$.

$$F(t) = m\frac{dv}{dt}$$

Separating the variables and integrating gives

$$m\int dv = \int F(t)\,dt.$$

Assuming you can solve the integral on the right-hand side, you then have v in terms of t.

Writing v as $\dfrac{ds}{dt}$, the displacement as a function of time can be found by integrating again.

The force is a function of displacement

When the force is a function, F(s), of displacement, you normally start from

$$\mathrm{F}(s) = mv\frac{\mathrm{d}v}{\mathrm{d}s}$$

then $$\int \mathrm{F}(s)\,\mathrm{d}s = m\int v\,\mathrm{d}v.$$

The force is a function of velocity

When the force is given as a function, F(v), of velocity, you have a choice. You can use

$$\mathrm{F}(v) = m\frac{\mathrm{d}v}{\mathrm{d}t}$$

or $$\mathrm{F}(v) = mv\frac{\mathrm{d}v}{\mathrm{d}s}$$

You can separate the variables in both forms; use the first if you are interested in behaviour over time and the second when you wish to involve displacement.

Variable force examples

The three examples that follow show the approaches used when the force is given respectively as a function of time, displacement and velocity.

When you are solving these problems, it is important to be clear about which direction is positive *before* writing down an equation of motion.

EXAMPLE 15.1

A crate of mass m is freely suspended at rest from a crane. When the operator begins to lift the crate further, the tension in the suspending cable increases uniformly from mg newtons to $1.2\,mg$ newtons over a period of 2 seconds.

(i) What is the tension in the cable t seconds after the lifting has begun ($t \leqslant 2$)?

(ii) What is the velocity after 2 seconds?

(iii) How far has the crate risen after 2 seconds?

Assume the situation may be modelled with air resistance and cable stretching ignored.

SOLUTION

When the crate is at rest it is in equilibrium and so the tension, T, in the cable equals the weight mg of the crate. After time $t = 0$, the tension increases, so there is a net upward force and the crate rises, see figure 15.1.

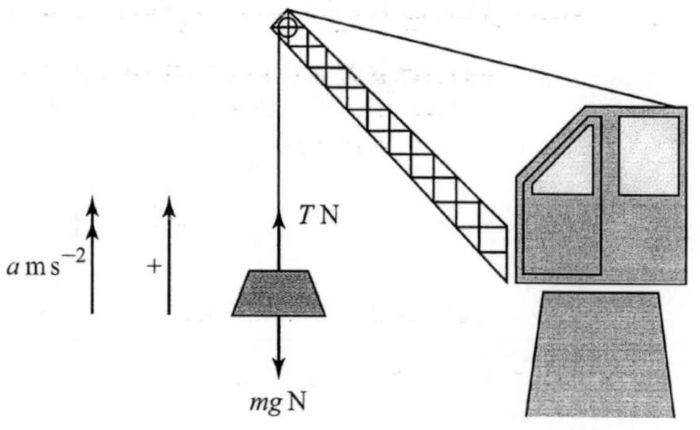

Figure 15.1

(i) The tension increases uniformly by $0.2mg$ newtons in 2 seconds, i.e. it increases by $0.1mg$ newtons per second, see figure 15.2.

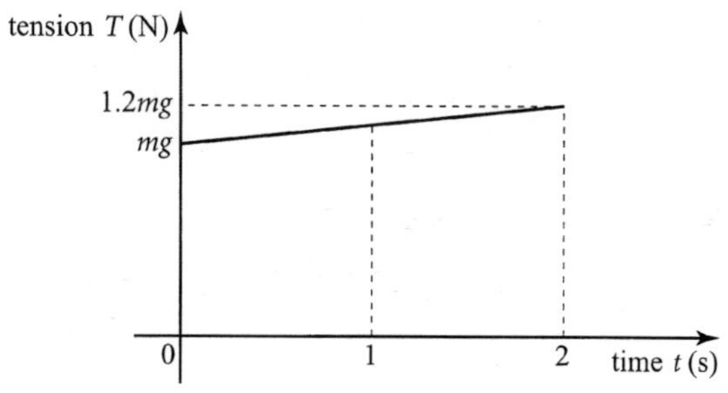

Figure 15.2

After t seconds, the tension is $T = mg + 0.1mgt$.

(ii) As the force is a function of time use $a = \frac{\mathrm{d}v}{\mathrm{d}t}$. Then at any moment in the 2-second period, $F = ma$ gives

$$(mg + 0.1mgt) - mg = m\frac{\mathrm{d}v}{\mathrm{d}t}$$

upwards is positive

$$\Rightarrow \quad \frac{\mathrm{d}v}{\mathrm{d}t} = 0.1gt$$

Integrating gives

$$v = 0.05gt^2 + k$$

where k is the constant of integration.

When $t = 0$, the crate has not quite begun to move, so $v = 0$. This gives $k = 0$ and $v = 0.05gt^2$.

When t is 2,

$$v = 0.05 \times 10 \times 4$$
$$= 2.$$

The velocity after 2 seconds is $2\,\mathrm{m\,s^{-1}}$.

(iii) To find the displacement s, write v as $\frac{ds}{dt}$ and integrate again.

$$\frac{ds}{dt} = 0.05gt^2$$

$$s = \int 0.05gt^2 dt$$

$$s = 0.05g \times \tfrac{1}{3}t^3 + c$$

When $t = 0$, $s = 0 \Rightarrow c = 0$.

When $t = 2$ and $g = 10$, $s = \frac{4}{3}$.

The crate moves $\frac{4}{3}$ m in 2 seconds.

? The displacement cannot be obtained by the formula $s = \frac{1}{2}(u + v)t$, which would give the answer 2 m. Why not?

EXAMPLE 15.2

A prototype of Jules Verne's projectile, mass m, is launched vertically upwards from the earth's surface but only just reaches a height of one tenth of the earth's radius before falling back. When the height, s, above the surface is small compared with the radius of the earth, R, the magnitude of the earth's gravitational force on the projectile may be modelled as $mg\left(1 - \frac{2s}{R}\right)$, where g is gravitational acceleration at the earth's surface.

Assuming all other forces can be neglected

(i) write down a differential equation of motion involving s and velocity, v

(ii) integrate this equation and hence obtain an expression for the loss of kinetic energy of the projectile between its launch and rising to a height s

(iii) show that the launch velocity is $0.3\sqrt{2gR}$.

SOLUTION

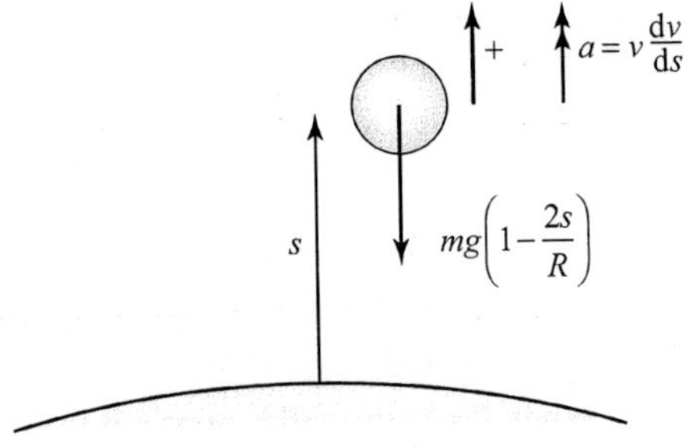

Figure 15.3

(i) Taking the upward direction as positive, the force on the projectile is $-mg\left(1-\frac{2s}{R}\right)$. The force is a function of s, so start from the equation of motion in the form

$$mv\frac{dv}{ds}=-mg\left(1-\frac{2s}{R}\right)$$

(ii) Separating the variables and integrating gives

$$\int mv\,dv=-\int mg\left(1-\frac{2s}{R}\right)ds$$

$$\Rightarrow \quad \tfrac{1}{2}mv^2=-mgs+\frac{mgs^2}{R}+k.$$

You would normally divide the equation by m, but it is useful to leave it in here in order to get kinetic energy directly from $\int mv\,dv$.

Writing v_0 for the launch velocity, $v=v_0$ when $s=0$, so $k=\tfrac{1}{2}mv_0^2$ and rearranging gives

$$\tfrac{1}{2}mv_0^2-\tfrac{1}{2}mv^2=mgs-\frac{mgs^2}{R}. \qquad ①$$

The left-hand side is the loss of kinetic energy, so

$$\text{loss of K.E.}=mgs-\frac{mgs^2}{R}.$$

(iii) Dividing equation ① by m and multiplying by 2 gives

$$v_0^2-v^2=2gs-\frac{2gs^2}{R}.$$

If the projectile just reaches a height $s=\frac{R}{10}$, then the velocity v is zero at that point.

Substituting $s=\frac{R}{10}$ and $v=0$ gives

$$v_0^2=2g\left(\frac{R}{10}\right)-\frac{2gR^2}{100R}$$

$$=\frac{18gR}{100}$$

$$\Rightarrow \quad v_0=\tfrac{3}{10}\sqrt{2gR}$$

So the launch velocity is $0.3\sqrt{2gR}$.

EXAMPLE 15.3

A body of mass 2 kg, initially at rest on a smooth horizontal plane, is subjected to a horizontal force of magnitude $\frac{1}{2v+1}$ N, where v is the velocity of the body ($v>0$).

(i) Find the time when the velocity is $1\,\mathrm{m\,s^{-1}}$.

(ii) Find the displacement when the velocity is $1\,\mathrm{m\,s^{-1}}$.

SOLUTION

(i) Using $F = ma = m\frac{dv}{dt}$

$$\frac{1}{2v+1} = 2\frac{dv}{dt}.$$

Write acceleration in $\frac{dv}{dt}$ form since time is required.

Separating the variables gives

$$\int dt = \int 2(2v+1)\,dv$$

$$\Rightarrow \quad t = 2v^2 + 2v + k.$$

When $t = 0$, $v = 0$ so $k = 0$ and therefore

$$t = 2v^2 + 2v$$

When $v = 1$, $t = 4$. That is, when the velocity is $1\,\mathrm{m\,s^{-1}}$, the time is 4 seconds.

(ii) Using $F = ma = mv\frac{dv}{ds}$

$$\frac{1}{2v+1} = 2v\frac{dv}{ds}$$

Write acceleration in $v\frac{dv}{ds}$ form since displacement is required.

Separating the variables gives

$$\int ds = \int 2v(2v+1)\,dv$$

$$\Rightarrow \quad s = \tfrac{4}{3}v^3 + v^2 + k.$$

When $s = 0$, $v = 0$ so $k = 0$ and therefore

$$s = \tfrac{4}{3}v^3 + v^2$$

When $v = 1$, $s = \frac{7}{3}$. When the velocity is $1\,\mathrm{m\,s^{-1}}$, the displacement is $2\frac{1}{3}$ m.

EXERCISE 15A

1 Each of the parts **(i)** to **(viii)** of this question assumes a body of mass 1 kg under the influence of a single force F N in a constant direction but with a variable magnitude given as a function of velocity, $v\,\mathrm{m\,s^{-1}}$, displacement, s m, or time, t seconds.

In each case, express $F = ma$ as a differential equation using either $a = \frac{dv}{dt}$ or $a = v\frac{dv}{ds}$ as appropriate. Then separate the variables and integrate, giving the result in the required form and leaving an arbitrary constant in the answer.

(i) $F = 2v$ — express s in terms of v

(ii) $F = 2v$ — express v in terms of t

(iii) $F = 2\sin 3t$ — express v in terms of t

(iv) $F = -v^2$ — express v in terms of t

(v) $F = -v^2$ — express v in terms of s

(vi) $F = -4s + 2$ — express v in terms of s

(vii) $F = -2v - 3v^2$ — express s in terms of v

(viii) $F = 1 + v^2$ — express s in terms of v

2 Each of the parts **(i)** to **(viii)** of this question assumes a body of mass 1 kg under the influence of a single force F N in a constant direction but with a variable magnitude given as a function of velocity, $v\,\mathrm{m\,s^{-1}}$, displacement, s m, or time, t seconds. The body is initially at rest at a point O.

In each case, write down the equation of motion and solve it to supply the required information.

(i) $F = 2t^2$ find v when $t = 2$

(ii) $F = -\dfrac{1}{(s+1)^2}$ find v when $s = -\frac{1}{9}$

(iii) $F = \dfrac{1}{s+3}$ find v when $s = 3$

(iv) $F = \dfrac{1}{v+1}$ find t when $v = 3$

(v) $F = 1 + v^2$ find t when $v = 1$

(vi) $F = 5 - 3v$ find t when $v = 1$

(vii) $F = 1 - v^2$ find t when $v = 0.5$

(Hint: Use partial fractions.)

(viii) $F = 1 - v^2$ find s when $v = 0.5$

3 A horse pulls a 500 kg cart from rest until the speed, v, is about $5\,\mathrm{m\,s^{-1}}$. Over this range of speeds, the magnitude of the force exerted by the horse can be modelled by $500(v+2)^{-1}$ N. Neglecting resistance,

(i) write down an expression for $v\dfrac{dv}{ds}$ in terms of v

(ii) show by integration that when the velocity is $3\,\mathrm{m\,s^{-1}}$, the cart has travelled 18 m

(iii) write down an expression for $\dfrac{dv}{dt}$ and integrate to show that the velocity is $3\,\mathrm{m\,s^{-1}}$ after 10.5 seconds

(iv) show that $v = -2 + \sqrt{4 + 2t}$

(v) integrate again to derive an expression for s in terms of t, and verify that after 10.5 seconds, the cart has travelled 18 m.

4 The acceleration of a particle moving in a straight line is $(x - 2.4)\,\mathrm{m\,s^{-2}}$ when its displacement from a fixed point O of the line is x m. The velocity of the particle is $v\,\mathrm{m\,s^{-1}}$, and it is given that $v = 2.5$ when $x = 0$. Find

(i) an expression for v in terms of x,

(ii) the minimum value of v.

[Cambridge AS and A Level Mathematics 9709, Paper 5 Q5 June 2005]

5 An object of mass 0.4 kg is projected vertically upwards from the ground, with an initial speed of $16\,\text{m}\,\text{s}^{-1}$. A resisting force of magnitude $0.1v$ newtons acts on the object during its ascent, where $v\,\text{m}\,\text{s}^{-1}$ is the speed of the object at time t s after it starts to move.

(i) Show that $\dfrac{\text{d}v}{\text{d}t} = -0.25(v + 40)$.

(ii) Find the value of t at the instant that the object reaches its maximum height.

[Cambridge AS and A Level Mathematics 9709, Paper 5 Q4 June 2006]

6 A particle P of mass 0.5 kg moves on a horizontal surface along the straight line OA, in the direction from O to A. The coefficient of friction between P and the surface is 0.08. Air resistance of magnitude $0.2v$ N opposes the motion, where $v\,\text{m}\,\text{s}^{-1}$ is the speed of P at time t s. The particle passes through O with speed $4\,\text{m}\,\text{s}^{-1}$ when $t = 0$.

(i) Show that $2.5\dfrac{\text{d}v}{\text{d}t} = -(v + 2)$ and hence find the value of t when $v = 0$.

(ii) Show that $\dfrac{\text{d}x}{\text{d}t} = 6\text{e}^{-0.4t} - 2$, where x m is the displacement of P from O at time t s, and hence find the distance OP when $v = 0$.

[Cambridge AS and A Level Mathematics 9709, Paper 5 Q7 June 2008]

7 A particle P starts from a fixed point O and moves in a straight line. When the displacement of P from O is x m, its velocity is $v\,\text{m}\,\text{s}^{-1}$ and its acceleration is $\dfrac{1}{x+2}\,\text{m}\,\text{s}^{-2}$.

(i) Given that $v = 2$ when $x = 0$, use integration to show that

$$v^2 = 2\ln\left(\tfrac{1}{2}x + 1\right) + 4.$$

(ii) Find the value of v when the acceleration of P is $\tfrac{1}{4}\,\text{m}\,\text{s}^{-2}$.

[Cambridge AS and A Level Mathematics 9709, Paper 5 Q3 June 2009]

8 A particle of mass 0.25 kg moves in a straight line on a smooth horizontal surface. A variable resisting force acts on the particle. At time t s the displacement of the particle from a point on the line is x m, and its velocity is $(8 - 2x)\,\text{m}\,\text{s}^{-1}$. It is given that $x = 0$ when $t = 0$.

(i) Find the acceleration of the particle in terms of x, and hence find the magnitude of the resisting force when $x = 1$.

(ii) Find an expression for x in terms of t.

(iii) Show that the particle is always less than 4 m from its initial position.

[Cambridge AS and A Level Mathematics 9709, Paper 5 Q7 November 2005]

9 A particle of mass 0.4 kg is released from rest and falls vertically. A resisting force of magnitude $0.08v$ N acts upwards on the particle during its descent, where $v\,\text{m s}^{-1}$ is the velocity of the particle at time t s after its release.

(i) Show that the acceleration of the particle is $(10 - 0.2v)\,\text{m s}^{-2}$.

(ii) Find the velocity of the particle when $t = 15$.

[Cambridge AS and A Level Mathematics 9709, Paper 5 Q4 November 2007]

10 A particle P of mass 0.5 kg moves along the x axis on a horizontal surface. When the displacement of P from the origin O is x m the velocity of P is $v\,\text{m s}^{-1}$ in the positive x direction. Two horizontal forces act on P; one force has magnitude $(1 + 0.3x^2)$ N and acts in the positive x direction, and the other force has magnitude $8e^{-x}$ N and acts in the negative x direction.

(i) Show that $v\dfrac{dv}{dx} = 2 + 0.6x^2 - 16e^{-x}$.

(ii) The velocity of P as it passes through O is $6\,\text{m s}^{-1}$. Find the velocity of P when $x = 3$.

[Cambridge AS and A Level Mathematics 9709, Paper 5 Q3 November 2008]

11 A particle P of mass 0.3 kg is projected vertically upwards from the ground with an initial speed of $20\,\text{m s}^{-1}$. When P is at height x m above the ground, its upward speed is $v\,\text{m s}^{-1}$. It is given that

$$3v - 90\ln(v + 30) + x = A,$$

where A is a constant.

(i) Differentiate this equation with respect to x and hence show that the acceleration of the particle is $-\frac{1}{3}(v + 30)\,\text{m s}^{-2}$.

(ii) Find, in terms of v, the resisting force acting on the particle.

(iii) Find the time taken for P to reach its maximum height.

[Cambridge AS and A Level Mathematics 9709, Paper 52 Q7 November 2009]

12 A particle P of mass 0.25 kg moves in a straight line on a smooth horizontal surface. P starts at the point O with speed $10\,\text{m s}^{-1}$ and moves towards a fixed point A on the line. At time t s the displacement of P from O is x m and the velocity of P is $v\,\text{m s}^{-1}$. A resistive force of magnitude $(5 - x)$ N acts on P in the direction towards O.

(i) Form a differential equation in v and x. By solving this differential equation, show that $v = 10 - 2x$.

(ii) Find x in terms of t, and hence show that the particle is always less than 5 m from O.

[Cambridge AS and A Level Mathematics 9709, Paper 51 Q7 June 2010]

KEY POINT

When a particle is moving along a line under a variable force F, Newton's second law gives a differential equation. It is generally solved by writing acceleration as

$\frac{dv}{dt}$	when F is given as a function of time, t
$v\frac{dv}{ds}$	when F is given as a function of displacement, s
$\frac{dv}{dt}$ or $v\frac{dv}{ds}$	when F is given as a function of velocity, v.

Answers

Neither University of Cambridge International Examinations nor OCR bear any responsibility for the example answers to questions from their past question papers which are contained in this publication.

Chapter 1

? (Page 3)

−4, 0, 5

(i) +4

(ii) −5

? (Page 4)

The marble is below the origin.

Exercise 1A (Page 5)

1 (i) +1 m

(ii) +2.25 m

2 (i) 3.5 m, 6 m, 6.9 m, 6 m, 3.5 m, 0 m

(ii) 0 m, 2.5 m, 3.4 m, 2.5 m, 0 m, −3.5 m

(iii) (a) 3.4 m

(b) 10.3 m

3 (i) 2 m, 0 m, −0.25 m, 0 m, 2 m, 6 m, 12 m

(ii)

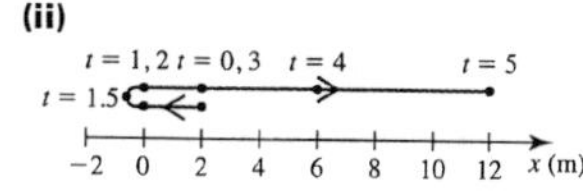

(iii) 10 m

(iv) 14.5 m

4 (i)

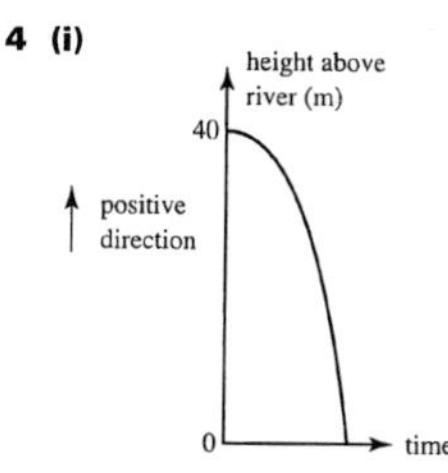

(ii)

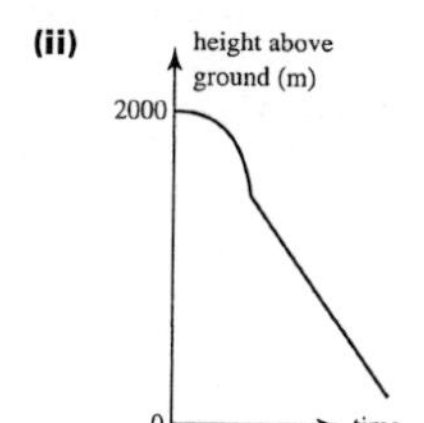

(iii)

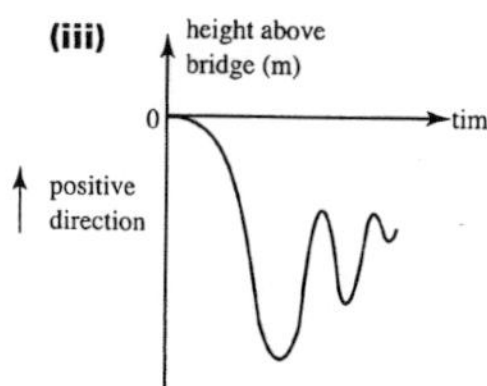

5 (i) The ride starts at $t = 0$. At A it changes direction and returns to pass its starting point at B continuing past to C where it changes direction again returning to its initial position at D.

(ii) An oscillating ride such as a swing boat.

? (Page 7)

10, 0, −10. The gradient represents the velocity.

? (Page 8)

The graph would curve where the gradient changes. Not over this period.

? (Page 9)

$+5\,\text{m s}^{-1}$, $0\,\text{m s}^{-1}$, $-5\,\text{m s}^{-1}$, $-6\,\text{m s}^{-1}$. The velocity decreases at a steady rate.

Exercise 1B (Page 10)

1

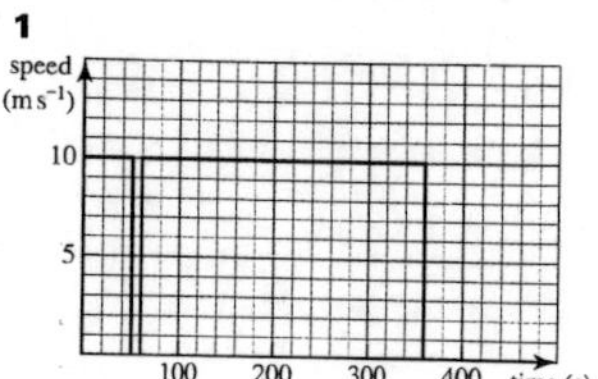

2 (i) The person is waiting at the bus stop.

(ii) It is faster.

(iii)

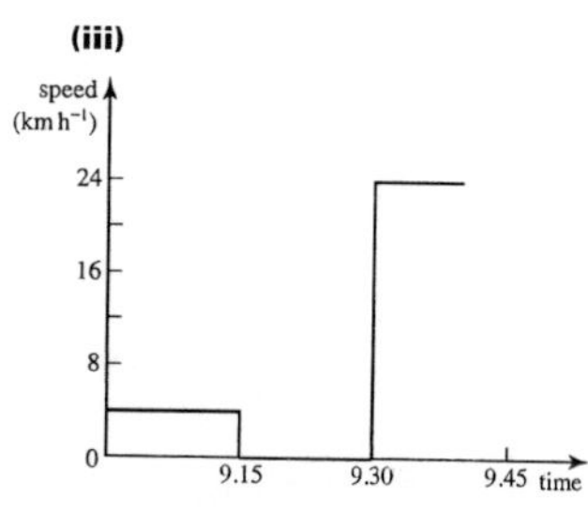

(iv) constant speed, infinite acceleration

3 (i) (a) 2 m, 8 m

(b) 6 m

(c) 6 m

(d) $2\,\text{m s}^{-1}$, $2\,\text{m s}^{-1}$

(e) $2\,\text{m s}^{-1}$

(f) $2\,\text{m s}^{-1}$

(ii) (a) 60 km, 0 km

(b) −60 km

(c) 60 km

(d) $-90\,\text{km h}^{-1}$, $90\,\text{km h}^{-1}$

(e) $-90\,\text{km h}^{-1}$

(f) $90\,\text{km h}^{-1}$

(iii) (a) 0 m, −10 m

(b) −10 m

(c) 50 m

(d) OA: $10\,\text{m s}^{-1}$, $10\,\text{m s}^{-1}$;
AB: $0\,\text{m s}^{-1}$, $0\,\text{m s}^{-1}$;
BC: $-15\,\text{m s}^{-1}$, $15\,\text{m s}^{-1}$

(e) $-1.67\,\text{m s}^{-1}$

(f) $8.33\,\text{m s}^{-1}$

(iv) (a) 0 km, 25 km

(b) 25 km

(c) 65 km

(d) AB: $-10\,\text{km h}^{-1}$, $10\,\text{km h}^{-1}$;
BC: $11.25\,\text{km h}^{-1}$, $11.25\,\text{km h}^{-1}$

(e) $4.167\,\text{km h}^{-1}$

(f) $10.83\,\text{km h}^{-1}$

4 $1238.7\,\text{km h}^{-1}$

❓ (Page 12)

(i) D

(ii) B, C, E

(iii) A

Exercise 1C (Page 12)

1 (i) (a) $+0.8\,\text{m s}^{-2}$

(b) $-1.4\,\text{m s}^{-2}$

(c) $+0.67\,\text{m s}^{-2}$

(d) 0

(e) $+0.5\,\text{m s}^{-2}$

(ii)

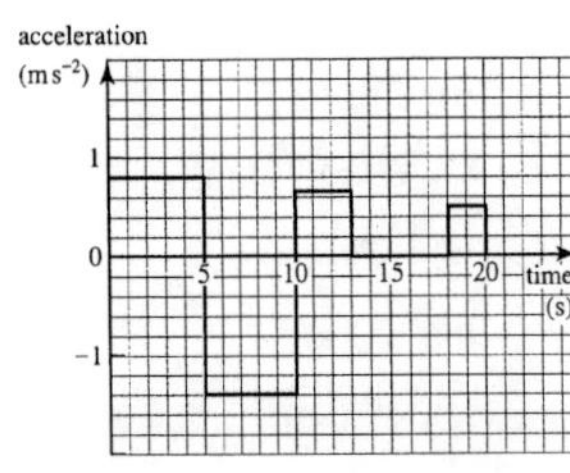

2 (i) 0 m, −16 m, −20 m, 0 m, 56 m

(ii)

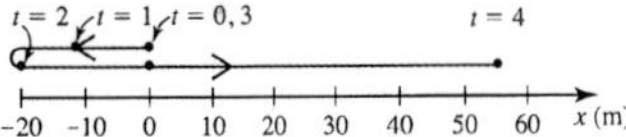

(iii)

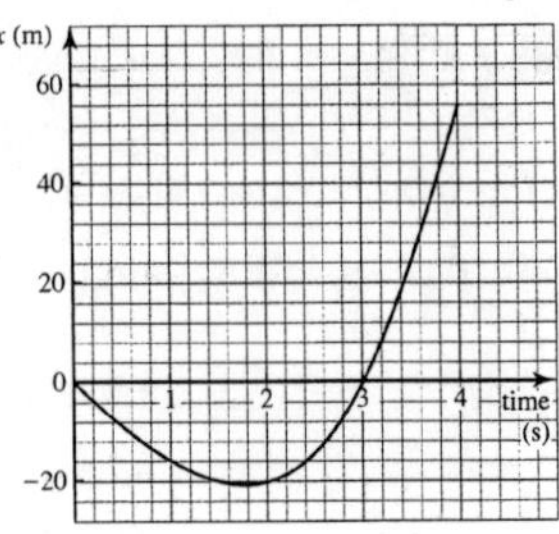

(iv) after 0 s (negative direction) and 3 s (positive direction)

3 (i) $32\,\text{km h}^{-1}$

(ii) $35.7\,\text{km h}^{-1}$

4 (i) (a) $56.25\,\text{km h}^{-1}$

(b) $97.02\,\text{km h}^{-1}$

(c) $46.15\,\text{km h}^{-1}$

(ii) The average speed is not equal to the mean value of the two speeds unless the same time is spent at the two speeds. In this case the ratio of distances must be 10 : 3.

5 (i)

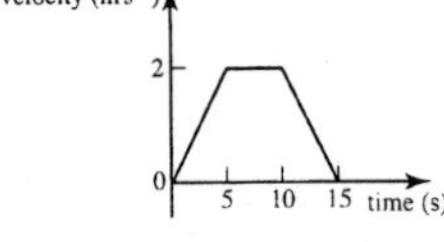

(ii)

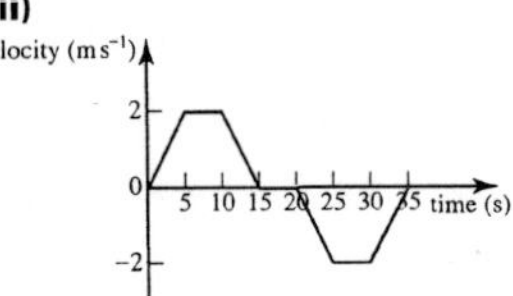

(iii) $+0.4\,\text{m s}^{-2}$, $0\,\text{m s}^{-2}$, $-0.4\,\text{m s}^{-2}$, $0\,\text{m s}^{-2}$, $-0.4\,\text{m s}^{-2}$, $0\,\text{m s}^{-2}$, $+0.4\,\text{m s}^{-2}$

(iv)

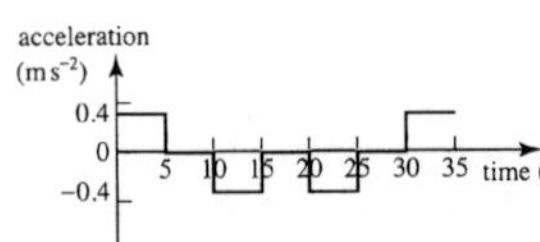

6 (i)

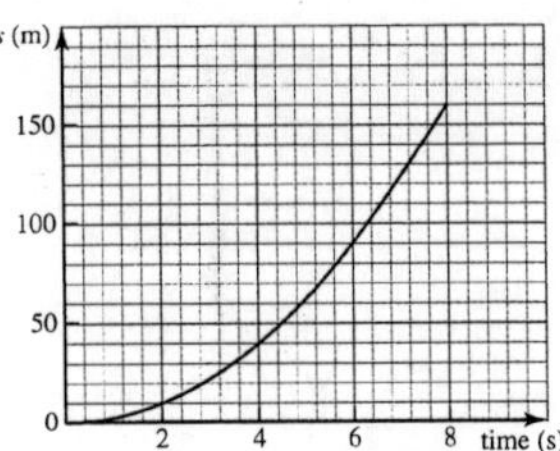

(ii) (a)

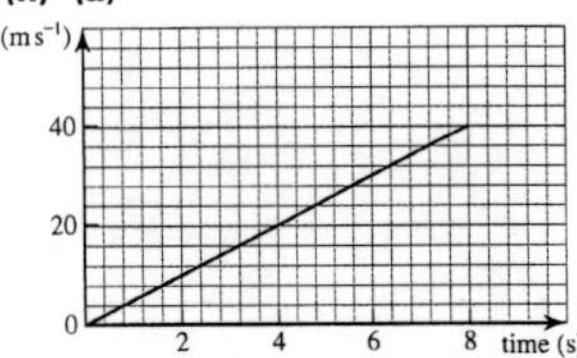

(b)

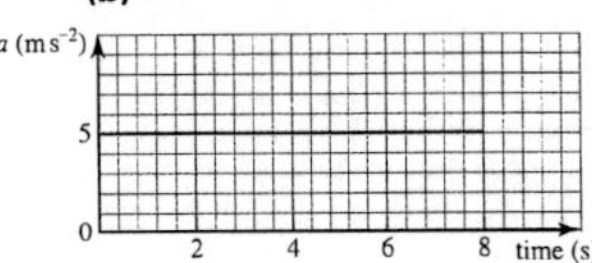

❓ (Page 14)

(i) 5

(ii) 20

(iii) 45

They are the same.

❓ (Page 14)

It represents the displacement.

❓ (Page 16)

Approx. 460 m

❓ (Page 16)

No, so long as the lengths of the parallel sides are unchanged the trapezium has the same area.

Exercise 1D (Page 17)

1 Car A

(i) $0.4\,\text{m s}^{-2}$, $0\,\text{m s}^{-2}$, $3\,\text{m s}^{-2}$

(ii) 62.5 m

(iii) $4.17\,\text{m s}^{-1}$

Car B

(i) $-1.375\,\text{m s}^{-2}$, $-0.5\,\text{m s}^{-2}$, $0\,\text{m s}^{-2}$, $2\,\text{m s}^{-2}$

(ii) 108 m

(iii) $3.6\,\text{m s}^{-1}$

2 (i) Enters the busy road at $10\,\text{m s}^{-1}$, accelerates to $30\,\text{m s}^{-1}$ and maintains this speed for about 150 s. Slows down to stop after a total of 400 s.

(ii) Approx. $0.4\,\text{m s}^{-2}$, $-0.4\,\text{m s}^{-2}$

(iii) Approx. 9.6 km, $24\,\text{m s}^{-1}$

3 (i)

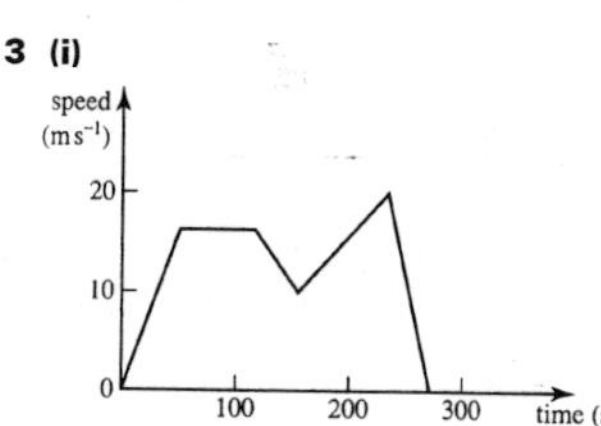

(ii) 3562.5 m

4 (i)

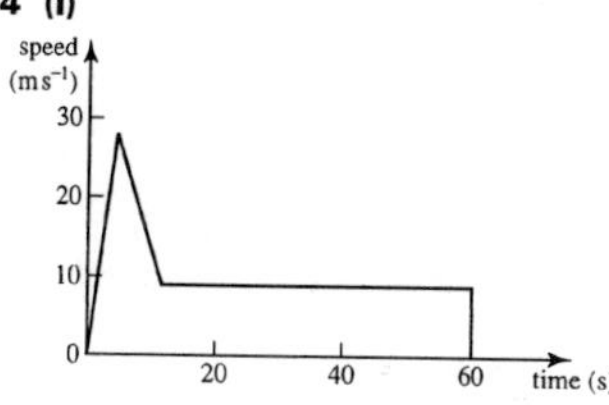

(ii) 558 m

5 (i)

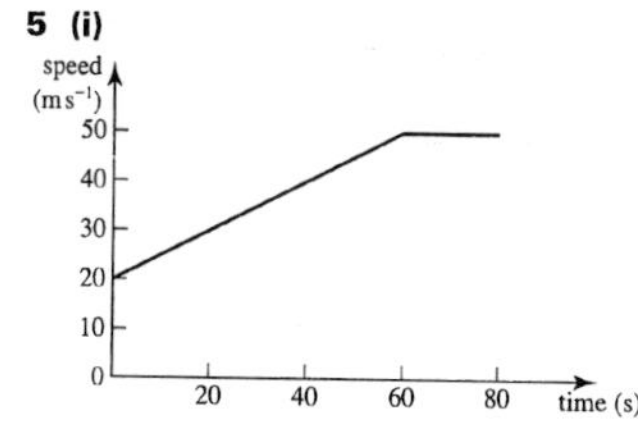

(ii) after 60 s

(iii) 6.6 km

(iv) $v = 20 + 0.5t$ for $0 \leqslant t \leqslant 60$, $v = 50$ for $t \geqslant 60$

6 (i)

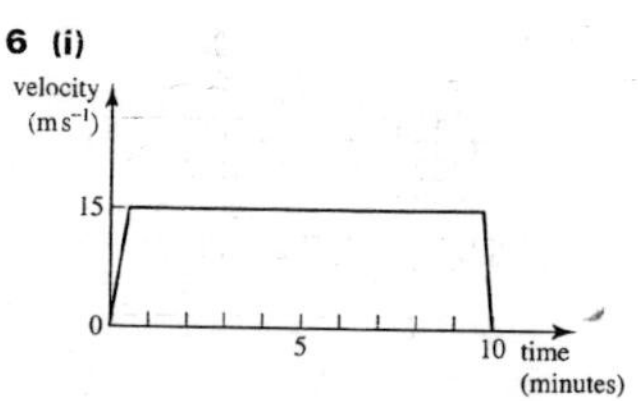

(ii) $15\,\text{m s}^{-1}$, $1\,\text{m s}^{-2}$, 8.66 km

7 (i) BC: constant deceleration, CD: stationary, DE: constant acceleration

(ii) $0.5\,\text{m s}^{-2}$, 2500 m

(iii) $0.2\,\text{m s}^{-2}$, 6250 m

(iv) 325 s

(v)

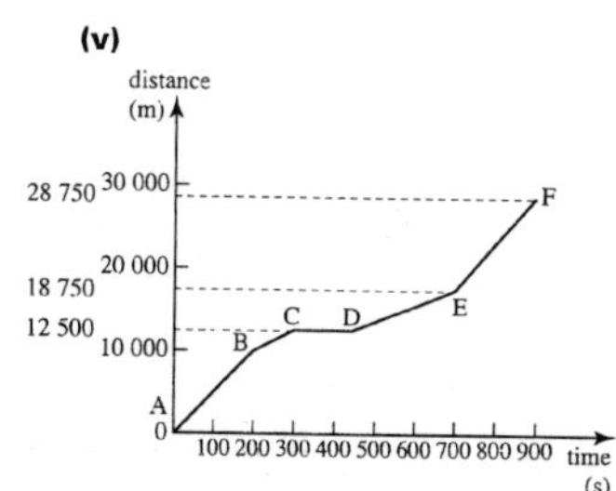

8 (i) $10\,\text{m s}^{-1}$, 0.7 s

(ii)

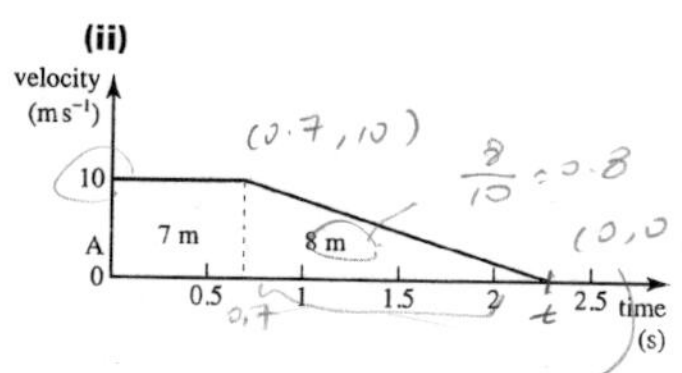

(iii) $6.25\,\text{m s}^{-2}$

(iv) 33.9 m

9 (i)

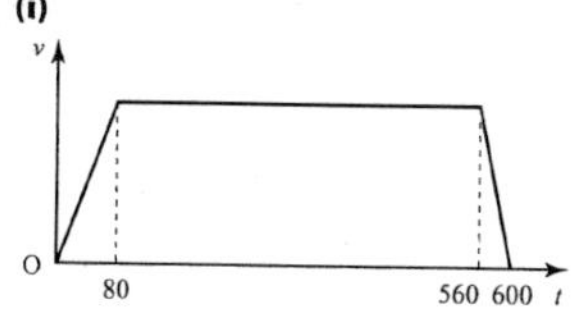

(ii) $10\,\text{m s}^{-1}$

(iii) 5400 m

(iv) $0.125\,\text{m s}^{-2}$

10 (i)

(ii) 25

(iii) 2920 m

11 (i) $0.09\,\text{m s}^{-2}$

(ii) 1.08 m

(iii) $0.72\,\text{m s}^{-1}$

Chapter 2

❓ (Page 22)

See text which follows.

❓ (Page 23)

It might be reasonable as much of the journey is on Interstate 5, a major road, but it would depend on the traffic.

❓ (Page 25)

For the fairground ride, $u = 4$, $v = 24$, $a = 2$, $t = 10$ and $s = 140$. The equations hold with these values.

❓ (Page 29)

$$s = \frac{1}{2}(2u + at) \times t$$

$$s = (u + \frac{1}{2}at) \times t$$

$$s = ut + \frac{1}{2}at^2$$

❓ (Page 29)

$$s = ut + \frac{1}{2}at^2$$

$$s = (v - at) \times t + \frac{1}{2}at^2$$

$$s = vt - at^2 + \frac{1}{2}at^2$$

$$s = vt - \frac{1}{2}at^2$$

Exercise 2A (Page 30)

1 (i) 22

(ii) 120

(iii) 0

(iv) −10

2 (i) $v^2 = u^2 + 2as$

(ii) $v = u + at$

(iii) $s = ut + \frac{1}{2}at^2$

(iv) $s = \frac{(u+v)}{2} \times t$

(v) $v^2 = u^2 + 2as$

(vi) $s = ut + \frac{1}{2}at^2$

(vii) $v^2 = u^2 + 2as$

(viii) $s = vt - \frac{1}{2}at^2$

3 (i) $10\,\text{m s}^{-1}$, $100\,\text{m s}^{-1}$

(ii) 5 m, 500 m

(iii) 2 s

Speed and distance after 10 s, both over-estimates.

4 $2.08\,\text{m s}^{-2}$, 150 m. Assume constant acceleration.

5 $4.5\,\text{m s}^{-2}$, 9 m

6 $-8\,\text{m s}^{-2}$, 3 s

7 (i) $s = 16t - 4t^2$, $v = 16 - 8t$

(ii) (a) 2 s

(b) 4 s

(iii)

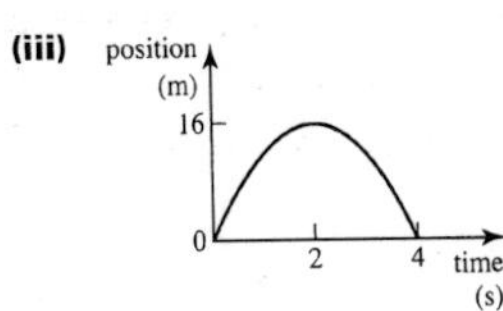

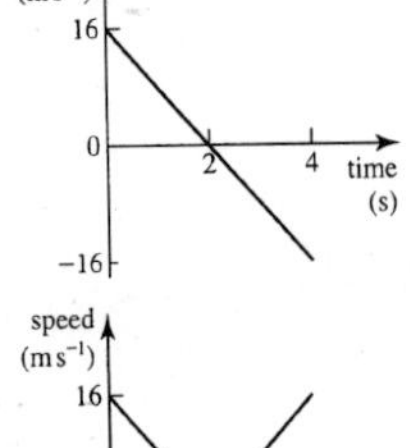

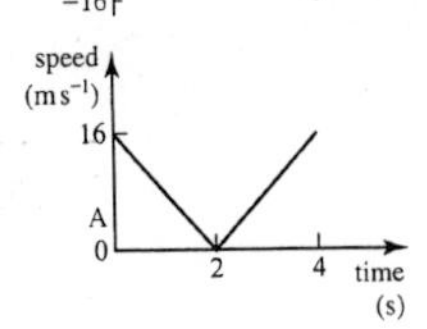

(Page 32)

$u = -15.0$, No

(Page 35)

$x = 5t^2$ and $1.25 - x = 5t - 5t^2$
so $1.25 = 5t$ as before.

(Page 35)

Because the velocity is not constant.

Exercise 2B (Page 35)

1 604.9 s, 9037 m or 9.04 km

2 (i) $v = 2 + 0.4t$

(ii) $s = 2t + 0.2t^2$

(iii) $18\,\text{m s}^{-1}$

3 No, he is 10 m behind when Sabina finishes.

4 (i) $h = 4 + 8t - 5t^2$

(ii) 2 s

(iii) $12\,\text{m s}^{-1}$

(iv) t greater, v less

5 (i) $12\,\text{m s}^{-1}$

(ii) 8.45 m

(iii) $13\,\text{m s}^{-1}$

(iv) 5.41 m

(v) underestimate

6 (i)

0 m s⁻¹
10 m s⁻²
15 m s⁻¹
h_b
h_s
30 m

(ii) $h_s = 15t - 5t^2$

(iii) $h_b = 30 - 5t^2$

(iv) $t = 2$ s

(v) 10 m

7 (i) $5.4\,\text{m s}^{-1}$

(ii) $-4.4\,\text{m s}^{-1}$

(iii) $1\,\text{m s}^{-1}$ increase

(iv) $9\,\text{m s}^{-1}$

(v) too fast

8 3 m

9 43.75 m

10 (ii) $u + 9a = 15.75$
or $u + 5a = 15.15$

(iii) 14.4

(iv) No, distance at constant acceleration is 166.5 m.

11 (i) $4\,\text{m s}^{-1}$

(ii) 6

(iii) 2 s

(iv) $\frac{4}{3}\,\text{m s}^{-2}$

12 6.8 m

Chapter 3

(Page 43)

The reaction between the person and the chair acts on the chair. The person's weight acts on the person only.

(Page 45)

Vertically up.

Exercise 3A (Page 46)

In these diagrams, W represents a weight, N a normal reaction with another surface, F a friction force, R air resistance and P another force.

1

2

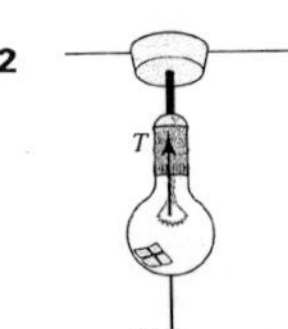

3

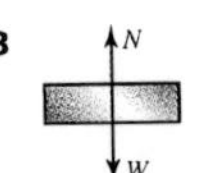

4 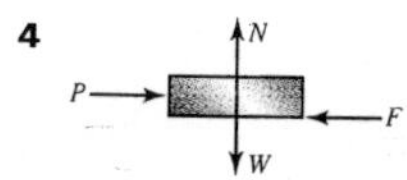

5 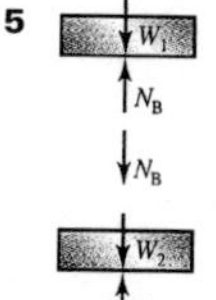

6 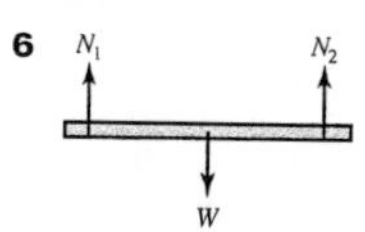

7 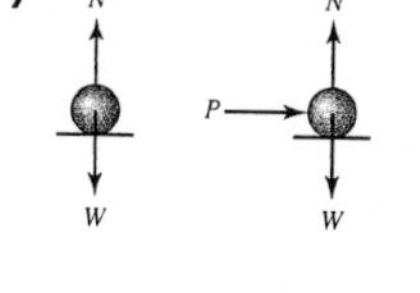

8 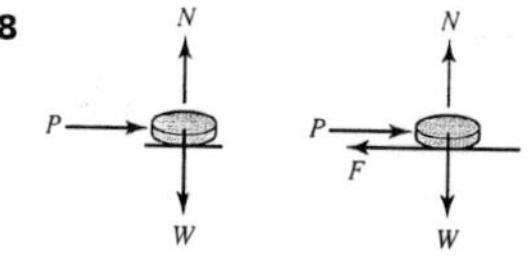

9 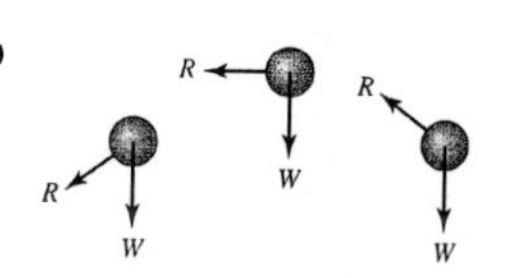

10 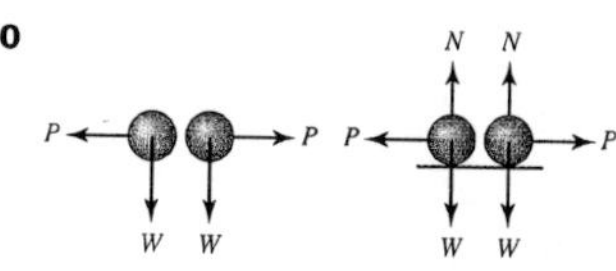

11 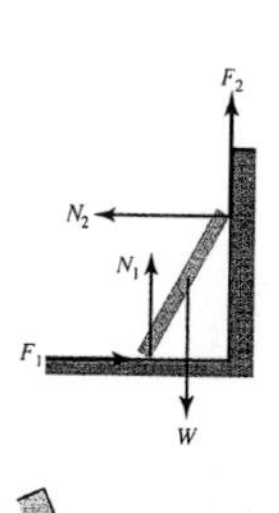

12 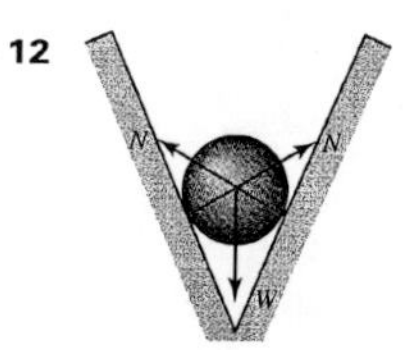

❓ (Page 47)

To provide forces when the velocity changes.

❓ (Page 48)

The friction force was insufficient to enable his car to change direction at the bend.

Exercise 3B (Page 49)

1 (i)

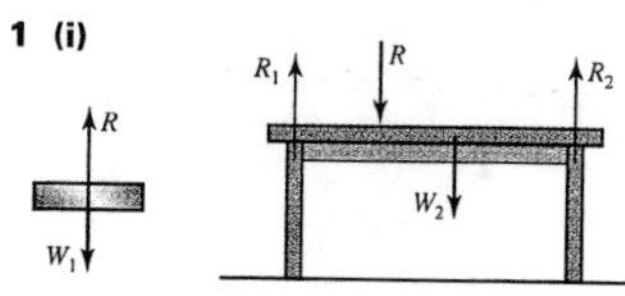

(ii) (a) $R = W_1$

(b) $R_1 + R_2 = W_2 + R$

2 (i) $R = W$, 0

(ii) $W > R$, $W - R$ down

(iii) $R > W$, $R - W$ up

3 (i) No

(ii) Yes

(iii) Yes

(iv) No

(v) Yes

(vi) Yes

(vii) Yes

(viii) No

4 Forces are required to give passengers the same acceleration as the car.

(i) A seat belt provides a backward force.

(ii) The seat provides a forward force on the body and the head rest is required to make the head move with the body.

❓ (Page 51)

Figure 3.18

Exercise 3C (Page 55)

1 (i) 150 N

(ii) 12 000 N = 12 kN

(iii) 0.5 N

2 (i) 60 kg

(ii) 1100 kg = 1.1 tonne

3 (i) 650 N

(ii) 650 N

4 112 N

5 (i) Both hit the ground together.

(ii) The balls take longer to hit the ground on the moon, but still do so together.

6 Answers for 60 kg

(ii) 600 N

(iii) 96 N

(iv) Its mass is 4 kg.

❓ (Page 55)

No. Scales which measure by balancing an object against fixed masses (weights).

Exercise 3D (Page 57)

In these diagrams, mg represents a weight, N a normal reaction with another surface, F a friction force, R air resistance, T a tension or thrust, D a driving force and P another force.

1 (i)

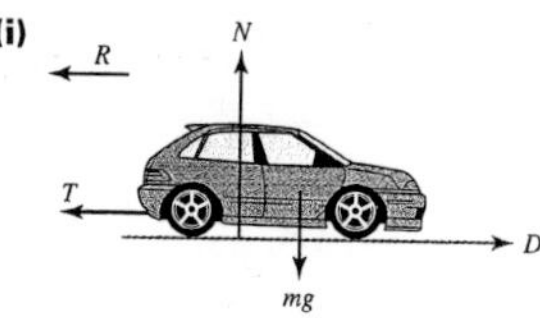

(ii)

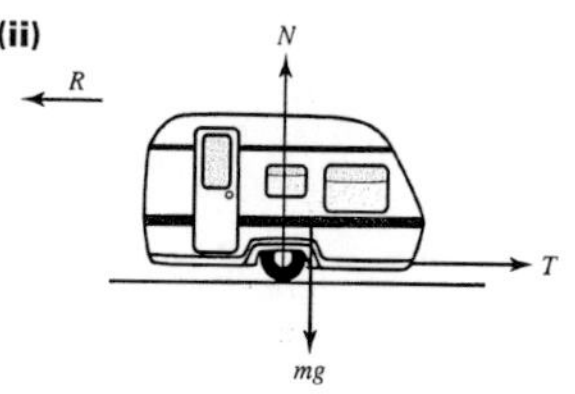

(iii)

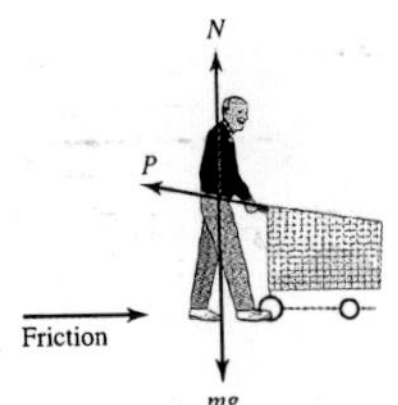

(iv) (a)

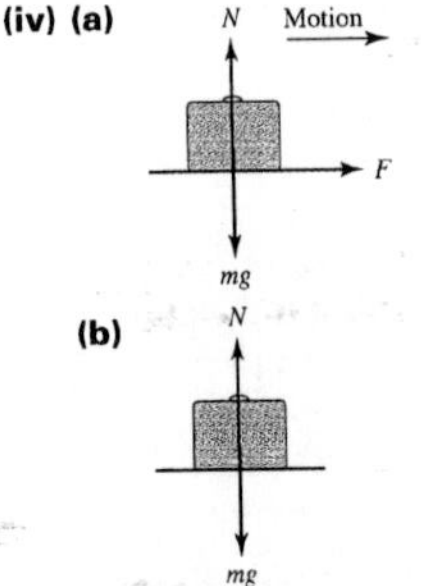

(b)

(v)

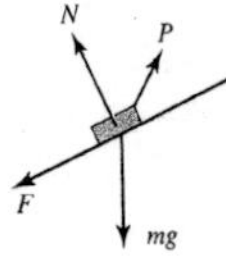

2 (i) Weight $5g = 50$ N down and reaction ($= 5g = 50$ N) up.

(ii) Weight $5g = 50$ N down, reaction with box above ($= 45g = 450$ N down) and reaction with ground ($= 50g = 500$ N up).

3 (i) $F_1 = 10$

(ii) $15 - F_2$ N

4 (i) towards the left

(ii)

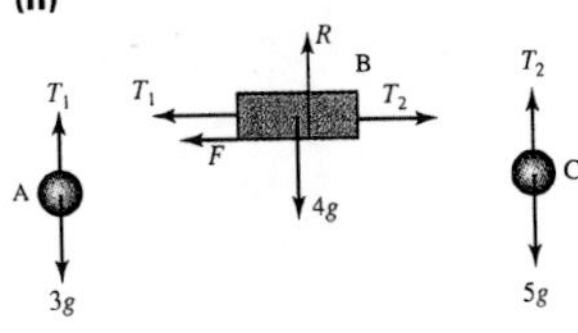

(iii) $3g\text{N} = 30\text{N}$, $5g\text{ N} = 50\text{N}$

(iv) $2g\text{N} = 20\text{N}$

(v) $T_1 - 3g\uparrow$, $T_2 - T_1 - F \rightarrow$, $5g - T_2 \downarrow$

5 All forces are in newtons

(i) greater

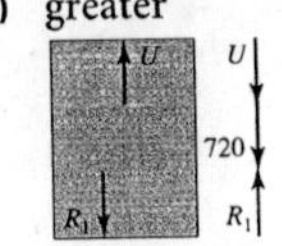

(ii) less

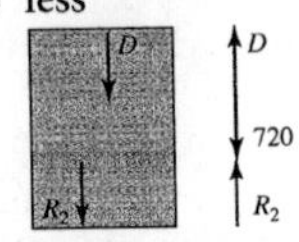

(iii) greater

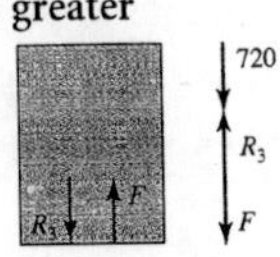

(iv) less

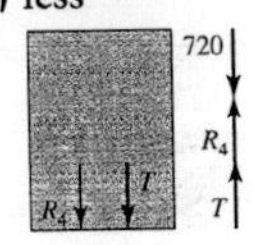

6 (i) 2400 N

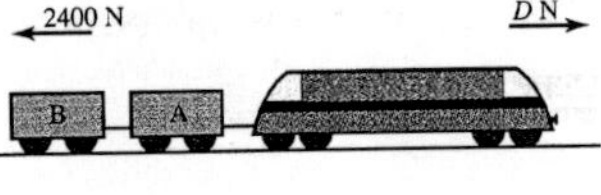

(ii) 2000 N

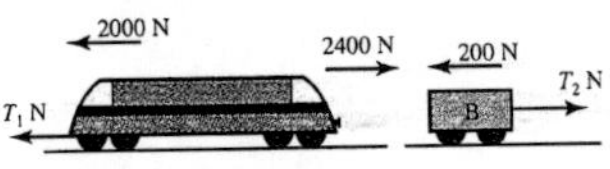

(iii) $T_1 = 400$

(iv) $T_2 = 200$

? (Page 60)

The air resistance seems to affect them more.

? (Page 60)

Sky divers and flying squirrels maximise air resistance by presenting a larger surface area in the direction of motion. Cyclists minimise air resistance by reducing the area.

? (Page 62)

Yes

? (Page 62)

Air resistance depends on velocity through the air. The velocities of a pair of cards in the experiment do not differ very much over such small heights.

Chapter 4

? (Page 64)

The pointer moves up and down as the force on the spring varies. Your weight would seem to change as the speed of the lift changed. You feel the reaction force between your hand and the book which varies as you move the book up and down.

Exercise 4A (Page 65)

1 (i) 800 N

(ii) 88 500 N

(iii) 0.0225 N

(iv) 840 000 N

(v) 8×10^{-20} N

(vi) 548.8 N

(vii) 8.75×10^{-5} N

(viii) 10^{30} N

2 (i) 200 kg

(ii) 50 kg

(iii) 10 000 kg

(iv) 1kg

3 (i) 7.76 N

(ii) 8 N

? (Page 67)

There is a resultant downward force because the weight is greater than the tension.

Exercise 4B (Page 89)

1 (i) $0.5\,\text{m s}^{-2}$

(ii) 25 m

2 **(i)** $1.67\,\text{m}\,\text{s}^{-2}$

(ii) 16.2 s

3 **(i)** 325 N

(ii) 1800 N

4 **(i)** 13 N

(ii) 90 m

(iii) 13 N

5 **(i)**

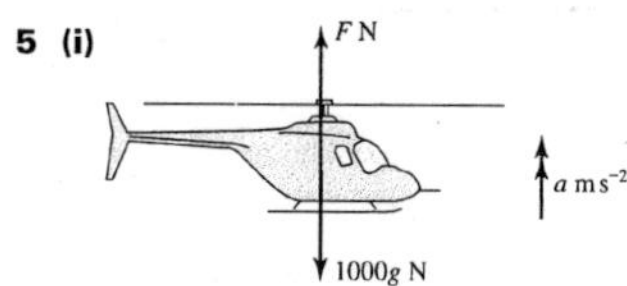

(ii) 11 500 N

6 **(i)** $400 - 250 = 12\,000a$,
$a = 0.0125\,\text{m}\,\text{s}^{-2}$

(ii) $0.5\,\text{m}\,\text{s}^{-1}$, 40 s

(iii) (a) 15 s

(b) 13.75 m

(c) 55 s

7 **(i)** $60\,\text{m}\,\text{s}^{-1}$

(ii) continues at $60\,\text{m}\,\text{s}^{-1}$

(iii) 1.25 N

(iv) the first by 655 km

8 **(i)** 7035 N, 7000 N, 6930 N, 2000 N

(ii) 795 kg

(iii) max $T < 9200$ N

9 **(i)** $8\,\text{m}\,\text{s}^{-2}$

(ii) $13.9\,\text{m}\,\text{s}^{-1}$ is just over $49\,\text{km}\,\text{h}^{-1}$

? (Page 71)

Your own weight acts on you and the tensions in the ropes with which you have contact; the other person's weight acts on them. The tension forces acting at the ends of the rope AB are equal and opposite. The accelerations of A and B are equal because they must always travel the same distance in each interval of time, assuming the rope does not stretch.

? (Page 72)

The tension in the rope joining A and B must be greater than B's weight because there must be a resultant force on B to produce an acceleration.

? (Page 75)

Using $v = u + at$ with $u = 0$ and maximum $a = 6$, the speed after 1 second would be $6\,\text{m}\,\text{s}^{-1}$ or about $22\,\text{km}\,\text{h}^{-1}$. Under the circumstances, a careful driver is unlikely to accelerate at this rate.

Alvin and his snowmobile and Bernard are two particles each moving in a straight line, otherwise Bernard could swing from side to side; contact between the ice and the rope is smooth, otherwise the tensions acting on Alvin and Bernard are different; the rope is light, otherwise its tension would be affected by its weight; the rope is of constant length, otherwise the accelerations would not be equal; there is no air resistance, otherwise the equations of motion would involve a force to allow for it.

Exercise 4C (Page 79)

1 **(i)**

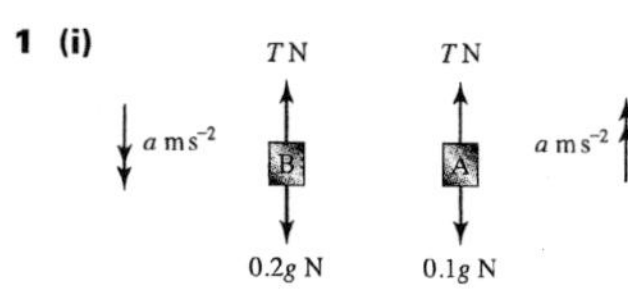

(ii) $T - 0.1g = 0.1a$,
$0.2g - T = 0.2a$

(iii) $3.33\,\text{m}\,\text{s}^{-2}$, 1.33 N

(iv) 1.10 s

2 **(i)**

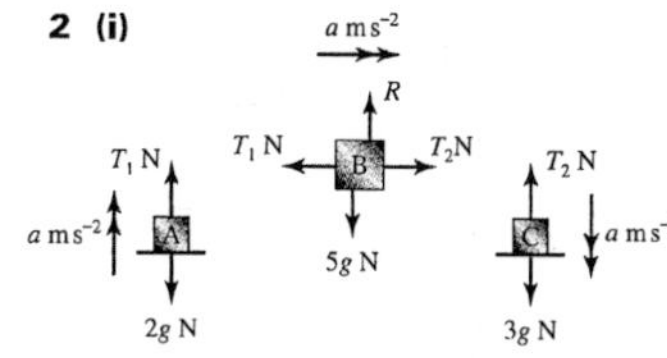

(ii) $T_1 - 2g = 2a$, $T_2 - T_1 = 5a$,
$3g - T_2 = 3a$

(iii) $1\,\text{m}\,\text{s}^{-2}$, 22 N, 27 N

(iv) 5 N

3 **(ii)** 750 N

RN TN

(iii) tension, 44 N

(iv) 170 N

4 **(i)** $0.625\,\text{m}\,\text{s}^{-2}$

(ii) 25 000 N

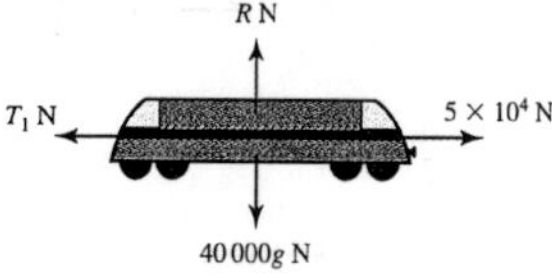

(iii) 12 500 N

(iv) reduced to 10 000 N

5 **(i)** $0.25\,\text{m}\,\text{s}^{-2}$

(ii) 9000 N, 6000 N

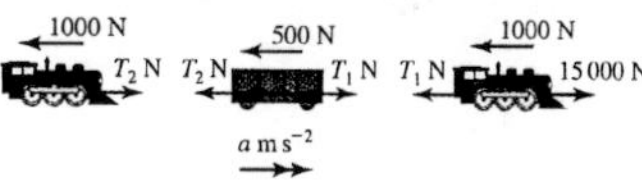

(iii) $0.25\,\text{m}\,\text{s}^{-2}$, tension 1500 N, thrust 1500 N. The second engine is now pushing rather than pulling back on the truck.

6 **(i)**

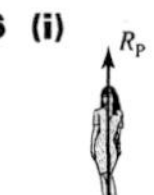

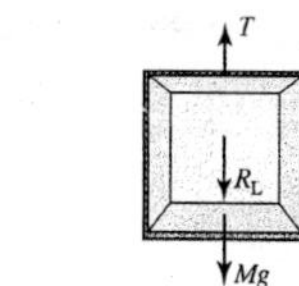

(ii) $R_P = R_L = 50g$, $T = 500g$

(iii) $T = 5400$, $R_P = R_L = 540$

7 **(i)**

R
a ms⁻²
70g

(ii) $1\,\text{m}\,\text{s}^{-2}$

(iii) stationary for 2 s, accelerates at $1\,\text{m}\,\text{s}^{-2}$ for 2 s, constant speed for 5 s, decelerates at $2\,\text{m}\,\text{s}^{-2}$ for 1 s, stationary for 2 s

(iv)

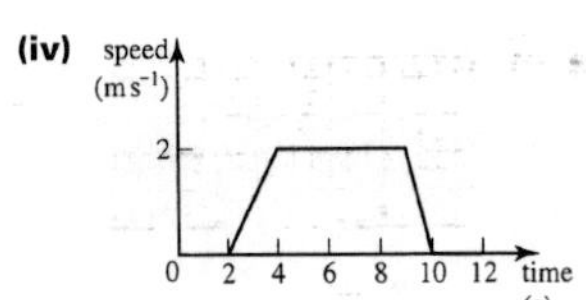

(v) 13 m

8 (i)

speed (m s⁻¹)

0 12.5 60 time (s)

(ii) $1\,\text{m s}^{-2}$

(iii) 20 s, $0.455\,\text{m s}^{-2}$

(iv) 62.3 kN

(v) 95.8 kg

9 (i) $5\,\text{m s}^{-2}$, 3 N

(ii) 0.6 s

10 (i) $2\,\text{m s}^{-2}$

(ii) 3.6 N

(iii) 0.3 kg

(iv) 0.792 m

11 (i) (a) $2.5\,\text{m s}^{-2}$

(b) 3.75 N

(ii) 0.3

12 (i) $1\,\text{m s}^{-2}$

(ii) (a) 3 m, 7 m

(b) $2\,\text{m s}^{-2}$

Chapter 5

❓ (Page 85)

See text which follows.

❓ (Page 87)

If the bird flies, say, 5 cm N, the wind would blow it 12 cm E and its resultant displacement would be 13 cm along DF. This would occur in any small interval of time.

❓ (Page 89)

–6

❓ (Page 90)

$\overrightarrow{AO} + \overrightarrow{OB} = -\mathbf{a} + \mathbf{b} = \mathbf{b} - \mathbf{a}$

Exercise 5A (Page 91)

1 (i) 6 m E, 2 m N

(ii) –6 m E, 0 m N

(iii) 6 m E, 4 m N

2 $\mathbf{a} = \begin{pmatrix} -2 \\ 0 \end{pmatrix}, \mathbf{b} = \begin{pmatrix} 0 \\ 1 \end{pmatrix}, \mathbf{c} = \begin{pmatrix} -3 \\ 0 \end{pmatrix},$

$\mathbf{d} = \begin{pmatrix} 0 \\ 3 \end{pmatrix}, \mathbf{e} = \begin{pmatrix} 2 \\ 0 \end{pmatrix}, \mathbf{f} = \begin{pmatrix} 1 \\ 1 \end{pmatrix},$

$\mathbf{g} = \begin{pmatrix} -2 \\ -1 \end{pmatrix}, \mathbf{h} = \begin{pmatrix} 1 \\ -2 \end{pmatrix}, \mathbf{k} = \begin{pmatrix} 1 \\ -1 \end{pmatrix}$

3 (4, –11)

4 (i) $\begin{pmatrix} 2 \\ 1 \end{pmatrix}$

(ii) $\begin{pmatrix} -10 \\ -24 \end{pmatrix}$

(iii) $\begin{pmatrix} 4 \\ -2 \end{pmatrix}$

(iv) $\begin{pmatrix} -3 \\ 22 \end{pmatrix}$

5 (i) $\begin{pmatrix} 1 \\ 2 \end{pmatrix}, \begin{pmatrix} 5 \\ 1 \end{pmatrix}, \begin{pmatrix} 7 \\ 8 \end{pmatrix}$

(ii) $\begin{pmatrix} 4 \\ -1 \end{pmatrix}, \begin{pmatrix} 2 \\ 7 \end{pmatrix}, \begin{pmatrix} -6 \\ -6 \end{pmatrix}$

(iii)

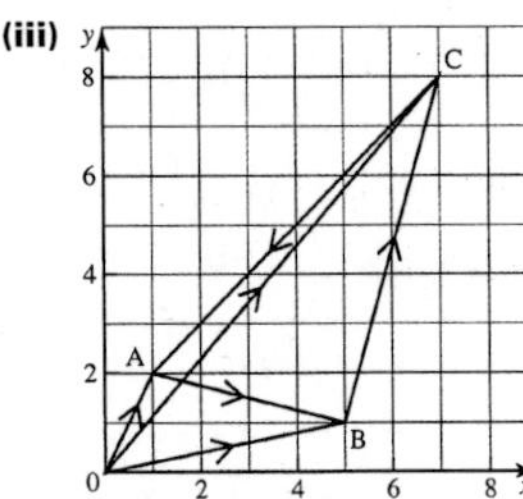

6 (i) $\begin{pmatrix} 0 \\ -3 \end{pmatrix}, \begin{pmatrix} 2 \\ 5 \end{pmatrix}, \begin{pmatrix} 3 \\ 9 \end{pmatrix}$

(ii) $\begin{pmatrix} 2 \\ 8 \end{pmatrix}, \begin{pmatrix} 1 \\ 4 \end{pmatrix}$

(iii) BC is parallel to AB

7 (i) $d = 9$

(ii) BC is equal and parallel to AD so ABCD is a parallelogram.

8 Acceleration $= \begin{pmatrix} 0 \\ -4 \end{pmatrix}\text{m s}^{-2}$,

Magnitude $= 4\,\text{m s}^{-2}$

Exercise 5B (Page 93)

1 (i) 10 at –53°

(ii) 4 at 180°

(iii) 2.24 at –117°

2 $\begin{pmatrix} 30 \\ 30 \end{pmatrix}$, 42.4 at 45°

3 $\begin{pmatrix} -1 \\ 2 \end{pmatrix}$, 2.24 at 117°

❓ (Page 94)

(i) $\sqrt{0.6^2 + 0.8^2} = 1$

(ii) (a) $\begin{pmatrix} 0.8 \\ 0.6 \end{pmatrix}$

(b) $\begin{pmatrix} 0.7071 \\ -0.7071 \end{pmatrix}$

Exercise 5C (Page 96)

1 (i) $\begin{pmatrix} 5.64 \\ 2.05 \end{pmatrix}$

(ii) $\begin{pmatrix} -5.36 \\ 4.50 \end{pmatrix}$

(iii) $\begin{pmatrix} 1.93 \\ -2.30 \end{pmatrix}$

(iv) $\begin{pmatrix} -1.45 \\ -2.51 \end{pmatrix}$

2 (i) $\begin{pmatrix} 113 \\ 65 \end{pmatrix}$

N
60°
130 km

(ii) $\begin{pmatrix} 192 \\ -161 \end{pmatrix}$

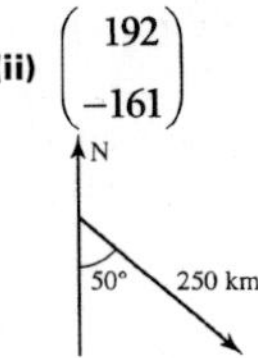

(iii) $\begin{pmatrix} -200 \\ -346 \end{pmatrix}$

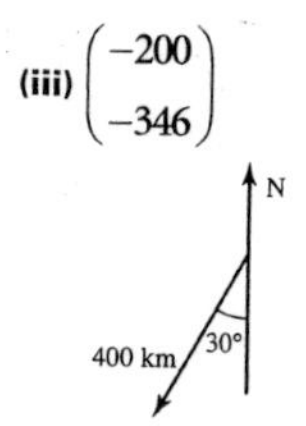

(iv) $\begin{pmatrix} -43 \\ 25 \end{pmatrix}$

3 $\begin{pmatrix} 2.83 \\ 2.83 \end{pmatrix}, \begin{pmatrix} 3 \\ 0 \end{pmatrix}$, 6.48 km h^{-1} at 064°

4 (i)

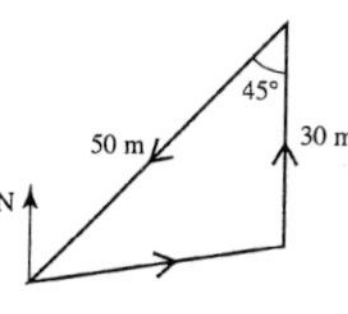

(ii) $\begin{pmatrix} 0 \\ 30 \end{pmatrix}, \begin{pmatrix} -35.4 \\ -35.4 \end{pmatrix}$

(iii) 081°

5 (i) (a) $\mathbf{p} = \begin{pmatrix} -0.92 \\ 2.54 \end{pmatrix}, \mathbf{q} = \begin{pmatrix} 2.30 \\ 1.93 \end{pmatrix}$,

$\mathbf{r} = \begin{pmatrix} 1.7 \\ -2.94 \end{pmatrix}, \mathbf{s} = \begin{pmatrix} -2.42 \\ -1.4 \end{pmatrix}$

(b) $\begin{pmatrix} 0.66 \\ 0.13 \end{pmatrix}$

(ii) (a) $\mathbf{t} = \begin{pmatrix} 1.35 \\ 2.34 \end{pmatrix}, \mathbf{u} = \begin{pmatrix} 2.68 \\ -1.55 \end{pmatrix}$,

$\mathbf{v} = \begin{pmatrix} -0.35 \\ -1.97 \end{pmatrix}, \mathbf{w} = \begin{pmatrix} -2 \\ 0 \end{pmatrix}$

(b) $\begin{pmatrix} 1.69 \\ -1.18 \end{pmatrix}$

6 (i) 1.45 km at 046°

(ii) 0 m

7 $\begin{pmatrix} 64.3 \\ 76.6 \end{pmatrix}, \begin{pmatrix} -153.2 \\ 128.6 \end{pmatrix}, \begin{pmatrix} -88.9 \\ 205.2 \end{pmatrix}$

8 $\begin{pmatrix} 5 \\ -1 \end{pmatrix}, \begin{pmatrix} 0.87 \\ 3.92 \end{pmatrix}, \begin{pmatrix} -3.21 \\ -4.83 \end{pmatrix}, \begin{pmatrix} 0 \\ -6 \end{pmatrix}$

9 079°, 5.1 km

Chapter 6

? (Page 99)

Yes if the cable makes small angles with the horizontal.

? (Page 100)

Parallel to the slope up the slope.

? (Page 101)

Start with AB and BC. Then draw a line in the right direction for CD and another perpendicular line through A. These lines meet at D.

? (Page 102)

(i) The sledge accelerates up the hill.

(ii) The sledge is stationary or moving with constant speed. (Forces in equilibrium.)

(iii) The acceleration is downhill.

Exercise 6A (Page 102)

1 (i)

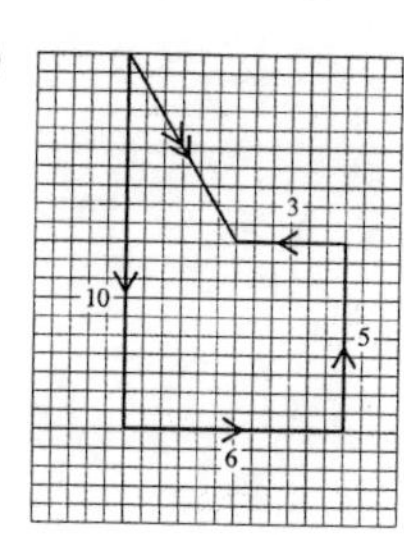

(iii) $\begin{pmatrix} 3 \\ -5 \end{pmatrix}$; 5.83 N, −59°

2 (i)

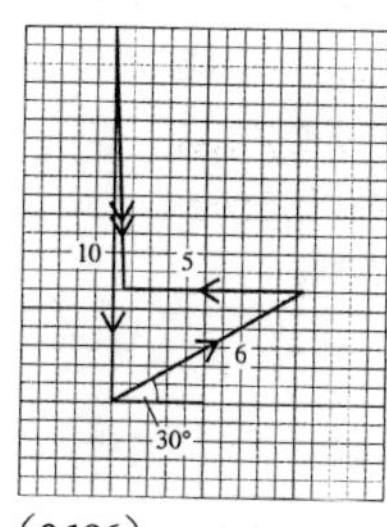

(iii) $\begin{pmatrix} 0.196 \\ -7 \end{pmatrix}$; 7.00 N, −88.4°

3 (i)

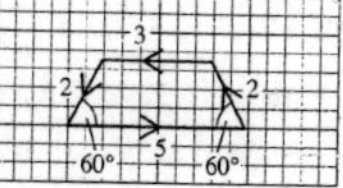

(iii) Equilibrium

4 (i)

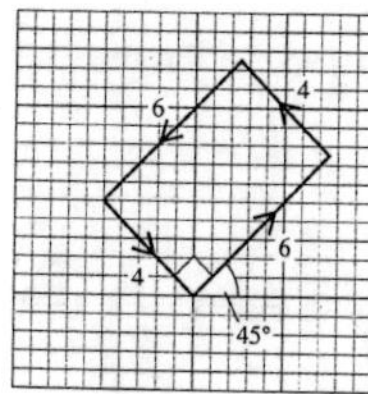

(iii) Equilibrium

5 (i)

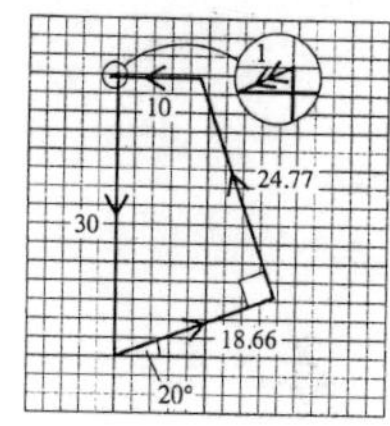

(iii) $\begin{pmatrix} -1 \\ 0 \end{pmatrix}$; 1 N down incline

6 (i)

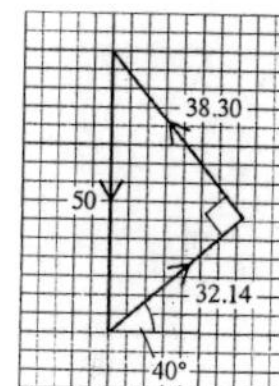

(iii) Equilibrium

7 (i) (a) 10.8 N

(b) 22.4 N

(ii) 64.2° anticlockwise from the x axis

? (Page 106)

Draw a vertical line to represent the weight, $10g$ N = 100 N. Then add the line of the force T_2 at 45° to the horizontal (note the length of this vector is unknown), and then the line of the force T_1 at 30° to the horizontal (60° to the vertical). C is the point at which these lines meet.

M1

Answers

? (Page 107)

The angles in the triangle are $180° - \alpha$ etc. The sine rule holds and $\sin(180° - \alpha) = \sin \alpha$ etc.

Exercise 6B (Page 109)

1 (i) 30 N, 36.9°; 65 N, 67.4°

(ii)

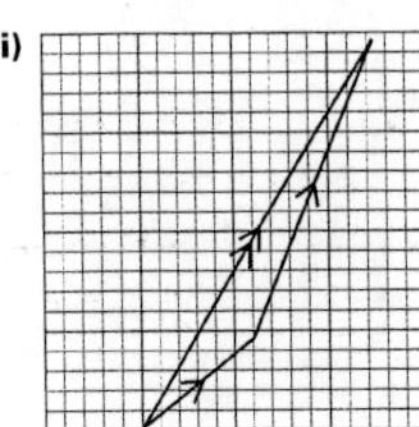

(iii) $\begin{pmatrix} 49 \\ 78 \end{pmatrix}$; 92.1 N, 57.9°

2 (i) T N, T N, C, D, 40°, 40°, $20\,000g$ N

(ii) $T\cos 40°$, $T\sin 40°$; $T\cos 40°$, $T\sin 40°$

(iv)

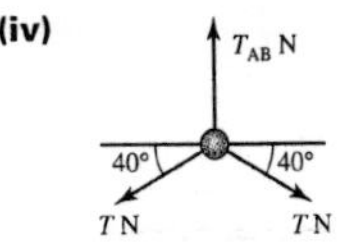

(v) 200 000 N

(vi) Resolve vertically for the whole system.

3 (i) (a)

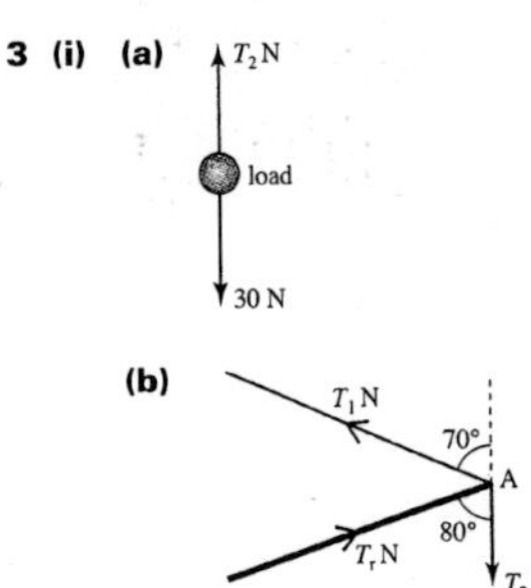

(ii) Rod: 56.4 N, compression, Cable 1: 59.1 N, tension

4 (i) 15.04 kg

(ii) Both read 10 kg

(iii) Both read 7.65 kg

(iv) Method A or C

5 (i) T N, θ, T_{AB} N, B, $1000g$ N

(ii) A force towards the right is required to balance the horizontal component of T.

(iii)

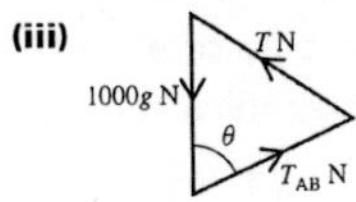

(iv) (a) 10 000 N, 14 142 N, 10 000 N

(b) 10 000 N, 10 000 N, 10 000 N

6 (ii) component form: $\begin{pmatrix} 56.1 \\ 61.2 \end{pmatrix}$

magnitude–direction form: 83 N, 47.5°

(iii) 30.8 N, −121°

7 (i) Cable 1: (5638, 2052); Cable 2: ($T_2 \cos 30°$, $T_2 \sin 30°$)

(ii) 4100 N

(iii) 9190 N

8 (i)

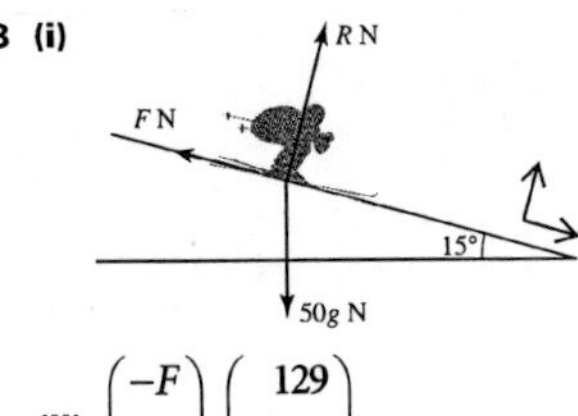

(ii) $\begin{pmatrix} -F \\ R \end{pmatrix}$, $\begin{pmatrix} 129 \\ -483 \end{pmatrix}$

(iii) 483 N, 129 N

(iv) 259 N

9 (i) R N, T N, F N, $5g$ N, 25°

(ii) $\begin{pmatrix} T \\ 0 \end{pmatrix}$, $\begin{pmatrix} -F \\ R \end{pmatrix}$, $\begin{pmatrix} -21.1 \\ -45.3 \end{pmatrix}$

(iii) $T = 30$, 8.87 N

(iv) 1.23 kg

10 (i) $\begin{pmatrix} 58 \\ 15.5 \end{pmatrix}$, $\begin{pmatrix} 59 \\ -10.4 \end{pmatrix}$

(ii) (a) 117 N

(b) 5.11 N

(iii) 97 N forwards

(iv) 3 N

11 (i) 11.2 N, 63.43°

(ii) A circle with centre A, radius 1m; No; two parallel forces and a third not parallel cannot form a triangle

12 $F = 28.3$, $G = 44.8$

13 $W_1 = 4.40$, $W_2 = 3.26$

? (Page 114)

Down the slope.

? (Page 117)

Anna and the sledge are a particle. There is no friction and the slope is straight. Friction would reduce both accelerations so Sam would not travel so far on either leg of his journey.

Exercise 6C (Page 119)

1 (i) $\begin{pmatrix} 1.5 \\ -1 \end{pmatrix}$ m s^{-2}

(ii) 1.8 m s^{-2}

2 (i) $\begin{pmatrix} 4 \\ 11 \end{pmatrix}$

(ii) $\begin{pmatrix} 8 \\ 8 \end{pmatrix}$, $\begin{pmatrix} 2 \\ 2 \end{pmatrix}$

3 (i) R N, T N, 30°, 10 N, $8g$ N

(ii) 11.55 N

(iii) 1 m s^{-2}

(iv) 0.4 s

4 (i)

(ii) 11.4 N, 29.7 N

(iii) $16.7\,\mathrm{m\,s^{-2}}$ at 69°

(iv) The fish swings sideways as it moves up towards Jones.

5 (i)

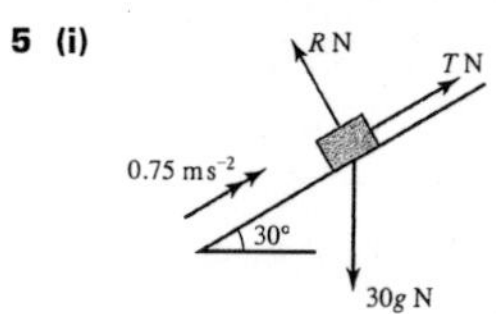

(ii) $\begin{pmatrix} T \\ 0 \end{pmatrix}, \begin{pmatrix} 0 \\ R \end{pmatrix}, \begin{pmatrix} -30g\sin 30° \\ -30g\cos 30° \end{pmatrix}$

(iii) 172.5 N

(iv) The crate slows down to a stop and then starts sliding down the slope.

6 (i)

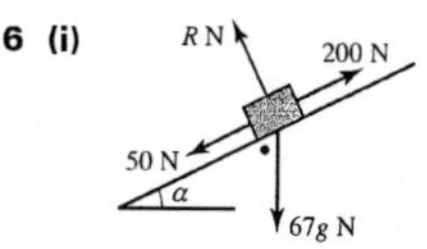

(ii) 12.9°

(iii) $0.85\,\mathrm{m\,s^{-2}}$, 35.6 m

(iv) $9.24\,\mathrm{m\,s^{-1}}$

7 (i)

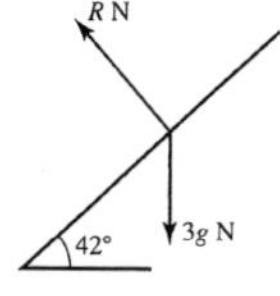

(ii) $6.69\,\mathrm{m\,s^{-2}}$

(iii) 1.73 s

(iv) 13.4 N

8 (i) The horizontal component of tension in the rope needs a balancing force.

(ii)

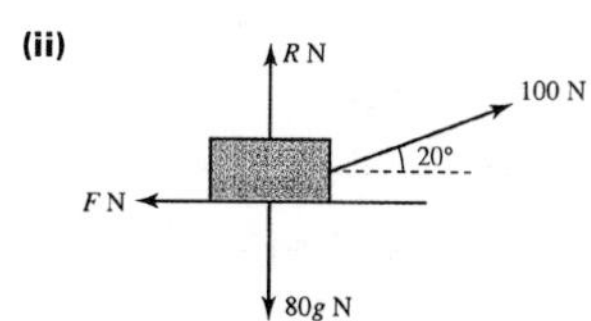

(iii) 94.0 N, 766 N

(iv) 128 N

(v) $0.144\,\mathrm{m\,s^{-2}}$

9 (i)

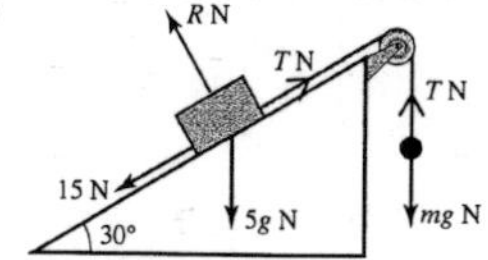

(ii) 4 kg, 40 N

(iii) $1.82\,\mathrm{m\,s^{-2}}$, 49.1 N

Chapter 7

Exercise 7A (Page 127)

1 (i) (a) $2-2t$

(b) 10, 2

(c) 1, 11

(ii) (a) $2t-4$

(b) 0, −4

(c) 2, −4

(iii) (a) $3t^2-10t$

(b) 4, 0

(c) 0, 4 and $3\frac{1}{3}$, −14.5

2 (i) (a) 4

(b) 3, 4

(ii) (a) $12t-2$

(b) 1, −2

(iii) (a) 7

(b) −5, 7

3 $v=4+t$, $a=1$

4 (i) (a) $v=15-10t$, $a=-10$

(b)

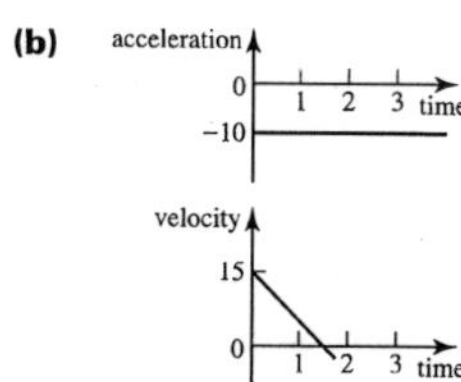

(c) The acceleration is the gradient of the velocity–time graph.

(d) The acceleration is constant; the velocity decreases at a constant rate.

(ii) (a) $v=18t^2-36t-6$, $a=36t-36$

(b)

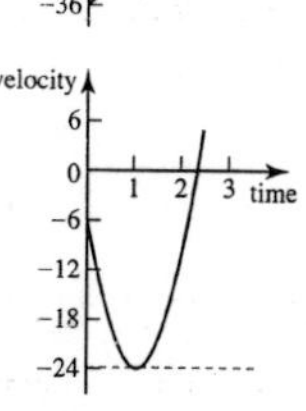

(c) The acceleration is the gradient of the velocity–time graph; velocity is at a minimum when the acceleration is 0.

(d) It starts in the negative direction. v is initially −6 and decreases to −24 before increasing rapidly to zero, where the object turns to move in the positive direction.

Exercise 7B (Page 131)

1 (i) $2t^2+3t$

(ii) $1.5t^4-\frac{2}{3}t^3+t+1$

(iii) $\frac{7}{3}t^3-5t+2$

2 (i)

(ii) 85 m

3 (i) When $t=6$

(ii) 972 m

4 (i) 4.47 s

(ii) 119 m

5 (i) $v=10t+\frac{3}{2}t^2-\frac{1}{3}t^3$, $x=5t^2+\frac{1}{2}t^3-\frac{1}{12}t^4$

(ii) $v=2+2t^2-\frac{2}{3}t^3$, $x=1+2t+\frac{2}{3}t^3-\frac{1}{6}t^4$

(iii) $v=-12+10t-3t^2$, $x=8-12t+5t^2-t^3$

❓ (Page 131)

Case (i); $s = ut + \frac{1}{2}at^2$; $v = u + at$; $a = 4$, $u = 3$.

In the other cases the acceleration is not constant.

❓ (Page 132)

Substituting $at = v - u$ in ② gives

$s = ut + \frac{1}{2}(v - u)t + s_0$

$\Rightarrow s = \frac{1}{2}(u + v)t + s_0$ ③;

$v - u = at$ and $v + u = 2(s - s_0)$

$\Rightarrow (v - u)(v + u) = at \times \frac{2}{t}(s - s_0)$

$\Rightarrow v^2 - u^2 = 2a(s - s_0)$ ④.

Substituting $u = v - at$ in ② gives

$s = vt - \frac{1}{2}at^2 + s_0$ ⑤.

Exercise 7C (Page 132)

1 (i) $15 - 10t$

(ii) 11.5 m, +5 m s⁻¹, 5 m s⁻¹; 11.5 m, −5 m s⁻¹, 5 m s⁻¹

(iii)

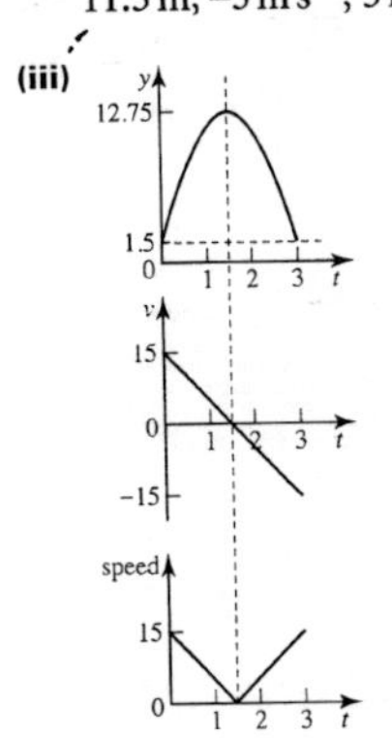

(iv) 3 s

(v) The expression does not equal the distance travelled because of changes in direction. The expression gives the displacement from the origin which equals 0.

2 (i) −3 m, −1 m s⁻¹, 1 m s⁻¹

(ii) (a) 1 s

(b) 2.15 s

(iii)

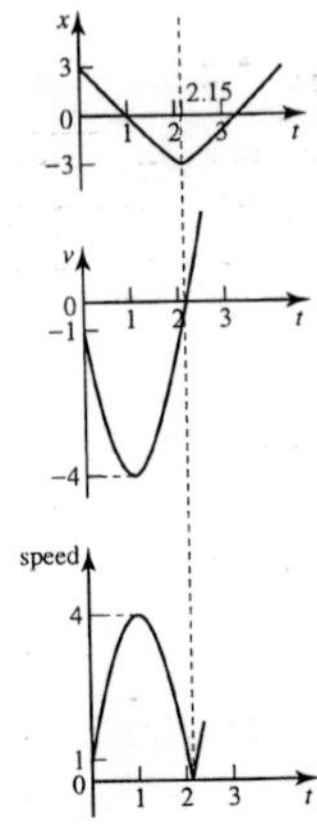

(iv) The object moves in a negative direction from 3 m to −3 m then moves in a positive direction with increasing speed.

3 2 s

4 (i) $v = 4 + 4t - t^2$,

$s = 4t + 2t^2 - \frac{1}{3}t^3$

(ii)

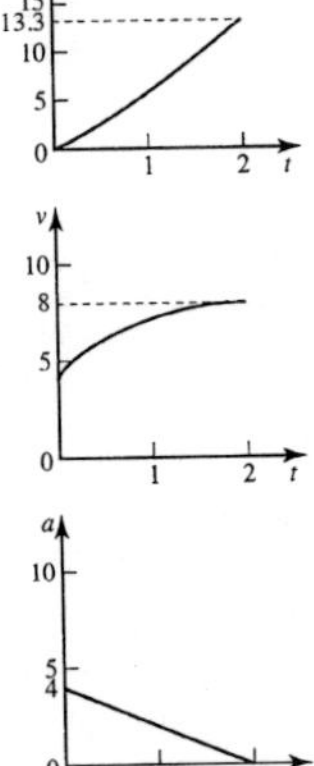

(iii) The object starts at the origin and moves in a positive direction with increasing speed reaching a maximum speed of 8 m s⁻¹ after 2 s.

5 (i) 0, 10.5, 18, 22.5, 24

(ii) The ball reaches the hole at 4 s.

(iii) $-3t + 12$ (m s⁻¹)

(iv) 0 m s⁻¹

(v) −3 m s⁻²

6 (i) Andrew 10 m s⁻¹, Elizabeth 9.6 m s⁻¹

(ii)

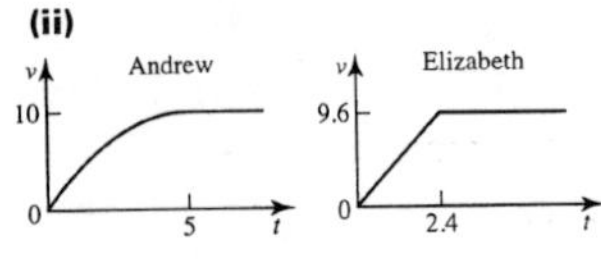

(iii) 11.52 m

(iv) 11.62 s

(v) Elizabeth by 0.05 s and 0.5 m

(vi) Andrew wins

7 (i)

v
85
5
0 10 20 30 40 50 60 70 80 90 t

Christine is in free fall until $t = 10$ s then the parachute opens and she slows down to terminal velocity of 5 m s⁻¹.

(ii) 1092 m

(iii) 8.5 m s⁻², $1.6t - 32$, 0 m s⁻², 16 m s⁻²

8 (i)

T N
1.6 m s⁻²
5g N

(iii) 870 N

(iv) max tension = 65 N, string breaks

9 (i) 40

(ii) $s = 0$ when $t = 0$ and 10

(iii) $25 - 5t$

(iv) 62.5 m

(v) In Michelle's model the velocity starts at 25 m s⁻¹ and then decreases. The teacher's model is better because the velocity starts at zero and ends at zero.

10 **(i)** **(a)** 112 cm

(b) 68 cm

(ii) $4t$, 16

(iii) $2t^2$, $16t - 32$

(iv) $111\frac{1}{9}$ cm, $\frac{8}{9}$ cm less

11 **(i)** $0.01t^3 + 1.25$

(ii) $3\,\mathrm{m\,s^{-1}}$

12 **(i)** 20 s

(ii) 80 s

(iii) $4\,\mathrm{m\,s^{-1}}$

(iv) 1170 m (to 3 s.f.)

13 **(ii)** $1\frac{2}{3}$

14 **(i)** 100 s, 200 m

(ii) **(a)** 0.0003

(b) $3\,\mathrm{m\,s^{-1}}$

(iii)

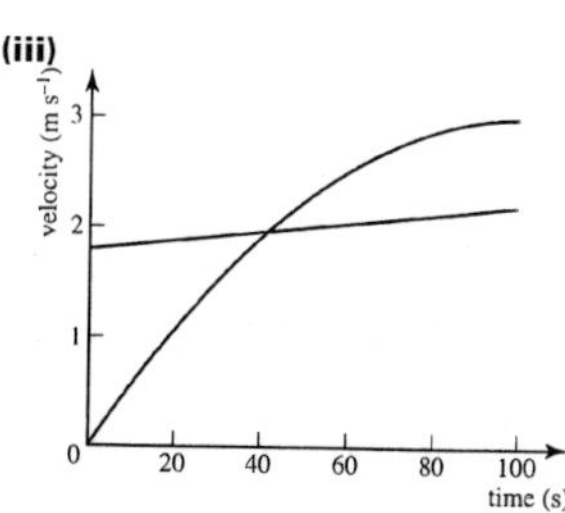

15 **(i)** $6\,\mathrm{m\,s^{-1}}$, 0.6

(ii) 13.9

(iii) 50 m

16 **(i)** $A = 4$

(ii) $450 - \dfrac{3375}{t}$

(iii) $5.4\,\mathrm{m\,s^{-1}}$

Chapter 8

(Page 138)

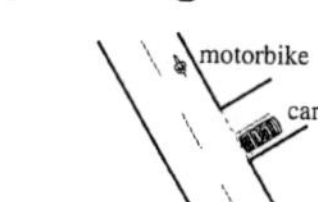

Assumptions: motorcycle is a particle, uniform frictional force with road, horizontal, linear motion with constant deceleration.

See also text which follows

(Page 141)

Downward slope would extend skid so u is an overestimate; opposite for upward slope. Air resistance would reduce skid so u is an underestimate. Smaller μ would extend skid so u is an overestimate.

(Page 141)

Friction is forwards when pedalling, backwards when freewheeling.

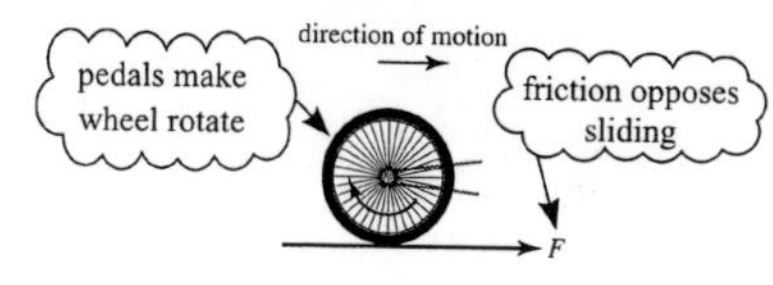

Exercise 8A (Page 145)

1

(i) 0.1

(ii) 0.05

2 **(i)** $F = 2g$

(ii) $2.5\,\mathrm{m\,s^{-2}}$

(iii) $F = 2g$

(iv) $2.25\,\mathrm{m\,s^{-2}}$

3 4.9 kN

4 **(i)** $1.02\,\mathrm{m\,s^{-2}}$

(ii) 0.102 N

(iii) 0.102

(iv) 49 m; independent of mass

5 0.8

6 **(i)** smoother contact

(ii) 0.2

(iii) 140 N

7 **(i)** $7.5\,\mathrm{m\,s^{-2}}$

(ii) $17.3\,\mathrm{m\,s^{-1}}$

(iii) 58.8 m

8 **(i)** 60 N

(ii) 63.8 N

9 **(i)** 0.577

(ii) 35°

(iii) 2.14

(iv) 50.2°

10 **(i)**

(ii) $4.51\,\mathrm{m\,s^{-2}}$

(iii) $5.20\,\mathrm{m\,s^{-1}}$

(iv) 5.42 m

11 **(i)** 0.194

(ii) $4.94\,\mathrm{m\,s^{-2}}$

(iii) $9.94\,\mathrm{m\,s^{-1}}$

(iv) $9.94\,\mathrm{m\,s^{-1}}$

12 **(i)**

$R = 28.4$
$F = 3.18$

(iii) 3.46 N, $4.05\,\mathrm{m\,s^{-2}}$

13 **(i)** **(a)** 37.9 N

(b) 37.2 N

(c) 37.5 N

(ii) $\dfrac{40}{\cos\alpha + 0.4\sin\alpha}$

(iii) 21.8°

14 0.346

15 **(i)** Mass of Q = 0.4 kg

(ii) 0.5 kg

(iii) 4.5 N

16 **(i)** 130, 50 N

(ii) 0.268

17 **(i)** 14.4 N, 75.2 N

(ii) 0.364

18 **(i)** $3200 - \frac{24}{25}X$

(ii) 1875

19 (i)

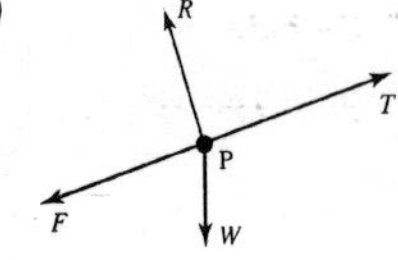

(iii) $0.1\,\mathrm{m\,s^{-2}}$

20 (i) 97.8 N

(ii) 28.3 N

21 (i) 0.546 N, 5.71 N, 0.0957

(ii) 2.18

22 (ii) 1.62

Chapter 9

❓ (Page 154)

The machine never stops (never loses energy).

No, it is an optical illusion.

Exercise 9A (Page 161)

1 (i) 2500 J

(ii) 40 000 J

(iii) 5.6×10^9 J

(iv) 3.7×10^{28} J

(v) 10^{-25} J

2 (i) 1000 J

(ii) 1070 J

(iii) 930 J

(iv) None

3 (i) 4320 J

(ii) 4320 J

(iii) 144 N

4 (i) 540 000 J, No

(ii) 3600 N

5 (i) 500 000 J

(ii) 6667 N

6 (i) (a) 5250 J

(b) −13 750 J

(ii) (a) 505 250 J

(b) 486 250 J

7 (i) 64 J

(ii) dissipated

(iii) 64 J

(iv) 400 N

(v) $89.4\,\mathrm{m\,s^{-1}}$

8 (i) 3.146×10^5 J

(ii) 8.28×10^3 N

(iii) dissipated as heat and sound

(iv) some of work is dissipated

9 $18.6\,\mathrm{m\,s^{-1}}$

10 (i) 240 N

(ii) 5.5 m

(iii) 1320 J

(iv) $0.5\,\mathrm{m\,s^{-2}}$, 270 N, 270 J

(v) 960 J, 90 J

❓ (Page 168)

More work cycling into the wind, less if at an angle, minimum if wind behind.

Exercise 9B (Page 170)

1 (i) 10 J

(ii) 96.4 J

(iii) −60 J

(iv) −60.1 J

2 (i) −28 J

(ii) 56 J

(iii) −12 J

3 18 J

4 23 760 J

5 (i) 157 000 J

(ii) 20 000 J

(iii) distance moved against gravity is 200 sin 5°

(iv) 138 000 J

6 (i) (a) 1500 J

(b) 280 J

(ii) (a) $15.6\,\mathrm{m\,s^{-1}}$

(b) $16.1\,\mathrm{m\,s^{-1}}$

7 (i) 2170 J

(ii) the same

8 (i) (a) 1740 J

(b) $8.8\,\mathrm{m\,s^{-1}}$

(ii) (a) unaltered

(b) decreased

9 (i) 1790 J

(ii) 1790 J, $8.45\,\mathrm{m\,s^{-1}}$

(iii) 50°

(iv) it is always perpendicular to the motion

10 (i) 160 J

(ii) $2(80 - x) = 160 - 10t^2$

(iii) $0 \leqslant t < 4$

(iv) $28.3\,\mathrm{m\,s^{-1}}$

(v) 60 m

11 (i) 35 000 J, 21 875 J

(ii) 262.5 N

(iii) 5061 N

12 (i) $9.8\,\mathrm{m\,s^{-2}}$

(ii) $1.47(10t - 4.9t^2)$, $0 \leqslant t \leqslant 2.04$

(iii) 5.1 m, $\begin{pmatrix} 10 \\ 0 \end{pmatrix}$

(iv) $\begin{pmatrix} 10 \\ 10 \end{pmatrix}$, $14.1\,\mathrm{m\,s^{-1}}$, 15 J

(v) No air resistance; No

13 (i) $110\,\mathrm{m\,s^{-1}}$

(ii) 117 N

(iii) The heavier (relatively less affected by resistance).

14 (i) 604 J

(ii) 774 000 J

(iii) 215 W

15 (i) 1000 J

(ii) 7500 J

(iii) 8000 J

(iv) 27.3

16 **(i)** 7100 kJ

(ii) 24 m

17 **(i)** 100 J

(ii) 5000 J

(iii) 50.4

18 **(i)** $12.2\,\mathrm{m\,s^{-1}}$

(ii) 4.9 J

19 2820 J

Exercise 9C (Page 177)

1 **(i)** 315 J

(ii) 37 800 J

(iii) 10.5 W

2 **(i)** 2400 J

(ii) 1200 W

(iii) 1920 W, 0 W, 2880 W

3 **(i)** 32 400 J

(ii) 16 200 J

(iii) 1620 J

(iv) 1350 N

(v) power = 1620 J

4 **(i)** 703 N

(ii) mass of car

5 576 N

6 250 kW

7 **(i)** 560 W

(ii) 168 000 J

8 **(ii)** 0.245 W

9 **(i)** $20\,\mathrm{m\,s^{-1}}$

(ii) $0.0125\,\mathrm{m\,s^{-2}}$

(iii) $25\,\mathrm{m\,s^{-1}}$

10 **(i)** $1.6 \times 10^7\,\mathrm{W}$

(ii) $0.0025\,\mathrm{m\,s^{-2}}$

(iii) $5.7\,\mathrm{m\,s^{-1}}$

11 **(i)** 16

(ii) 320 N

(iii) 6400 W

(iv) **(a)** $1.98\,\mathrm{m\,s^{-2}}$

(b) $1.10\,\mathrm{m\,s^{-2}}$

12 **(i)** 10 000 N, 9945 N

(ii) 3045 N

(iii) 50.8 kW

(iv) $94.6\,\mathrm{km\,h^{-1}}$

13 **(i)** 62 500 J

(ii) 521 W

(iii) 47.2 s

14 **(i)** 300 000 J

(ii) 6970 N

(iv) 3.23 m

15 $20\,\mathrm{m\,s^{-1}}$, $30\,\mathrm{m\,s^{-1}}$, 250 000 J

16 $0.845\,\mathrm{m\,s^{-2}}$

17 **(i)** $1.25\,\mathrm{m\,s^{-2}}$

(ii) 590 m

Chapter 10

? (Page 186)

(i) the vertical component of velocity is zero

(ii) $y = 0$

? (Page 187)

1 Yes for a parabolic path.

$u_y - gt = 0$ when $t = \frac{u_y}{g}$

and $u_y t - \frac{1}{2}gt^2 = 0$ when $t = 2\frac{u_y}{g}$.

2 The balls and the bullet can be modelled as projectiles when there is no spin or wind and air resistance is negligible. Also a rocket with no power. The air affects the motion of the others.

Exercise 10A (Page 188)

1 **(i)** **(a)**

(b) $u_x = 8.2$

$u_y = 5.7$

(c) $v_x = 8.2$

$v_y = 5.7 - 10t$

(d) $x = 8.2t$

$y = 5.7t - 5t^2$

(ii) **(a)**

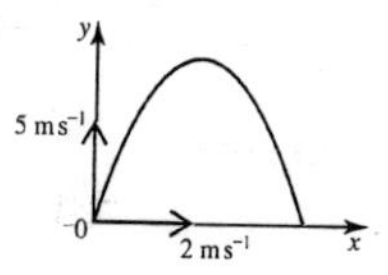

(b) $u_x = 2$

$u_y = 5$

(c) $v_x = 2$

$v_y = 5 - 10t$

(d) $x = 2t$

$y = 5t - 5t^2$

(iii) **(a)**

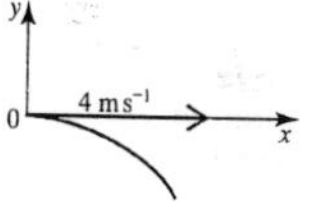

(b) $u_x = 4$

$u_y = 0$

(c) $v_x = 4$

$v_y = -10t$

(d) $x = 4t$

$y = -5t^2$

(iv) **(a)**

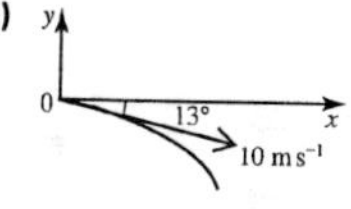

(b) $u_x = 9.7$

$u_y = -2.2$

(c) $v_x = 9.7$

$v_y = -2.2 - 10t$

(d) $x = 9.7t$

$y = -2.2t - 5t^2$

(v) **(a)**

(b) $u_x = U\cos\alpha$

$u_y = U\sin\alpha$

(c) $v_x = U\cos\alpha$

$v_y = U\sin\alpha - gt$

(d) $x = Ut\cos\alpha$

$y = Ut\sin\alpha - \frac{1}{2}gt^2$

(vi) (a)

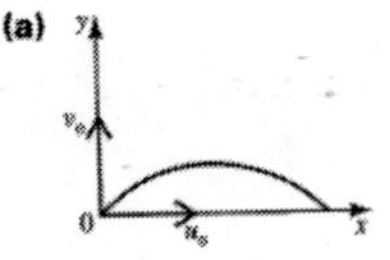

(b) $u_x = u_0$

$u_y = v_0$

(c) $v_x = u_0$

$v_y = v_0 - gt$

(d) $x = u_0 t$

$y = v_0 t - \frac{1}{2}gt^2$

2 (i) (a) 1.5 s

(b) 11 m

(ii) (a) 0.5 s

(b) 1.25 m

3 (i) (a) 4 s

(b) 80 m

(ii) (a) 0.88 s

(b) 2.17 m

Exercise 10B (Page 191)

1 (i) (a)

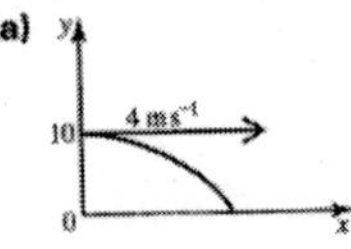

(b) $\begin{pmatrix} 4 \\ -10t \end{pmatrix}$

(c) $\begin{pmatrix} 4t \\ 10 - 5t^2 \end{pmatrix}$

(ii) (a)

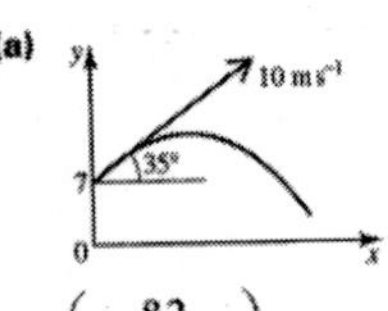

(b) $\begin{pmatrix} 8.2 \\ 5.7 - 10t \end{pmatrix}$

(c) $\begin{pmatrix} 8.2t \\ 7 + 5.7t - 5t^2 \end{pmatrix}$

(iii) (a)

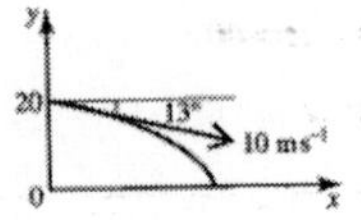

(b) $\begin{pmatrix} 9.7 \\ -2.2 - 10t \end{pmatrix}$

(c) $\begin{pmatrix} 9.7t \\ 20 - 2.2t - 5t^2 \end{pmatrix}$

(iv) (a)

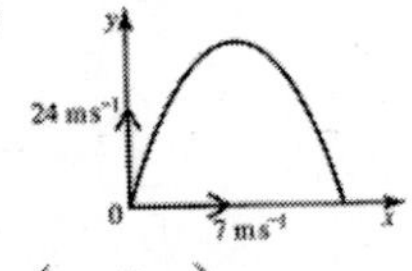

(b) $\begin{pmatrix} 7 \\ 24 - 10t \end{pmatrix}$

(c) $\begin{pmatrix} 7t \\ 24t - 5t^2 \end{pmatrix}$

(v) (a)

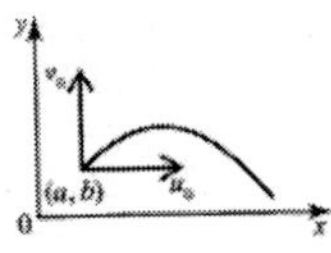

(b) $\begin{pmatrix} u_0 \\ v_0 - gt \end{pmatrix}$

(c) $\begin{pmatrix} a + u_0 t \\ b + v_0 t + \frac{1}{2}gt^2 \end{pmatrix}$

2 (i) (a) 1.47 s

(b) 26 m

(ii) (a) 0.3 s

(b) 10.45 m

3 (i) 2.8 m

(ii) 2.8 m

(iii)

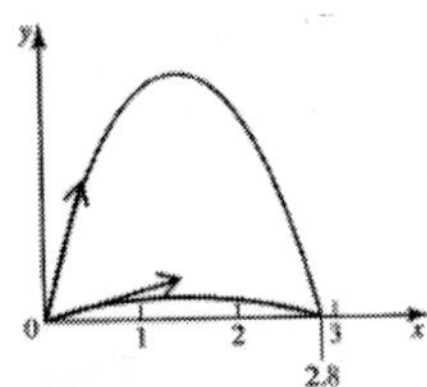

? (Page 196)

They land together because *u*, *s* and *a* in the vertical direction are the same for both.

Exercise 10C (Page 196)

1 (i) 17.3, 10 m s⁻¹

(ii) 0, –10 m s⁻²

(iii) 35.3 m

(iv) 1 s

(v) 5 m

2 (i) 41, 28.7 m s⁻¹

(ii)

t	0	1	2	3	4	5	6
x	0	41	82	123	164	205	246
y	0	24	38	42	36	21	–4.3

(iii)

(iv) 42 m, 239.7 m

(v) The ball is a particle, no spin, no air resistance so acceleration = *g*

3 (i) 17.2, 8 m s⁻¹

(ii) 1.61 s

(iii) 28.8 m

(iv) 0.84 s

(v) 3.4 m

(vi) 2.58 m, No

4 (i) 10.3, 14.7 m s⁻¹

(ii) 2.91 s

(iii) Into the goal

(iv) No

5 (i) 44.7 s

(ii) 13.4 km

(iii) 539 m s⁻¹

(iv) 56.1°

6 (i) 0.47 s

(ii) 0.63 s

(iii) 25.7 m s⁻¹

(iv) 29.1 m s⁻¹

7 (i) Yes, the range is 68.9 m.

(ii) $33.0\,\mathrm{m\,s^{-1}}$

8 (i) (a) 34.6 m

(b) 39.4 m

(c) 40 m

(d) 39.4 m

(e) 34.6 m

(ii) $80 \sin\alpha \cos\alpha = 80 \cos(90° - \alpha) \sin(90° - \alpha)$

(iii) 57.9°

(iv) +30 cm, 31 cm; lower angle slightly more accurate.

9 (i) $26.3\,\mathrm{m\,s^{-1}}$

(ii) $27.6\,\mathrm{m\,s^{-1}}$

(iii) $26.4 < u < 27.2$

10 25.2 m

11 (i) 1.74 s

(ii) 3.5 m, Juliet's window

(iii) $9.12\,\mathrm{m\,s^{-1}}$

12 (i) 3.2 m, vertical component of velocity is always $\leqslant 8\,\mathrm{m\,s^{-1}}$

(ii) 5.5 m

(iii) 52°

13 (i) 20

(ii) 32 m

(iii) 3.2 m

14 (i) $7 - 5t^2$

(ii) (a) $Vt\cos\theta$

(b) $Vt\sin\theta - 5t^2$

? (Page 202)

$\begin{pmatrix}20\\30\end{pmatrix}\mathrm{m\,s^{-1}}$, (0, 6), $10\,\mathrm{m\,s^{-2}}$

Exercise 10D (Page 202)

1 (i) $y = \frac{5}{16}x^2$

(ii) $y = 6 + 0.4x - 0.2x^2$

(iii) $y = -14 + 17x - 5x^2$

(iv) $y = 5.8 + 2.4x - 0.2x^2$

(v) $y = 2x - \dfrac{gx^2}{2u^2}$

2 (i) $x = 40t$

(iii)

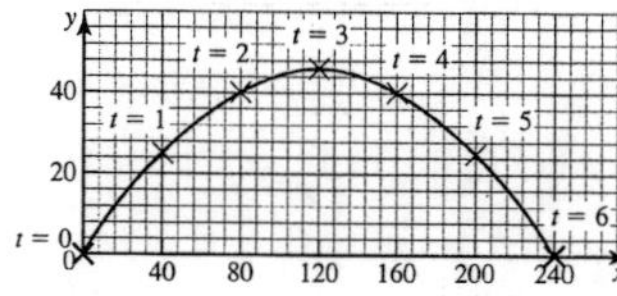

3 (i) $y = \frac{3}{4}x - \frac{1}{320}x^2$

(ii)

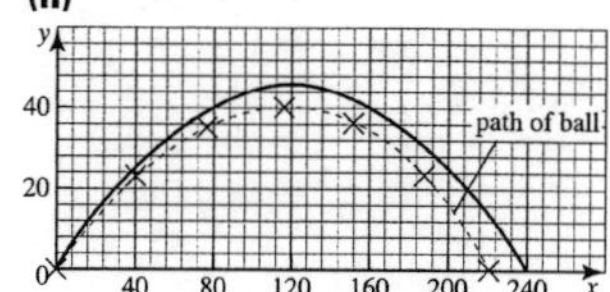

Air resistance would reduce x.

(iii) Yes, horizontal acceleration $= -0.5\,\mathrm{m\,s^{-2}}$

? (Page 205)

The projectile is a particle and there is no air resistance or wind. The particle is projected from the origin.

Exercise 10E (Page 205)

1 (i) $(21.2t, 21.2t - 5t^2)$

(iii) 8.9

(iv) 29.7 or 61.2

2 (ii) 6.9 m

3 (ii) 24.1 m

(iii) Yes, $y = 2.4$ m

4 (i) $y = 1 + 0.7x - \dfrac{7.45x^2}{u^2}$

(ii) $u > 7.8\,\mathrm{m\,s^{-1}}$

(iii) $u < 8.5\,\mathrm{m\,s^{-1}}$

5 (i) $u = 20\,\mathrm{m\,s^{-1}}, \theta = 45°$

(ii) $y = x - 0.025x^2$

6 (i) Speed $= 13.5\,\mathrm{m\,s^{-1}}$, $T = 0.721$

(ii) 2.85° to the horizontal

7 (i) $x = \frac{1}{2}vt$, $y = \dfrac{\sqrt{3}}{2}vt - 5t^2$

(ii) 29.7

(iii) 55.3° downward from the horizontal

8 (ii) When $\tan\theta = \frac{3}{4}$, OA = 38.4 m

When $\tan\theta = \frac{17}{4}$, OA = 17.8 m

(iii)

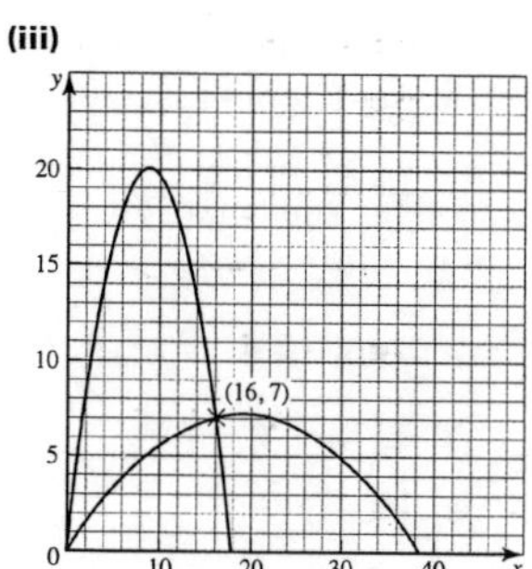

Experiment (Page 207)

1 $g\sin\theta$ where θ is the angle between the table and the horizontal

2 Not according to the simplest model

3 Yes, at angles α and $(90° - \alpha)$

4 45°

5 a parabola

Chapter 11

? (Page 213)

The tool shown in figure 11.9(i) works with one hand but has less leverage (moment). See also text.

? (Page 219)

(i) $P + Q$ line of action parallel to P and Q and in same direction; distance from O is $a + \dfrac{bQ}{P+Q}$ (between P and Q)

(ii) $P - Q$ line of action parallel to P and Q and in direction of the larger; distance from O is $a - \dfrac{bQ}{P-Q}$ (to the left of P for $P > Q$)

❓ (Page 219)

You produce equal and opposite couples using friction between one hand and the lid and between the other hand and the jar so that they turn in opposite directions. Pressing increases the normal reactions and hence the maximum friction possible.

Exercise 11A (Page 219)

1 (i) 15 Nm

(ii) −22 Nm

(iii) 18 Nm

(iv) −28 Nm

2 (i) 2.1 Nm

(ii) 6.16 Nm

(iii) 0.1 Nm

(iv) 0.73 Nm

3 29.2 N, 20.8 N

4 (i) 1250 N, 1250 N

(ii) 1479 N, 1821 N

5 96.5 N, 138.5 N

6 (i) 55 kg

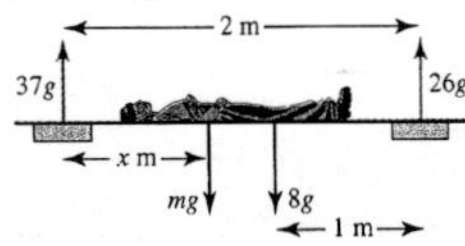

(ii) 0.8 m

7 (i) $P = 27.5g$, $Q = 147.5g$

(ii) $P = 2.5g$, $Q = 172.5g$

(iii) If child is less than 0.95 m from the adult, $P < 0$ so the bench tips unless A is anchored to the ground.

(iv) The bench tips if A is not anchored.

8 (i) $15g$N, $30g$N

(ii) $90g$N, $5g$N

(iii) zero

(iv) $\frac{2}{3}$ m

9 (i) $0.5g(30 - x)$ kN, $0.5g(20 + x)$ kN

(ii) its centre of mass

(iii) constant $15g$kN each

10 (i) $35g$N, $75g$N

(ii) no

(iii) 36 kg

❓ (Page 226)

No, the system is symmetrical providing the rod is uniform.

Exercise 11B (Page 228)

1 (i) 6 Nm

(ii) −10.7 Nm

(iii) 23 Nm

(iv) 0

(v) −4.24 Nm

(vi) 4.24 Nm

2 David and Hannah (by radius × 0.027 Nm)

3 (i) 5915 kg

(ii) $4532 \sec\theta$ kg

4 (i) 43.3 N

(ii) 28.1 N

(iii) 30.7 N

5 (i) $T\cos 30°$, $T\sin 30°$

(iii) 30 Nm

(v) 8.04 N, 15.36 N

(vi) (a) 33.7°

(b) 3.23 m

6 (i) 1434 N

(ii) 651 N, 1644 N

(iv) (a) jib stays put, $T = 0$

(b) A drops

7 (i)

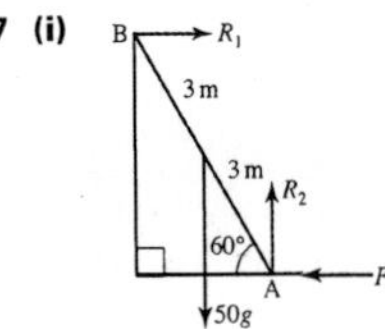

(ii) 0

(iii) 144 N

(iv) 144 N, $\mu \leqslant 0.289$

8 (i)

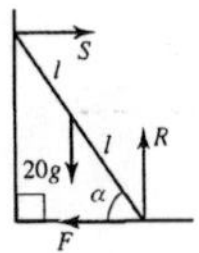

(a) (ii) 57.7 N, 57.7 N, 200 N

(iii) 0.29

(b) (ii) 100 N, 100 N, 200 N

(iii) 0.5

9 (i)

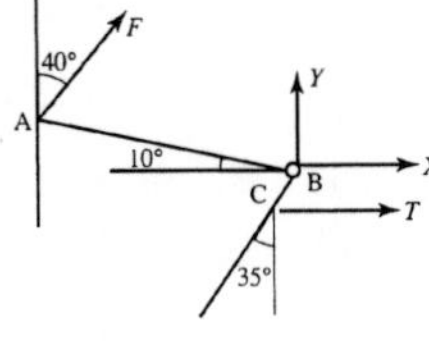

(ii) 162 N

(iii) 61.7 N

10 (ii) 1 600 000 Nm (to 3 s.f.)

(iii) 6830 N, 3830 N

11 (i)

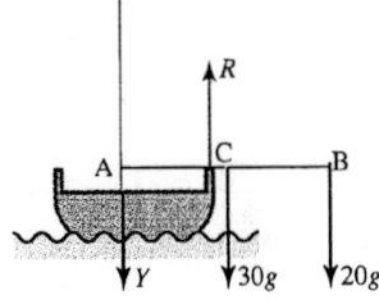

(ii) $80g = 800$ N

(iii) $30g = 300$ N vertically down

(iv) $17.5g = 175$ N

12 (i)

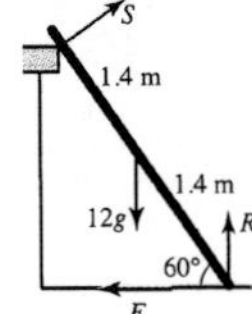

(ii) 26.0 N, 105 N

(iii) 0.51

(iv) 2.25 m

13 (i)

YN RN

A 1.5 m 1 m 1.5 m 1 m B

1000 N 1500 N 1000 N

(ii) 3596 N

(iii) 3497 N at 28.15° above the horizontal

14 (i) 960 N

(ii) $X = 269$, $Y = 522$

Chapter 12

? (Page 235)

Yes, centre of mass vertically below P.

? (Page 237)

$4 \times 1\frac{2}{15} + 5 \times \frac{2}{15} = 6 \times \frac{13}{15}$

Exercise 12A (Page 239)

1 (i) 0.2 m

(ii) 0

(iii) −0.72 m

(iv) 1.19 m

(v) +0.275 m

(vi) 0.36 m

(vii) −0.92 m

(viii) 0.47 m

2 2.18 m from 20 kg child

3 4.2 cm (towards the 60 kg mass)

4 4680 km (to 3 s.f.)

5 0.92 m

6 3.33 mm from centre

7 2.95 cm

8 1.99 kg

9 42 kg

10 $\frac{m_2 l}{(m_1 + m_2)}$ from m_1 end

11 (i) 3.35 m from centre line, tips over

(ii) 4.55 tonnes

(iii) $L(l - d) < Md + C(a + d)$, $C(a - d) < Md$

(iv) $\frac{2Mad}{(l - d)(a - d)}$

Exercise 12B (Page 247)

1 (i) (2.3, −0.3)

(ii) (0, 1.75)

(iii) $(\frac{1}{24}, \frac{1}{6})$

(iv) (−2.7, −1.5)

2 $(5, 6\frac{1}{3})$

3 (i) (20, 60)

(ii) (30, 65)

(iii) (30, 60)

4 23 cm

5 (i) (5, 2)

(ii) (3, 6)

(iii) $\left(4, \frac{20}{\pi}\right)$

(iv) $\left(4, \frac{30}{\pi}\right)$

6 (i) (1.5, −1.5)

(ii) (1.5, −2.05)

7 (i) 0.2 cm below O

(ii) 9.1°

8 (i) $(0.5a, 1.2a)$

(ii) 3.9°

(iii) $2m$

9 (i) 2.25

(ii) 0.56 m

(iii) 0.40, $(\frac{1}{2}, 1\frac{1}{2})$

10 (i) 16 N

(ii) 12.8 N, 30.4 N

11 (ii) 15.3

12 (i) $\frac{1}{3}$ m (towards C)

(ii) Tension $= \frac{8}{9} W$, force at C $= \frac{1}{9} W$

13 (i) 39.9 N

(ii) $\theta = 47.5$, tension = 18.5 N

14 (ii) 30°

? (Page 253)

It is likely to topple. Toppling depends on relative mass of upstairs and downstairs passengers.

? (Page 254)

1st slide, 2nd topple.

? (Page 255)

R cannot act outside the surfaces in contact so there is a resultant moment about the edge E.

? (Page 257)

Yes, when $\mu = 0.5$ and $\alpha = 26.6°$

Exercise 12C (Page 257)

1 (i) $2.8g$ N

(ii) $3.5g$ N

(iii) slide

2 (i) $P = 2\mu g$

(ii) $0.6g$ N

(iii) $\mu < 0.3$

(iv) $\mu > 0.3$

3 It slides.

4 (a) (i) 22.5 N

(ii) 22.0 N

(iii) topples

(b) (i) 22.8 N

(ii) 25.9 N

(iii) slides. 63.4°, 22.4 N

5 (i) 14.0°

(ii) 18.4°

(iii) sliding

6 (i) (a) 50 by 20

(b) 20 by 10

(ii) The shortest side is perpendicular to the plane of the slope for maximum likelihood of sliding.

(iii) (a) $\mu < 0.2$

(b) $\mu > 5$

7 (i) stays put

(ii) topples

8 (i) $W\cos\alpha$

(iii) 137 Nm; $P > 137$

(v) $\mu < 0.211$

9 (i) (28, 60)

(ii) (52, 60)

(iii) (64, 60)

(iv) 40 cm

10 (i) 0.5 m from ground

(ii) 16.7°

(iv) 19.4 kg

11 (i) (a) (10, 2.5)

(b) (12.5, 5)

(c) (15, 7.5)

(d) (17.5, 10)

(e) (20, 12.5)

(ii) 5

(iii) $(9+n, 2.5n)$, 11

(iv) 102.5 cm

12 (i) 5.25

13 The prism falls on the face containing BC.

14 (i) 7.5

15 (i) 48

(ii) 39.8

Chapter 13

? (Page 266)

Forces which pull them towards the centre of the circle.

Gravity pulls it in.

It moves off at a tangent.

No

? (Page 268)

They are often given in radians per second and one turn is 2π radians.

Exercise 13A (Page 269)

1 (i) $8.2\ \text{rad s}^{-1}$

(ii) $4.7\ \text{rad s}^{-1}$

(iii) $3.5\ \text{rad s}^{-1}$

2 2865 rpm

3 (i) (a) 0.033 rpm

(b) $0.0035\ \text{rad s}^{-1}$

(ii) $0.24\ \text{m s}^{-1}$

4 $32.5\ \text{rad s}^{-1}$

5 (i) $50\ \text{rad s}^{-1}$

(iii) $150\ \text{rad s}^{-1}$

6 (i) 3820 rpm

(ii) 2080 rpm (to 3 s.f.)

7 (i) $1.99 \times 10^{-7}\ \text{rad s}^{-1}$

(ii) $7.27 \times 10^{-5}\ \text{rad s}^{-1}$

(iii) $465\ \text{m s}^{-1}$

(iv) about $290\ \text{m s}^{-1}$ at latitude 51.5°

8 2.29:1

9 (i) 61.7 J

(ii) points on a large object would travel with different speeds

10 (i) big: $2.09 \times 10^{-3}\ \text{m s}^{-1}$, small: $1.16 \times 10^{-4}\ \text{m s}^{-1}$

(ii) 18:1 the radius is also involved

11 (i) 4.91 m

(ii) $12.3\ \text{m s}^{-1}$

12 (i) no

(ii)

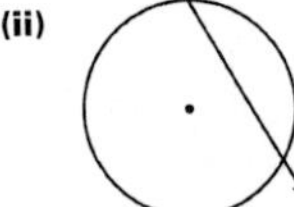

? (Page 274)

(i) $0\ \text{m s}^{-1}$ so $u = r\omega$

(ii) $\frac{u}{r}$

(iii) $2u\ \text{m s}^{-1}$ forwards and $\sqrt{2}u\ \text{m s}^{-1}$ at 45° to the forward direction.

Activity 13.1 (Page 280)

When you turn fast enough, the coin flies off the card. When the card is tilted it stays put at higher angular speeds.

? (Page 281)

See text that follows.

? (Page 283)

1 Resolve perpendicular to the slope

$R - mg\cos\alpha = m\frac{v^2}{r}\sin\alpha.$

Then parallel to the slope. No slipping down if

$mg\sin\alpha - F = m\frac{v^2}{r}\cos\alpha$, and $F < \mu R$. No slipping up if

$mg\sin\alpha + F = m\frac{v^2}{r}\cos\alpha$, and $F < \mu R$.

Substitute for F and R and rearrange.

2 The bend is safe at $48\ \text{km h}^{-1}$. In general bends are safer for larger r. So long as $\mu > \tan\alpha$, there is no lower limit for v and then α can be increased in order to increase the upper limit.

Exercise 13B (Page 285)

1 (i) Neither, both have $\omega = 0.75\ \text{rad s}^{-1}$.

(ii) No because they have the same angular speed.

(iii) $13.5\ \text{m s}^{-2}$, $11.25\ \text{m s}^{-2}$

(iv) towards the centre for circular motion

2 (i) (a) neither slips

(b) B slips, A doesn't

(c) both slip

(ii) A slips first, radius matters, mass doesn't matter

3 (i) accelerates because direction changes

(ii) $11.25\ \text{m s}^{-2}$

(iii) 2250 N

(iv) No, outside wheels go faster so force is greater

4 (i) True for fixed seats

(ii) False, as $v = r\omega$ so speed depends on r.

(iii) False, as $a = r\omega^2$ so acceleration depends on r.

5 (i) (a) $0.5\,\text{rad}\,\text{s}^{-1}$

(b) $1\,\text{m}\,\text{s}^{-2}$

(c) 60 N towards centre

(ii) skater is particle

6 B has greater force because greater acceleration

7 (i) (a) $\dfrac{(2\times10^7)}{\sqrt{r}}$ (3 s.f.)

(b) $\pi r^{3/2}\times10^{-7}\,\text{s}$

(ii) $T^2 = \pi^2\times10^{-14}r^3$

(iii) $4.23\times10^7\,\text{m}$

8 (i) $5.72\times10^{-3}\,\text{m}\,\text{s}^{-2}$

(ii) $1.77\times10^{30}\,\text{kg}$

(iii) any planet in this orbit would have the same period whatever its mass

9 (i) vertical force required to balance weight

(ii) $12.57\,\text{rad}\,\text{s}^{-1}$

(iii)

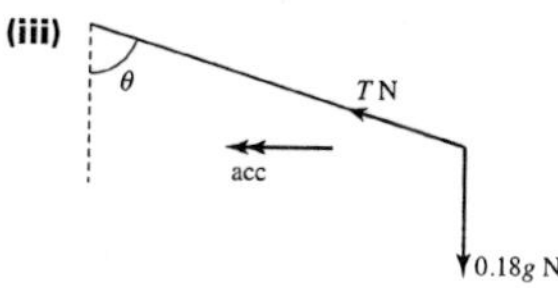

(iv) $T\sin\theta = mr\omega^2$
$= 0.18\times0.8\sin\theta\times(4\pi)^2$;
$T\cos\theta = mg = 1.8$

(vi) 22.7 N

10 (i) 263 N

(ii) The weight of the sphere is very small compared with the tension in the string.

(iii) $23.1\,\text{m}\,\text{s}^{-1}$

11 (i) 6400 N

(a) The car is about to skid on a bend.

(b) The car is accelerating or braking and is about to skid.

(ii) $111\,\text{km}\,\text{h}^{-1}$ ($31\,\text{m}\,\text{s}^{-1}$)

(iii)

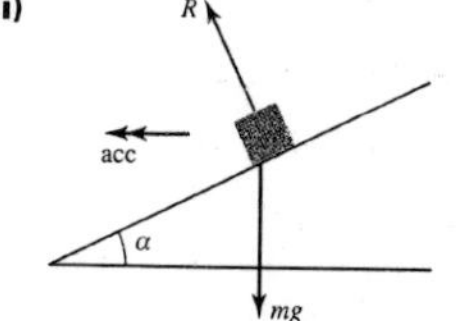

(iv) $R\sin\alpha = \dfrac{mv^2}{r}$, $R\cos\alpha = mg$

(vi) 10.6° or 0.185 rad

12 (i) $72\,\text{km}\,\text{h}^{-1}$ ($20\,\text{m}\,\text{s}^{-1}$)

(ii) $6.6\,\text{km}\,\text{h}^{-1}$ ($1.84\,\text{m}\,\text{s}^{-1}$) faster

13 (i) $\pi\,\text{rad}\,\text{s}^{-1}$, $5\pi\,\text{m}\,\text{s}^{-1}$

(iii) 40.5 rpm

(iv) 100 m

14 (ii) $\cos\alpha > \cos\beta \Rightarrow \alpha < \beta$
so $x > 2a - x$
$x > a$

15 (i) 35

(ii) 1.83 N

(iii) $2.83\,\text{m}\,\text{s}^{-1}$

16 $R = 1.10$, $S = 0.784$

17 (ii) 3.93

18 (ii) $2.31\,\text{m}\,\text{s}^{-1}$

(iii) 1.09 s

19 (i) $0.735\,\text{rad}\,\text{s}^{-1}$

(ii) (a) 400 N

(b) $10.3\,\text{m}\,\text{s}^{-1}$

20 (i) $T = 3$

(ii) $3\,\text{m}\,\text{s}^{-1}$

Chapter 14

❓ (Page 295)

Use energy considerations. See the investigation at the end of the chapter.

Exercise 14A (Page 299)

1 (i)

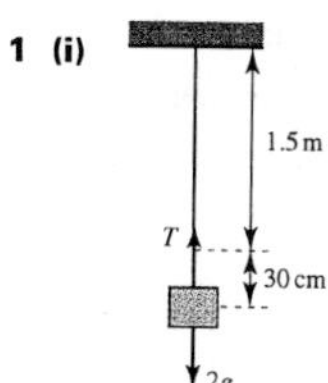

(ii) 100 N

2 (i) 20 N

(ii) 20 N

(iii) Tension required to double the length is the same. There is a 20 N force at the fixed end in part (i).

3 (i) 0.03 N

(ii) 0.0375 m

(iii) 0.08

4 (i) 5 N

(ii) 75 N

(iii) 0.625 kg

5 (i) 30 N

(ii) 3

(iii) Spring becomes fully compressed with fewer than seven blocks.

6 $\dfrac{\lambda a}{\lambda - ma\omega^2}$

Exercise 14B (Page 303)

1 (i) 60 N

(ii) 0.04 m

(iii) 2400 N; could be two strings together

2 (i) 12 000 N

(ii) 12 000 N

(iii) 0.05 m

(iv) 0.02 m

3 (i) 7000 N

(ii) 7000 N

(iii) 4000 N

4 (i) 198 N

(ii) 654 N

(iii) 118%

(iv) 51.6 m

5 (i) $(0.3 - x)$ m

(ii) $16x$ N, $25(0.3 - x)$ N

(iii) 0.183

6 (i) $2.2 - h$ m, $h - 1.2$ m

(ii) $44 - 20h$ N, $30h - 36$ N

(iii) 1.2

(iv) 20 N, 0 N

7 (i) $\frac{l_0}{\lambda} mg\sin\alpha$

(ii) (a) $\frac{l_0}{\lambda} mg(\mu\cos\alpha + \sin\alpha)$

(b) $\frac{l_0}{\lambda} mg(\sin\alpha - \mu\cos\alpha)$

8 (ii) 0.313 m

(iii) An elastic string is unlikely to pass smoothly over a peg.

9 1.3 m

10 (i) 1.33 N

(ii) 16

Exercise 14C (Page 311)

1 (i) 0.1 J

(ii) 0.001 J

(iii) 0.4 J

(iv) 0 J

2 (i) 4 J

(ii) 0.25 J

(iii) 1 J

(iv) 0.0625 J

3 (i) 0.75 J

(ii) 5.48 m s^{-1}

4 (i) 0.006 67 J

(ii) 0.577 m s^{-1}

5 (i) 5×10^4 J

(ii) 7.07 m s^{-1}

(iii) 1.29 m

(iv) 7.07×10^4 N, 35.4 m s^{-2}

(v) Truck moves back along the rail at 5 m s^{-1} if other forces ignored.

6 0.433 m, 0.067 m

7 (i) B is in equilibrium

(ii) 0.8 m, 6 N

(iii) 2.7 J

(iv) 10 m s^{-1}

8 (i) 0.8 J

(ii) 0.1

9 (i) 26 N, 7 N

(ii) 20 m s^{-2}

(iii) 0.933 m

10 (i) 12 N, 24 N

(iii) 7.5 m s^{-2}

(iv) 0.5

Exercise 14D (Page 316)

1 (i) $5h^2$ J

(ii) $0.2gh$ J

(iii) 0.4

2 $2l_0$

3 $l_0\left(1 + \frac{mg}{\lambda}\right)$

4 (ii) 0.13

5 0.463 m

6 (i) 1 J

(iii) 0.2 m

7 (i) 0.1098 m

(ii) 0.149 m

9 (i) $\cos\theta = \frac{g}{\omega^2 l}$

(ii) $\cos\theta = \frac{g(\lambda - m\omega^2 l_0)}{\omega^2 l_0 \lambda}$

10 (i) 0.8 m

(ii) $98(x - 0.5)$ N, $98(2 - x)$ N

(iii) 57.3 J

(iv) 0.81 m s^{-1}

11 (ii) 8 m

(iii) 6.71 m s^{-1}

12 (i) Tension = 6.25 N

(ii) 4.90 m s^{-1}

? (Page 319)

20.3 m s^{-2} for a person weighing 90 kg. A lighter person would feel a greater deceleration, approximately 27 m s^{-2} for somebody weighing 60 kg.

? (Page 320)

The jumper would slow down more quickly at the end.

Chapter 15

? (Page 321)

The forces are air resistance and the gravitational forces due to the earth and the moon.

? (Page 326)

Because the acceleration is not constant.

Exercise 15A (Page 328)

1 In the following answers, k is an arbitrary constant.

(i) $s = \frac{1}{2}v + k$

(ii) $v = ke^{2t}$

(iii) $v = -\frac{2}{3}\cos 3t + k$

(iv) $v = \frac{1}{t + k}$

(v) $v = ke^{-s}$

(vi) $v = \sqrt{4s(1 - s) + k}$

(vii) $s = -\frac{1}{3}\ln k(3v + 2)$

(viii) $s = \frac{1}{2}\ln(1 + v^2) + k$

2 (i) $\frac{16}{3}$

(ii) $-\frac{1}{2}$

(iii) $\sqrt{2\ln 2}$ or 1.177

(iv) $7\frac{1}{2}$

(v) $\frac{\pi}{4}$ or 0.785

(vi) $\frac{1}{3}\ln\frac{5}{2}$ or 0.305

(vii) $\frac{1}{2}\ln 3$ or 0.549

(viii) $\frac{1}{2}\ln\frac{4}{3}$ or 0.144

3 (i) $v\frac{dv}{ds} = \frac{1}{v + 2}$

(iii) $\frac{dv}{dt} = \frac{1}{v + 2}$

(v) $s = -2t + \frac{1}{3}(4 + 2t)^{3/2} - \frac{8}{3}$

4 (i) $v = \sqrt{x^2 - 4.8x + 6.25}$

(ii) 0.7

5 (ii) 1.35

6 (i) $t = 2.75$

(ii) OP = 4.51 m

7 (ii) 2.32

8 (i) Acceleration = $4x - 16$
Resisting force when $x = 1$ is 3 N

(ii) $x = 4(1 - e^{-2t})$

9 (ii) 47.5 m s^{-1}

10 (ii) 5.33 m s^{-1}

11 (ii) $0.1v$ N

(iii) 1.53 s

12 (i) $0.25v\dfrac{dv}{dx} = -(5 - x)$

(ii) $x = 5(1 - e^{-2t})$

Index

Page numbers are in black in *Mechanics 1*. Page numbers are in blue in *Mechanics 2*.